Dragonfly Genera of the New World

An illustrated and annotated key to the Anisoptera

Dragonfly Genera of
An Illustrated and

the New World
Annotated Key to the Anisoptera

Rosser W. Garrison,
Natalia von Ellenrieder, and
Jerry A. Louton

The Johns Hopkins University Press
Baltimore

The Johns Hopkins University Press
2715 North Charles Street
Baltimore, Maryland 21218-4363
www.press.jhu.edu

Library of the Congress Cataloging-in Publication Data

Garrison, Rosser W.
    Dragonfly Genera of the New World: an illustrated and
    annotated key to the Anisoptera/
    Rosser, W. Garrison, Natalia von Ellenrieder,
    Jerry A. Louton

p.cm.
Includes bibliographical references and index.
ISBN 0-8018-8446-2 (alk. paper)
1. Dragonflies — North America — Classification. 2.
Dragonflies — South America — Classification. I.
Ellenrieder, Natalia von. II. Louton, Jerry A. III. Title.
QL520.2.A1G37 2006
595.7'33097–dc22   2006041810

A catalog record for this book is available from the British
Library

Title page illustration *Stylurus plagiatus* (Gomphidae) male
– by R.W. Garrison.

Front cover above: *Erythemis peruviana* (Libellulinae)
male – photographed by R.W. Garrison in Venezuela, Río
Caroni; below: *Fylgia amazonica* (Libellulinae) male
– photographed by R.W. Garrison in Brazil, Rondônia State

To our families, friends, and all dragonfly lovers

Contents

Preface

During the past 30 years, dragonflies and damselflies (Order Odonata) have become one of the most studied groups of insects. Their relatively large size, striking coloration, interesting biology, and relative fewness in numbers of species (about 5,600 species) compared with other orders of insects are factors contributing to their popularity. Dragonfly larvae generally have high habitat fidelity, and can thus be used as indicators of the health of aquatic ecosystems. Adults are uually active only during daylight hours and may be collected by aerial netting.

Despite their popularity, there has virtually been no means of easily identifying species from the richest area of biodiversity: the neotropical region. This area houses about 1,650 species in 195 genera, or about 30% of the world's total (Garrison, 1991; Bridges, 1994; Tsuda, 2000). The closest areas in number of species are continental Southeast Asia with 959 species and Indonesia with 673 species (Tsuda, 2000). Within the past eight years a comprehensive account of the Odonata from the United States and Canada has been published (Westfall and May, 1996; Needham, Westfall, and May, 2000), resulting in an increase in interest by students and the general public. In contrast, for Mexico and Central America there are only the less comprehensive keys of Calvert (1905–1908) and a modified, up-to-date volume (Förster, 2001) for Central America, and nothing equivalent has ever been published for the vast South American continent. A significant amount of research in the neotropical Odonata by Latin American students has only occurred during the past fifteen to twenty years. However, all students of the order have had to rely on scattered publications that treat regional fauna or review particular groups of Odonata, or have access to a synoptic collection in order to successfully identify material.

It is surprising that such a small and well-known order of insects has lagged behind others as regards taxonomic knowledge, as is the case in South America. This is most likely due to the few numbers of active researchers in the past as well as to the paucity of material in collections; odonates are fragile and, except for a few groups such as the magnificent helicopter damselflies (Pseudostigmatidae), are not attractive to the general collector since preserved material often undergoes significant postmortem discoloration. Because there is no comprehensive treatment allowing for the identification of neotropical dragonflies, an illustrated key should greatly facilitate and accelerate taxonomic and ecological studies of the order in this rich geographic region. We believe that this book will be a reliable reference source for identification of dragonflies for limnologists, ecologists, and other biologists relatively unfamiliar with these insects, and will encourage Latin American workers to study their dragonfly faunas.

This work attempts to provide keys to adults, supported by abundant illustrations, to genera of dragonflies (suborder Anisoptera) from anywhere in the New World. We are well aware that our goal of allowing the nonspecialists or those little familiar with entomology to properly place specimens correctly to genus using our keys will not always be realized. Future studies of vastly unexplored areas, especially in South America, will augment species and genera rendering our keys out-of-date. Still, our book should allow biologists to determine the existence of undescribed taxa by comparison with the keys, illustrations, and references to literature, which would otherwise not be possible. All the keys produced for this book using the DELTA program are open-ended and should allow for the inclusion of newly described genera. Almost all illustrations and wing scans are from preserved material we have personally examined. Readers will find a strong bias in supportive illustrations toward the Meso- and South American fauna. Because of the relative disparity between the knowledge of North American versus Meso- and South American faunas, we have felt justified in concentrating our efforts on the less known neotropical region.

We have steered a conservative course with regard to recognition of families and genera. Several authors (Carle, 1982a, 1986, 1995; Bechly, 1996; Lohmann, 1996; Jarzembowski and Nel, 1996; Trueman, 1996; von Ellenrieder, 2002), based on phylogenetic analyses, have shown the existing classification of the order to be partially artificial, and some of them have proposed new — and in some cases radical — reclassifications. However, phylogenetic relationships within the order are not yet fully resolved, and further research will most likely help build stronger hypotheses of relationships reflecting a natural classification. Since our purpose in this book is the identification of taxa, we have refrained from deviating from most of the currently recognized families and genera, although we provide references and comments regarding phylogenetic relationships of genera when known.

Although we provide brief summaries and references on habits and other aspects of biology for each genus, we have refrained from including sections on the general biology and ecology of Odonata in this volume. A standard reference for this is Tillyard (1917), and the up-dated and thorough volume by Corbet (1999), as well as a shorter but very good reference by Miller (1995), will provide the interested student with information on the biology of Odonata in general.

No effort was spared in trying to examine as many species as possible for every described genus. To this end, we have received invaluable help from many of our colleagues and institutions who have assisted us with needed specimens, literature, and advice. Chief among those are Drs. Janira M. Costa, Frederico A. Lencioni, Jürg De Marmels, Enrique González-Soriano, Rodolfo Novelo-Gutiérrez, Carlos Esquivel, Javier Muzón, Dennis R. Paulson, Michael L. May, Thomas W. Donnelly, Oliver S. Flint, Jr., Mark O'Brian, William J. Mauffray, Kenneth J. Tennessen, Jerrell J. Daigle, Sidney W. Dunkle, Andrew C. Rehn, Richard Hoebecke, Jérôme Constant, Stephen Brooks, and David Goodger. We are deeply indebted to Drs. Dennis R. Paulson, Sidney W. Dunkle, Kenneth J. Tennessen, and Jürg De Marmels for their valuable comments on the manuscript and to Dennis R. Paulson for his generous sharing of unpublished information. We thank the Smithsonian Institution for providing RWG and NVE a travel grant to make this study more complete.

ABBREVIATIONS

The following are used in the text:
FW = fore wing
HW = hind wing
S1, S2, etc = first, second, etc. abdominal segment
L = larva described
* = species examined
** = examined from photographs
S, N, E, W, SW, NE, etc. = South, North, East, West, Southwest, Northeast, etc.

Terminology and abbreviations of wing veins and fields are explained and illustrated in table 1 (page 2) and figures 6–7 (pages 3–4).

Acronyms for collections used throughout the text are spelled out on pages 7–8.

Dragonfly Genera of the New World

1. Introduction

What is a dragonfly?

Dragonflies and damselflies belong to the insect order Odonata. As in other insects, their body is divided into three regions: head, thorax and abdomen. The hemispherical or dumbbell-shaped **head** (Figs. 1–3) has large compound eyes which cover the entire sides, and the antennae are small and setaceous. The head

Fig. 13) and other accessory structures (**hamules** and **genital lobes**) are found. The **caudal appendages** are at the tip of the abdomen and are used by the male to hold the female by its head during copulation. Before copula ensues, the male curves his abdomen to charge the vesica spermalis with sperm, which is produced by the gonads that open, as in other insects, on the venter of the ninth segment. After the male grasps the female, she bends her abdomen placing her genital opening in contact with the charged vesica spermalis so that the characteristic wheel position or copula takes place. Females almost always lay their eggs in the water, where the larvae develop.

This volume provides keys to the 124 genera of New World Anisoptera or dragonflies; the suborder Zygoptera (damselflies) will be covered in a forthcoming volume. Dragonflies are easily differentiated from damselflies by their largely hemispherical head,

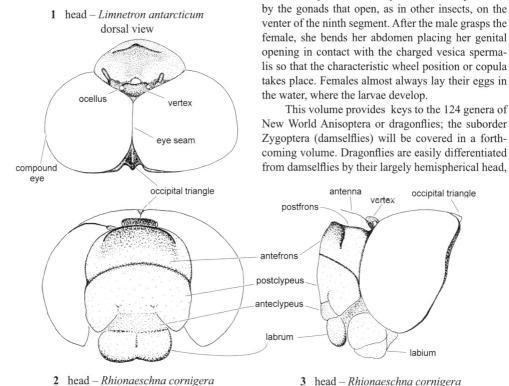

1   head – *Limnetron antarcticum*
dorsal view

ocellus   vertex

eye seam

compound eye

occipital triangle

antenna   vertex   occipital triangle
postfrons

antefrons

postclypeus

anteclypeus

labrum

labium

2   head – *Rhionaeschna cornigera*
frontal view (left antenna missing)

3   head – *Rhionaeschna cornigera*
lateral view

is attached to the thorax by a narrow, membranous neck. The thorax includes a small anterior **prothorax**, divided into three lobes (Fig. 4), followed by the meso- and metathorax which are fused into a box-like **pterothorax** (Fig. 5). The pterothorax in odonates is unique in having very large lateral (pleural) areas which have pushed the two pairs of membranous wings (Figs. 6–7) to the back and the legs to the front, so that the latter form a basket used for capturing their prey in flight. The **abdomen** is more or less cylindrical (Figs. 8–11), with terga covering dorsum and sides of the segments. The sterna consist of narrow, longitudinal ventral sclerites– the membranous pleura are normally not seen externally. Unique for Odonata is the secondary genitalia of the males, as well as their copulatory process. On the venter of second and third segments, males have a **genital fossa** (Fig. 12), where the copulatory organ (**vesica spermalis**,

differential wing shape and venation between fore and hind wing (base of hind wing broader than in fore wing), and in males, by the presence of a pair of cerci and a well developed epiproct. In damselflies, the head is transversely dumbbell-shaped with widely separated compound eyes, the fore and hind wings are similar, and in males, the pair of cerci are accompanied by a pair of well developed paraprocts. In the field, damselflies can be told apart from dragonflies by their widely spaced eyes, generally smaller size and wings usually held over their backs in their resting position; dragonflies are usually larger and perch with their wings outspread (*Zenithoptera* is an exception). Damselflies tend to fly close to the water surface while dragonflies are often the dominant aerial predators at pond or stream environments, and are often seen hawking over fields, trails, or defending a territory at streams or ponds.

4 prothorax – *Erythrodiplax castanea*
lateral view

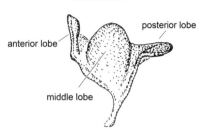

anterior lobe

posterior lobe

middle lobe

5 pterothorax – *Tachopteryx thoreyi*
lateral view

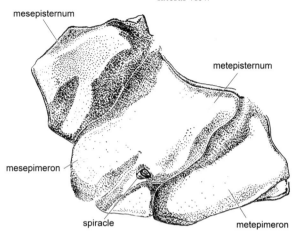

mesepisternum

metepisternum

mesepimeron

spiracle

metepimeron

Methodology

Two seminal works greatly influenced the form that our volume takes: the key to the Zygoptera of the world (Munz, 1919) and the key to New World Libellulidae (Borror, 1945). Although valuable, both keys suffer from being out of date, long, and contorted. For example, we initially used the characters and states from Borror's key to test the usability of the DELTA (DEscription Language for TAxonomy, Dallwitz, Paine, and Zurcher, 2000) key-generating feature and found that the number of couplets was reduced by two-thirds with no loss of resolving power and greatly increased usability. This program has several advantages over conventional key-making. It generates most parsimonious keys using recognizable

Vein name	Comstock and Needham (1898-1899)	Tillyard and Fraser (1938-1940)	Carle (1982a)	*Riek and Kukalová-Peck (1984)
Costa	C	C	CA	C
Subcosta	Sc	Sc	CP	ScP
Radius anterior	R_1	R_1	RA	RA
Radius posterior, first branch	M1	R_2	RP_1	RP_1
Radius posterior, second branch	M2	R_3	RP_2	RP_2
Radius posterior, third branch	M3	R_4	MP	RP_{3-4}
Intercalar vein 1	$M1_a$	IR_2	-	IR_1
Intercalar vein 2	RS	IR_3	MA	IR_2
Media anterior	M4	MA	CuA	MA
Media posterior	Cu1	CuP	CuP	MP
Cubitus anterior	Cu2	1A	A1	CuA
Anal anterior and posterior	A1-A3	A	A2/A3	AA/AP

Table 1. Odonata venational terminologies. * = used in this volume

character states, and is open-ended, *i.e.* it allows for easy modification of the keys should some character prove to be incorrect, or if additional characters or new taxa are added.

The process of constructing keys to families and genera was as follows:

1. A character matrix for all families/genera was made using diagnostic characters from examined specimens and/or descriptions and illustrations. We examined several species representing each genus in order to encompass intrageneric variability, and we were able to examine at least one representative of each genus except for six - *Racenaeschna* (Aeshnidae), *Anomalophlebia*, *Brasiliogomphus*, *Praeviogomphus* (Gomphidae), *Lauromacromia* and *Santosia* (Corduliinae).

2. Character states were incorporated into the standard data DELTA files. Most of the useful characters in the data set were unordered multistate characters. Integer and real data could not be gap-coded effectively because of high overall intrageneric (both within and among species) variability. Specification files (specs, tokey, and key) for key generation used DELTA system defaults except that "confirmatory characters" was reduced to "2" when key couplets became too cumbersome. Character reliabilities were scored as a compromise between clarity and reliability in order to prevent more difficult to see characters from appearing early in the keys.

3. Keys were generated and checked for readability. They were tested and corrected iteratively until no additional variation was found. When keys were in-

complete, new characters were found and added to the data scts until all genera were resolved.

4. All diagnostic characters were illustrated using a camera lucida, or from edited scans (wings), or scanning electron images. In order to facilitate the use of the keys, we have tried to place all figures on the same or opposite page. We use wing terminology from Riek and Kukalová-Peck (1984) as modified by Bechly (1996) as this is presently considered most correct (Rehn, 2003). Table 1 shows the equivalent wing venation systems that have been used for the last 150 years.

5. The main body of the text includes extensive annotations for each genus as follows: 1. Generotype designation; 2. Total number of world species followed by number of New World species; 3. Synonymic list of all species including reference to larval description ("**L**"); 4. References to taxonomic revisions; 5. General distribution; 6. Short description; 7. Unique characters; 8. Status of classification; 9. Potential for discovery of new species; and 10. Habitat and general notes including adult and larval biology. Each generic account includes a thumbnail polygon area distribution map, and illustrations of the taxonomically important caudal appendages, wings, and further diagnostic characters. Supplementary illustrations are given for speciose or diverse genera.

6. Certain genera could not be easily placed in the key because they are poorly known and/or not separable from other genera. Others probably represent artificial groups. To treat these taxa, we have occasionally strayed from using DELTA and annotated a few cou-

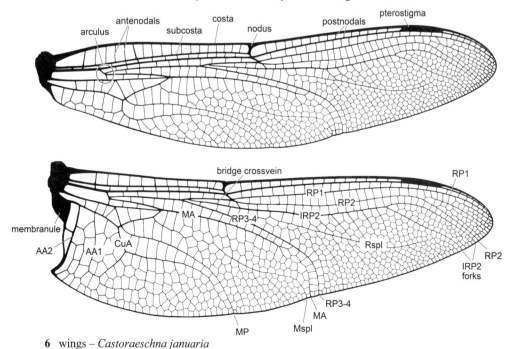

6 wings – *Castoraeschna januaria*

7 wings – *Orthemis attenuata*

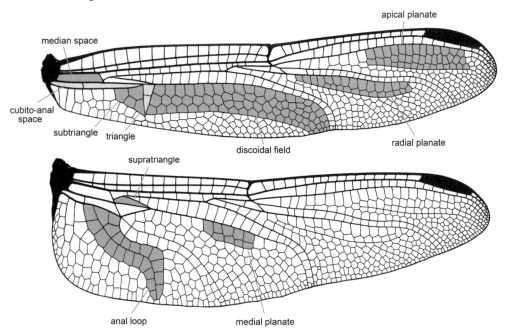

plets in order to accommodate their placement in the keys. We have not formally synonymized any existing genus, although we realize that such actions may need to be taken in the future.

Collection and preservation of specimens for study

The adult odonate fauna of the United States and Canada is so well known that it is possible to identify many species in the field with a good pair of binoculars (Dunkle, 2000). Several illustrated regional guides have and are currently being published, which allow relatively easy identification for some species from photographs. However, the neotropical fauna is still poorly known, and discovery of new species and range extensions are fairly common. Therefore identification of dragonflies from photographs is not possible or can be misleading, and it is essential to document discoveries with properly preserved voucher material, which can be then carefully studied in the laboratory.

Although adult Odonata are most often collected near water, they can be found almost anywhere. Many recent field guides (Dunkle, 1989; Carpenter, 1991; Biggs, 2000; 2004; Curry, 2001; Cannings, 2002; Nikula, Sones, Stokes, and Stokes, 2002; Nikula, Loose, and Burne, 2003; Legler, Legler, and Westover, 2003; Manolis, 2003; Mead, 2003; Abbott, 2005) discuss habitat selection for most of the nearctic fauna. Less is known for the neotropical species and, in general, new species will likely be found by searching in poorly explored areas and habitats. For example, some gomphids, which are ordinarily rarely seen, can sometimes be common along agricultural stubble bordering strips of forest along stream habitats, and some crepuscular species of *Gynacantha* (Aeshnidae) have been reported to be attracted to Mercury vapor lights at night.

The following habitats may prove useful for tropical sampling of adults and larvae for biodiversity studies: 1) lakes, ponds lagoons; 2) rivers, large streams; 3) springs, seeps, trickles, under wet leaf litter in dried-out depressions; and 4) phytotelmata (water contained in treeholes, bamboo internodes, leaf axils, *Heliconia* inflorescences, fallen nut-husks and palm spathes). Necessary equipment for collecting dragonflies includes an aerial net and a means of temporarily storing specimens. Other useful equipment may include binoculars, a waterproof notebook, and a GPS (Geographic Positioning System device) for recording geographic coordinates. Most adults can be collected with a net, although some species occurring in confined understory areas may actually be collected by hand. A telescopic (2-foot extended to 6-foot) 18-inch aerial net has the advantage of being maneuverable in confined (understory) areas and can easily be packed in suitcases for transport. The well-known procedure of collecting aerial insects is addressed in most standard entomological textbooks. Take time to collect and rear larvae, if possible. This almost always adds taxa and is practical if full-term last instar larvae are placed in small wire-mesh pillow-cages and returned to shallow water for a few days. Here, we add some caveats that may prove useful for dragonfly collection:

Try to collect a series of specimens rather than just one, since more than one cryptic (sibling) species may be present. Likelihood of new discoveries is high and sufficient documentation of individual variation can only be assessed through a series. It also ensures that supplementary material may be available for other researchers and institutions. If specimens are plentiful, try to preserve a few of both sexes in 95 percent ethyl alcohol. They can be used to accurately describe color and they may also be useful for molecular analysis. Photographs of specimens in the field can be used to describe perching habits and accurate body and eye color.

Always take a GPS reading at each collection locality. These data are especially valuable in remote areas and will also allow for data capture in GIS (Geographic Information Systems) programs. Keep complete and clear field notes.

Try to keep specimens alive once captured, since that allows them to void their remaining excrement, and helps ensure that their natural colors are preserved until they are fixed. Place only one specimen per envelope and keep them cool and dry. Envelopes can be placed in large shirt pockets or pouches. Teneral specimens are fragile and preserve poorly. If they must be collected, they should be kept alive in vials thus preventing morphological distortion of body parts that would ensue if placed in field envelopes. Rearing of larval specimens is an excellent way of obtaining material for descriptions of life history. Both exuviae and adults can be preserved in alcohol.

Equipment needed to preserve odonates includes a large, wide-mouth jar, pure acetone, glassine envelopes (9 cm x 5 1/2 cm is best), and disposable insulin syringes. Acetone is a highly flammable solvent and should be used with extreme caution. Never use acetone near an open flame and always work in a well-ventilated area. Acetone is available in hardware stores, but finger nail polish remover may be used as a temporary substitute (acetone is difficult to obtain or even illegal in Pan-Andean countries), later resoaking the specimens in pure acetone.

Specimens should be preserved when alive or freshly killed. We use a syringe to inject live specimens with acetone. The needle can be inserted between the legs or at the side of one leg. Be careful not inject too much acetone, or the abdomen may burst. If syringes are not available, specimens can go directly into an acetone bath as described below. Proper injection of specimens will often help by expanding the taxonomically important caudal appendages allowing for easy observation under a microscope. If possible, use a small insect pin to excise the vesica spermalis, since this structure may aid in the identification of some groups.

Injected specimens are placed in glassine envelopes with wings folded over the back, and steeped in an acetone bath in a jar. Specimens should be left in the acetone bath for 1–2 days. They are then removed from the acetone and from the envelopes and placed on absorbent toweling to dry for about three hours.

Acetoned specimens are often wonderfully preserved: delicate yellows and greens are well-preserved, although the latter may turn yellow. It is best to experiment and see how well (or how poorly) various specimens are preserved, since acetone often destroys pruinosity. The acetone extracts the body fats; some specimens (members of the subfamily Libellulinae, for example) have a naturally high fat content. Such specimens may need to go into two or three clean acetone baths. The bath should be changed when it becomes yellow, or when its efficacy at removing grease diminishes. Specimens with shiny (greased) wings may be cleaned up similarly using an acetone bath.

We keep preserved specimens in clear plastic envelopes with a 3" x 5" index card on which complete collection data are typed. Specimens are filed on end like library index cards. However, dried acetoned specimens can also be pinned and placed into a collection box or drawer.

Using the keys

This work is intended to provide an illustrated and annotated key to all 124 anisopteran genera of the New World (North, Middle and South America). Anyone with general knowledge on keying out insects and with access to a stereoscopic microscope should be able to identify any New World anisopteran to genus. Please note that some of the keys contain a third alternative (*n"*) in addition to the usual two.

We have endeavored to make the keys as user friendly as possible. Almost all characters are illustrated from voucher material. The greatest impediment to accurately identifying material is improperly collected and preserved material. Field collected specimens often become laterally compressed (crushed), hiding or distorting structures such as the female vulvar lamina, cerci, and structures within the male genital pocket. Poorly preserved specimens may need to be relaxed or softened in warm water (and afterwards dehydrated in acetone again) so that their important taxonomic characters can be seen. For example, it may be necessary to gently spread the cerci in males so that their ventral surfaces can be seen or the anterior hamules and vesica spermalis extruded with an insect pin or watch forceps so that their structures can be viewed. Structures of the vesica may be difficult to resolve unless they are adequately illuminated.

8 male abdomen – *Rhionaeschna brasiliensis*   dorsal view

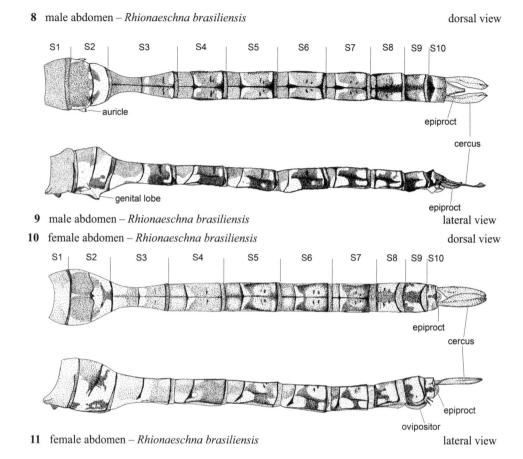

9 male abdomen – *Rhionaeschna brasiliensis*   lateral view
10 female abdomen – *Rhionaeschna brasiliensis*   dorsal view

11 female abdomen – *Rhionaeschna brasiliensis*   lateral view

Specimens will be encountered that will not place correctly. Isolated females may be especially difficult to identify. If you are unable to confidently reach a choice, check your specimen against the generic account and illustrations. Distribution summaries for some geographically restricted genera may also aid in properly placing a specimen to genus, however, be aware that the known distribution of some Latin American genera is poorly known. Once you have made a generic identification, consult the list of papers relevant to the genus in question for species identification.

Finally, do not assume that every specimen will be assignable to an existing genus. While we have tried to include intrageneric variation by examining as many species as possible, there may be exceptions which we have unintentionally excluded or there may be the possibility of finding a new species not belonging to any described genus.

One final warning: Some photographs of odonates on the web have been used to document range extensions or new species. We strongly suggest that voucher specimens be collected and confirmed before these records are introduced into the literature.

Where to go from here

We hope that our keys will stimulate further interest in understanding the biodiversity of this group in the neotropics. Reference to revisions and/or papers treating adults of various species will be the next step in identifying collected material; the generic accounts will point the way to groups which are in need of revision. The larvae of only a few of the neotropical species are known and most of these from isolated descriptions. Taxonomic revisions of the immature stages of neotropical species are urgently needed so that aquatic ecologists can use odonate larvae as subjects in their research.

Main repositories for New World Odonata collections

Several of these collections have been and will continue to be indispensable for students of the order in comparing material to authoritatively identified specimens. Descriptions of new taxa often involve comparison with primary type material and a few institutions have published type catalogs which we reference here.

12 male genital fossa – *Limnetron antarcticum*
ventral view

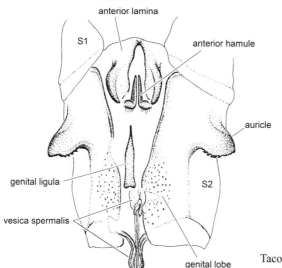

13 male vesica spermalis
Gomphomacromia paradoxa
lateral view

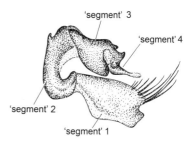

European Collections:

BMNH: British Museum of Natural History, Dept. of Entomology, British Museum (Natural History), Cromwell Road, London SW7 5BD, ENGLAND, U.K. Large collection, mostly pinned and of old material, but containing type material described by Calvert (1901-1906) in the Biologia Centrali Americana, and by other authorities. Type catalogs: Kimmins, 1966; 1969a; 1969b; 1970.

FNS: Forschungsinstitut und Naturmuseum Senckenberg, Senckenberganlage 25, D-60326, Frankfurt-am-Main, GERMANY. Contains the papered collection of Friedrich Ris who worked on Odonata from around the world.

IRSN: Institut Royal des Sciences Naturelles de Belgique, 29, rue Vautier, B-1000 Bruxelles, BELGIUM. Very old collection containing the Odonata of Baron Edmund de Selys-Longchamps, considered the "Father of Odonatology" who described approximately one-fifth of the world's Odonata. Almost all of the collection is pinned and is rich in type material although no listing has ever been made.

RNHL: Rijksmuseum van Natuurlijke Historie, Department of Entomology, Rijksmuseum van Natuurlijke Historie, Raamsteeg 2, Postbus 9517, NL 2300 RA Leiden, THE NETHERLANDS. Contains the collections of Jean Belle and Dirk Geijskes who collected for many years in Surinam. The collection is especially rich in neotropical Gomphidae. Type catalog: Lieftinck, 1971.

North American Collections:

DRP (registered with FSCA): Dennis R. Paulson personal collection, Slater Museum of Natural History, University of Puget Sound, 1500 North Warner,

Tacoma, WA 98416, U.S.A. A large, well-curated, data-based collection mostly of New World Odonata put together by an experienced odonatologist over the last forty years.

FSCA: Florida State Collection of Arthropods, 1911 SW 34th Street, Gainesville, FL 32608-1268, U.S.A. One of the largest holdings of New World Odonata in the United States, collected over the past sixty years. Collection includes over 150 primary types, listed on line at: <http://www.afn.org/~iori/odontype.html>.

MCZ: Museum of Comparative Zoology, Harvard University, Cambridge, Massachusetts 02138, U.S.A. Very rich, old, pinned collection containing type material of Herman Burmeister, Toussaint de Charpentier, Herman Hagen, and others. Listing of over 550 types (some with images) on line at: <http://mcz-28168.oeb.harvard.edu/mcz/>.

RWG (registered with USNM): Rosser W. Garrison personal collection, Plant Pest Diagnostics, California Department of Food and Agriculture, 3294 Meadowview Road, Sacramento, CA 95832-1448, USA. Well-curated collection, especially rich in New World material with emphasis on neotropical fauna.

TWD (registered. with FSCA): Thomas W. Donnelly personal collection, 2091 Partridge Lane, Binghamton NY 13903, U.S.A. Well-curated, worldwide collection containing numerous New World specimens collected over the past 50 years.

UMMZ: University of Michigan, Museum of Zoology, 1109 Geddes Avenue, Ann Arbor, MI 48109-1079, U.S.A. One of the largest holdings of Odonata in the world, containing the collections of F. Förster, C.H. Kennedy, E.B. Williamson, and others. Particularly rich in neotropical material though vast amounts of this material await study. Type catalog: Garrison, von Ellenrieder, and O'Brien (2003).

UNAM: Instituto de Biología, Departamento de Zoología, Apartado Postal 70-153, MX-04510, D. F. Large, well-curated, papered collection of Odonata from Mexico put together mostly by E. González-Soriano.

USNM: National Museum of Natural History, Smithsonian Institution, Washington, D.C. 20560, U.S.A. A large well-curated world-wide collection containing numerous neotropical material collected by Oliver S. Flint, Jr. over the past 40 years. Type catalog: Flint (1991).

South American Collections:

ABMM: Angelo B.M. Machado personal collection, Departamento de Zoologia, Instituto de Ciências Biológicas, Universidade Federal de Minas Gerais, Caixa Postal 486, BR-31270-901, Belo Horizonte - Minas Gerais, BRAZIL. Private collection including types of Brazilian material.

MIZA: Museo del Instituto de Zoología Agrícola "Francisco Fernández Yépez", Maracay, VENEZUELA. Particularly rich collection of papered Venezuelan Odonata including collected material by the late J. Rácenis and by Jürg De Marmels.

MLP: Museo de Ciencias Naturales de La Plata, Departamento de Entomología, Paseo del Bosque s/n, 1900 La Plata, ARGENTINA. Rich collection including types of Argentine species.

MNRJ: Museu Nacional, Deptamento de Entomología, Museu Nacional, Univ. Federal Río de Janeiro, Quinta da Boa Vista, BR-20942-040, Río de Janeiro, RJ, BRAZIL. One of the largest Odonata collections in South America containing the collections of Santos and J.M. Costa. Type catalog: Costa and Mascarenhas (1998).

2. Key to families

1. Eyes on top of head separated by space equal to distance between lateral ocelli (Fig. 14) **2**

1'. Eyes on top of head meeting at a single point or separated by space much shorter than distance between lateral ocelli (Fig. 15) [with the exception of *Diastatops*, see Page 235, Fig. 1449] **3**

1". Eyes on top of head meeting for a considerable distance, forming an eye seam (Fig. 16) **5**

2(1). HW pterostigma starting at distal 1/3 of postnodal portion of wing or further distally (Fig. 17); female ovipositor modified as vulvar laminae (Fig. 19); prementum entire (Fig. 21) ....
.... **Gomphidae** (Page 65)

2'. HW pterostigma starting at about 1/2 of postnodal portion of wing (Fig. 18); female ovipositor sawlike for endophytic oviposition (Fig. 20); prementum cleft (Fig. 22) .... **Petaluridae** (Page 13)

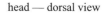
head — dorsal view

eyes widely separated
14 *Perigomphus pallidistylus*

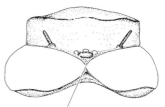
eyes close together or touching
15 *Neopetalia punctata*

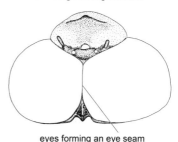
eyes forming an eye seam
16 *Limnetron antarcticum*

postnodal portion of HW

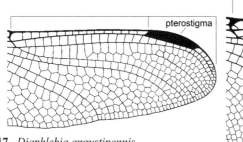
pterostigma

17 *Diaphlebia angustipennis*

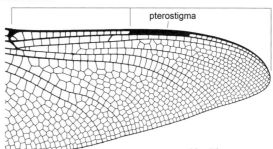
pterostigma

18 *Phenes raptor*

distal end of female abdomen

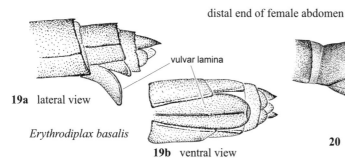
vulvar lamina

19a lateral view

Erythrodiplax basalis

19b ventral view

sawlike ovipositor

20 *Gynacantha bartai*
lateral view

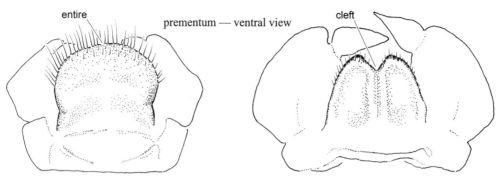

entire   prementum — ventral view   cleft

21 *Peruviogomphus moyobambus*   **22** *Phenes raptor*

3(1'). Pterostigma bracevein present (Fig. 23); wings with costal series of 5–8 red or reddish-brown spots (Fig. 23); male mesotibial spines gradually tapering distally (Fig. 25) **4**

3'. Pterostigma bracevein absent (Fig. 24); wings lacking costal series of 5–8 red or reddish-brown spots (Fig. 24); male mesotibial spines approximately quadrangular (Fig. 26) .... **Cordulegastridae** (Page 135)

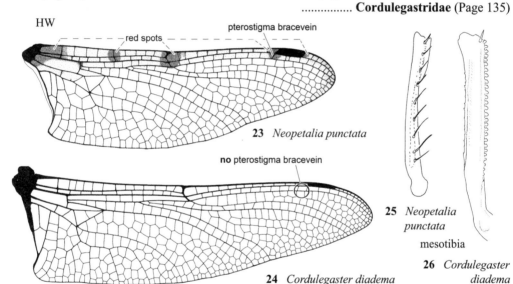

HW   red spots   pterostigma bracevein

23 *Neopetalia punctata*

no pterostigma bracevein

25 *Neopetalia punctata*

mesotibia

26 *Cordulegaster diadema*

24 *Cordulegaster diadema*

4(3). Abdominal terga 5–8 in male and 2–8 in female with ventro-apical tufts of black hairs (Fig. 27); male posterior hamule tip bilobed (Fig. 28); male anterior lamina lacking elongate median cleft (Fig. 28); vertex forming a prominent tubercle between ocelli (Fig. 30) .... **Neopetaliidae** (Page 133)

4'. Abdominal terga 5–8 in male and 2–8 in female lacking apical tufts of black hairs; male posterior hamule tip not bilobed (Fig. 29); male anterior lamina with elongate median cleft (Fig. 29); vertex approximately flat between ocelli (Fig. 31) .... **Austropetaliidae** (Page 19)

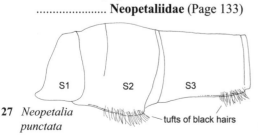

S1   S2   S3

27 *Neopetalia punctata*   tufts of black hairs

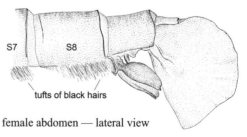

S7   S8

tufts of black hairs

female abdomen — lateral view

male genital fossa — ventral view

28 *Neopetalia punctata*   **29** *Phyllopetalia pudu*

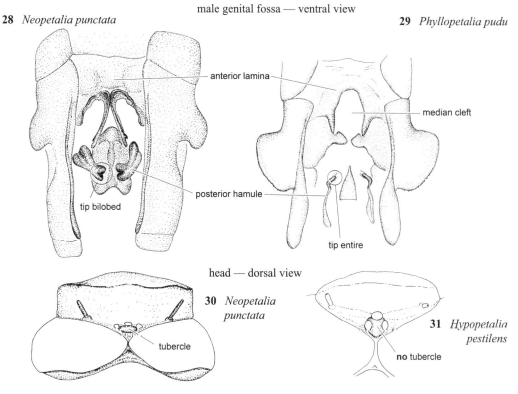

anterior lamina

median cleft

posterior hamule

tip bilobed

tip entire

head — dorsal view

30 *Neopetalia punctata*

tubercle

31 *Hypopetalia pestilens*

no tubercle

5(1"). HW triangle at about same distance from arculus (**B**) as FW triangle (**A**) (Fig. 32); FW triangle longitudinally elongated (Fig. 32); male anterior lamina with elongate median cleft (as in Fig. 29); female ovipositor sawlike for endophytic oviposition (Fig. 20) ..... **Aeshnidae** (Page 25)

5'. HW triangle closer to arculus (**B**) than HW triangle (**A**) (Fig. 33); FW triangle transversely elongated (Fig. 33); male anterior lamina lacking elongate median cleft (as in Fig. 28); female ovipositor modified as vulvar laminae (Fig. 19) .... **Libellulidae** (Page 139)

HW — antenodal portion

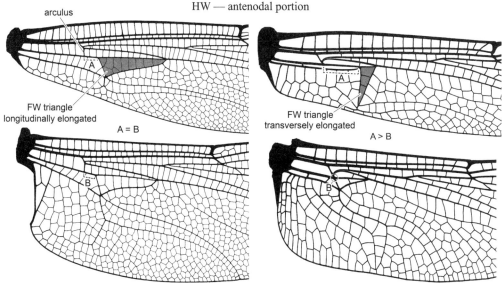

arculus

A

FW triangle longitudinally elongated

A = B

FW triangle transversely elongated

A > B

B

B

32 *Neuraeschna calverti*   **33** *Libellula herculea*

3. Petaluridae

Nearctic, austral and palearctic regions: 11 spp. in 5 genera.

New World: 3 spp. in 3 genera (all endemic).

Diagnostic characters: Prementum cleft; labial palps with spatulate end hooks (Carle, 1995; Fig. 49); eyes separated by more than distance between lateral ocelli; occiput rounded, tumid posteriorly (Needham, Westfall, and May, 2000; Fig. 60); pterostigma relatively long and narrow, occupying 12–15 % of wing length (Bechly, 1996), situated at middle third of postnodal portion of wing (Figs. 53, 57, 63); triangles crossed; supratriangles free; FW subtriangles with 3 crossveins, of HW free; no basal subcostal cross-veins; sectors of arculus separated (Figs. 53, 57, 63); bladelike gonapophyses in females, forming a well developed endophytic ovipositor (Figs. 48, 66); larvae with tibial spurs and apical burrowing hooks on legs, and abdominal segments with paired lateral tubercles.

Status of classification: Petaluridae is recognized as a monophyletic group composed of a few relictual species. Its placement within the Anisoptera is controversial; Lohmann (1996) considered this family the sister group of the Gomphidae and Libelluloidea; Carle (1996) and Bechly (1996) the sister group of the remaining anisopteran families.

Key to males

1. Rear of occiput with a pair of small horns (Fig. 34); ventral portion of mesepimeron with a spine (Fig. 37); brace vein proximal to pterostigma (Fig. 40); epiproct long, narrow and upturned beyond level of cerci (Fig. 44); distribution: Chile and SW Argentina (Map 1) .... *Phenes* (Page 15)

1'. Rear of occiput entire (Fig. 36); ventral portion of mesepimeron rounded (Fig. 38); brace vein at proximal end of pterostigma (Fig. 41); epiproct short, broad, not extending beyond cerci (Figs. 45–46); distribution: N America (Maps 2–3) **2**

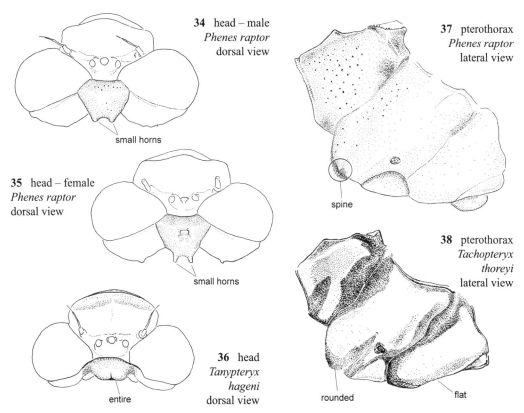

34 head – male
Phenes raptor
dorsal view

small horns

37 pterothorax
Phenes raptor
lateral view

spine

35 head – female
Phenes raptor
dorsal view

small horns

38 pterothorax
Tachopteryx thoreyi
lateral view

36 head
Tanypteryx hageni
dorsal view

entire

rounded

flat

39 pterothorax – *Tanypteryx hageni*

42 epiproct – *Tanypteryx hageni*

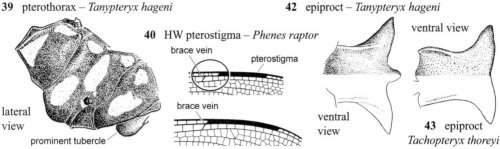

40 HW pterostigma – *Phenes raptor*

brace vein

pterostigma

brace vein

lateral
view

prominent tubercle

ventral view

ventral
view

43 epiproct
Tachopteryx thoreyi

41 HW pterostigma – *Tachopteryx thoreyi*

2(1). Epiproct with a pair of small spines on its dorsal surface (Fig. 45); venter of thorax with a reduced tubercle or flat (Fig. 38); epiproct bifid (Fig. 43); distribution: E United States (Map 2) .... ***Tachopteryx*** (Page 16)

2'. Epiproct lacking a pair of small spines on its dorsal surface (Fig. 46); venter of thorax with a prominent rounded tubercle (Fig. 39); epiproct trifid (Fig. 42); distribution: W United States and SW Canada (Map 3) .... ***Tanypteryx*** (Page 17)

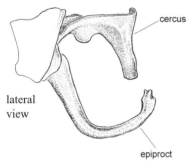

cercus

lateral
view

epiproct

44 caudal appendages – *Phenes raptor*

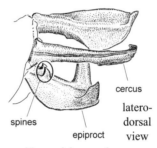

cercus

spines

epiproct

latero-
dorsal
view

45 caudal appendages
Tachopteryx thoreyi

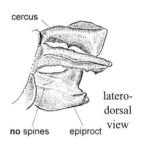

cercus

no spines   epiproct

latero-
dorsal
view

46 caudal appendages
Tanypteryx hageni

Key to females

1. Rear of occiput with a pair of small horns (Fig. 35); ventral portion of mesepimeron with a spine (as in Fig. 37); brace vein proximal to pterostigma (as in Fig. 40); distribution: Chile and SW Argentina (Map 1) .... ***Phenes*** (Page 15)

1'. Rear of occiput entire (as in Fig. 36); ventral portion of mesepimeron rounded (as in Figs. 38–39); brace vein at proximal end of pterostigma (as in Fig. 41); distribution: N America (Maps 2–3) **2**

2(1). S10 directed dorsally (Fig. 47); venter of thorax with a prominent rounded tubercle (as in Fig. 39); distribution: W United States and SW Canada (Map 3) .... ***Tanypteryx*** (Page 17)

2'. S10 in the same plane as previous segments (as in Fig. 48); venter of thorax with a reduced tubercle or flat (as in Fig. 38); distribution: E United States (Map 2) .... ***Tachopteryx*** (Page 16)

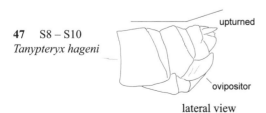

47 S8 – S10
Tanypteryx hageni

upturned

ovipositor

lateral view

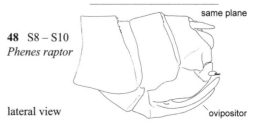

48 S8 – S10
Phenes raptor

same plane

lateral view

ovipositor

Phenes Rambur, 1842: 175.
[♂ pp. 13, couplet 1; ♀ pp. 14, couplet 1]
Type species: *Phenes raptor* Rambur, 1842 [by monotypy]
1 species:

raptor Rambur, 1842* – **L** [Schmidt, 1941b; Needham and Bullock, 1943]
  syn. *raptor centralis* Jurzitza, 1989

References: Fraser, 1933.
Distribution: Central and S Chile and SW Argentina.

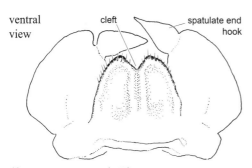

Map 1. Distribution of *Phenes* sp.

Large petalurids (55-88 mm); pterothorax largely pale brown or yellow to gray with an irregular small dark brown spot anterior to upper end of lateral sutures (Fig. 52); abdomen largely pale brown to gray with irregular bands of brown on posterior 1/2 of segments; frontal portion of postfrons smoothly rounded (Fig. 51). Wings (Fig. 53) hyaline, 2 Cu-A crossveins. Male cercus bent ventrally at a right angle, with a subbasal tooth (Fig. 54); anterior hamule small, posterior hamule long and arcuate. Female S10 in the same plane as previous segments (Fig. 55). **Unique characters**: Rear of occiput with a pair of small horns, and with a medial conical tubercle on dorsal surface in female (Figs. 50–51); lobes of anteclypeus acutely pointed (Fig. 51); mesepisternum coarsely denticulate and ventral portion of mesepimeron with a spine (Fig. 52); anal loop not defined, brace vein proximal to pterostigma (Fig. 53); male epiproct long, narrow and upturned beyond level of cerci (Fig. 54).
Status of classification: Easily recognized species.
Potential for new species: Very unlikely.
Habitat: Adults fly over streams in narrow shady canyons, and frequent sunny clearings in forests (Joseph, 1928; Svihla, 1960a). Larvae have been found living in moist debris under damp pieces of wood on the forest ground (Garrison and Muzón, 1995).

ventral view   cleft   spatulate end   hook

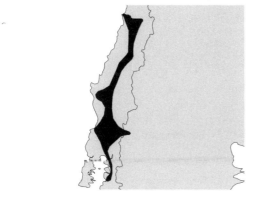

49   prementum – male *Phenes raptor*

50   head – male *Phenes raptor*

dorsal view

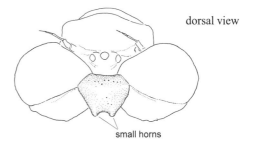

small horns

lateral view

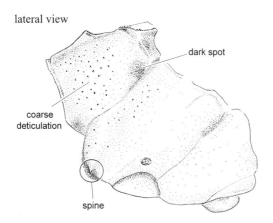

dark spot

coarse deticulation

spine

52   pterothorax – *Phenes raptor*

medial conical tubercle

frontal view

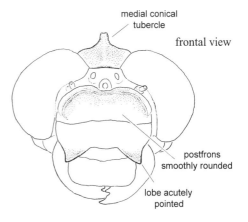

postfrons smoothly rounded

lobe acutely pointed

51   head – female *Phenes raptor*

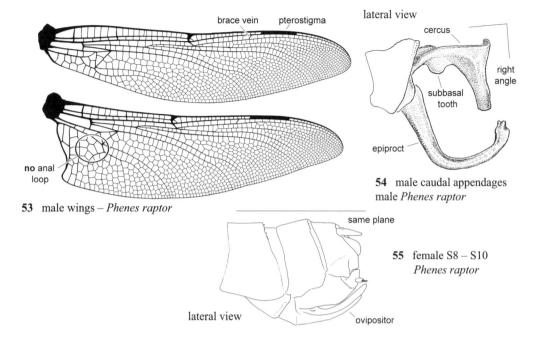

brace vein   pterostigma

lateral view
cercus
right angle
subbasal tooth
epiproct

no anal loop

53 male wings – *Phenes raptor*

54 male caudal appendages
male *Phenes raptor*

same plane

55 female S8 – S10
Phenes raptor

lateral view   ovipositor

Tachopteryx Uhler *in* Selys, 1859: 551 (25 reprint).
[♂ pp. 14, couplet 2; ♀ pp. 14, couplet 2]
Type species: *Uropetala thoreyi* Hagen *in* Selys, 1859
[by monotypy]
1 species:

thoreyi (Hagen *in* Selys, 1858)* – **L** [Williamson, 1901; Kennedy, 1917]

References: Fraser, 1933; Needham, Westfall, and May, 2000.
Distribution: E United States.

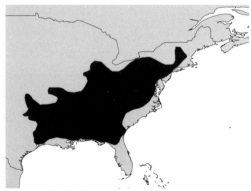

Map 2. Distribution of *Tachopteryx* sp.

Large petalurids (71-80 mm); pterothorax largely pale brown or gray with dark brown stripes along sutures (Fig. 56); abdomen largely pale brown to gray with irregular bands of brown on posterior 1/2 of segments; lobes of anteclypeus rounded (as in Fig. 61); frontal portion of postfrons with a transverse sinus

(shared with *Tanypteryx*, as in Fig. 61); occiput entire (shared with *Tanypteryx*, as in Fig. 60). Wings (Fig. 57) hyaline, 1–3 Cu-A crossveins. Male cercus linear, with a sub-basal tooth (Fig. 59); anterior hamule small, posterior hamule long and arcuate. Female S10 in the same plane as previous segments (as in Fig. 55). **Unique characters**: Male epiproct trifid (Fig. 58) and with a pair of small spines on its dorsal surface (Fig. 59).
Status of classification: Easily recognized species.
Potential for new species: Very unlikely.
Habitat: Larvae found under leaves on wet soil in surface burrows where ground is kept perpetually moist by ground water seepage (Louton, 1982b). Males perch in sunny spots on tree trunks, where they defend one-day territories (Dunkle, 2000). Females oviposit among roots of grasses in decaying vegetable matter above water surface (Needham, Westfall, and May, 2000).

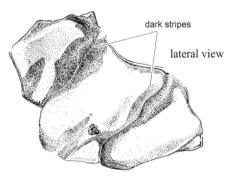

dark stripes

lateral view

56 pterothorax – *Tachopteryx thoreyi*

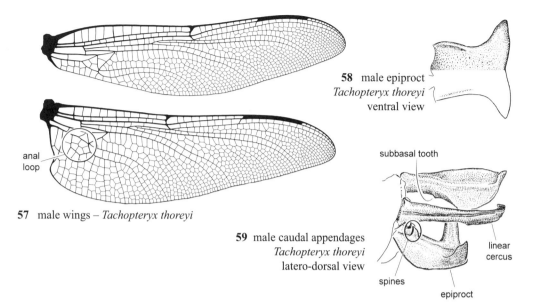

57 male wings – *Tachopteryx thoreyi*

anal loop

58 male epiproct
Tachopteryx thoreyi
ventral view

59 male caudal appendages
Tachopteryx thoreyi
latero-dorsal view

subbasal tooth

linear cercus

spines

epiproct

Tanypteryx Kennedy, 1917: 507.

[♂ pp. 14, couplet 2; ♀ pp. 14, couplet 2]

Type species: *Tachopteryx hageni* Selys, 1879 [by original designation]
2 species, 1 New World species:

hageni (Selys, 1879)* – **L** [Svihla, 1958]

References: Kennedy, 1917; Fraser, 1933, Needham, Westfall, and May, 2000.
Distribution: W United States and SW Canada.
[NOTE: There is a record of a single female from Missoula, Montana, that is considered doubtful by Miller and Gustafson, 1996.]

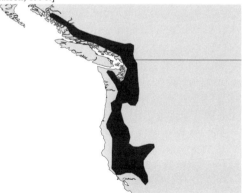

Map 3. Distribution of *Tanypteryx* sp.

Medium petalurids (54–59 mm), largely black with yellow spots on pterothorax and abdomen; lobes of anteclypeus rounded; frontal portion of postfrons with a transverse sinus (shared with *Tachopteryx*, Fig. 61); occiput entire (shared with *Tachopteryx*, Fig. 60). Wings (Fig 63) hyaline, 1–3 Cu-A crossveins. Male cercus flattened, with medio-ventral and postero-ventral teeth (Fig. 64); anterior hamule small, posterior hamule long and arcuate. **Unique characters**: Venter of thorax with a prominent rounded tubercle (Fig. 62); male epiproct trifid (Fig. 65); female S10 directed dorsally (Fig. 66).
Status of classification: Easily recognized species.
Potential for new species: Very unlikely.
Habitat: Mossy bogs, where adults perch on trees and other vertical surfaces. Larvae live in 'L'-shaped burrows in permanent seepage areas located on hillsides (Svihla, 1960b).

60 head –*Tanypteryx hageni*

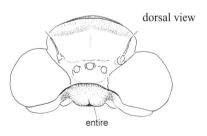

dorsal view

entire

61 head – *Tanypteryx hageni*

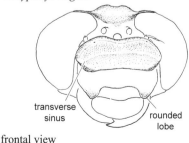

transverse sinus

rounded lobe

frontal view

62 pterothorax – *Tanypteryx hageni*

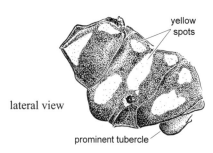

yellow
spots

lateral view

prominent tubercle

63 male wings – *Tanypteryx hageni*

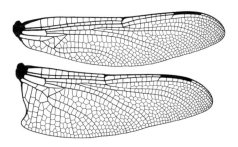

64 male caudal appendages – *Tanypteryx hageni*

flattened cercus

ventral view

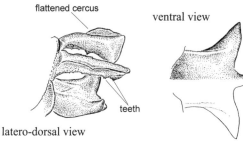

teeth

latero-dorsal view

65 male epiproct
Tanypteryx hageni

66 female S8 – S10 – *Tanypteryx hageni*
lateral view

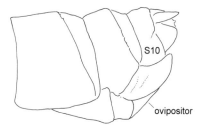

S10

ovipositor

4. Austropetaliidae

Austral region: 10 spp. in 4 genera, in Australia, Tasmania and SW South America.

New World: 7 spp. in 2 genera (all endemic).

Diagnostic characters: Prementum cleft; eyes meeting on top of head at a single point, vertex approximately flat between ocelli (Fig. 75); wings with costal series of 5–8 red or reddish-brown spots (Figs. 67–68); leaf like lateral expansions of terga S7–8 in adults (Figs. 94–96); vesica spermalis with pendulous distal segment and postero-ventrally directed paired flagella (Figs. 81, 92); and distally widened larval labrum; massive ventro-lateral development of larval occipital ridge; dorsally excrescent femora; obsolete transverse larval abdominal muscles; extensively granulate larval body surface; and well developed lateral lobes on larval S1–10 (Carle, 1996).

Status of classification: Austropetaliids were linked for many years to the libelluloid *Neopetalia punctata* (Hagen) within family Neopetaliidae, which they resemble superficially in wing color pattern, until Carle and Louton (1994) made the aeshnoid lineage of Austropetaliidae evident and placed them in their own family. There are no doubts as to the close association of austropetaliids to aeshnids (Carle and Louton, 1994; Bechly, 1996; Carle, 1996), both sharing anterior lamina with elongate anterior cleft, anterior hamules directed medially, vestigial posterior hamules, abdomen bearing dorso-longitudinal carinae, and proventricular lobes of larval gizzard small and mound-like with 8 or fewer clustered teeth. Molecular data also support this relationship (Misof, Rickert, Buckley, Fleck, and Sauer, 2001).

Key to males

1. FW costal area with 7 reddish spots, subtriangle with 2–3 cells (Fig. 67); dorsum of pterothorax with coarse tubercles, pterothorax with 3 lateral whitish spots (Fig. 69); auricles about as long (**A**) as wide (**B**) (Fig. 71) ....
.... ***Hypopetalia*** (Page 20)

1'. FW costal area with 5–6 reddish spots, subtriangle with 1 cell (Fig. 68); dorsum of pterothorax smooth, pterothorax with dorsal and lateral yellowish stripes (Fig. 70); auricles about 1 1/2–2 times as wide (**B**) as long (**A**) (Fig. 72) ....
.... ***Phyllopetalia*** (Page 22)

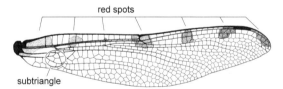

67 FW – *Hypopetalia pestilens*

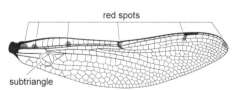

68 FW – *Phyllopetalia apicalis*

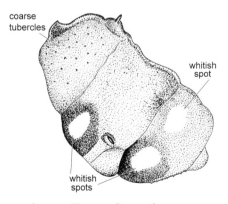

69 pterothorax – *Hypopetalia pestilens*
lateral view

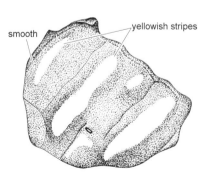

70 pterothorax – *Phyllopetalia apicalis*
lateral view

71 genital fossa – *Hypopetalia pestilens*
ventral view

72 genital fossa – *Phyllopetalia pudu*
ventral view

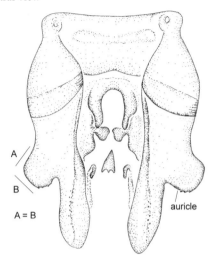

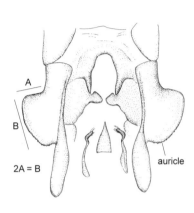

Key to females

1. FW costal area with 7 reddish spots, subtriangle with 2-3 cells (as in Fig. 67); dorsum of pterothorax with coarse tubercles, pterothorax with 3 lateral whitish spots (as in Fig. 69); sternum of S10 smooth (Fig. 73) ....
.... ***Hypopetalia*** (Page 20)

1'. FW costal area with 5-6 reddish spots, subtriangle with 1 cell (as in Fig. 68); dorsum of pterothorax smooth; pterothorax with dorsal and lateral yellowish stripes (as in Fig. 70); sternum of S10 with denticles (Fig. 74) ....
.... ***Phyllopetalia*** (Page 22)

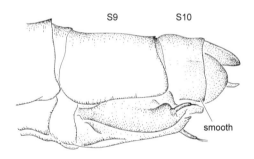

73 S9–10 – *Hypopetalia pestilens*
ventro-lateral view

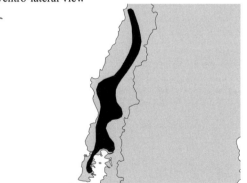

74 S9–10 – *Phyllopetalia apollo*
ventro-lateral view

Hypopetalia McLachlan, 1870: 170.
[♂ pp. 19, couplet 1; ♀ pp. 20, couplet 1]
Type species: *Hypopetalia pestilens* McLachlan, 1870 [by monotypy]
1 species:

pestilens McLachlan, 1870* – **L** [Schmidt, 1941b]

References: Schmidt, 1941b; Carle, 1996.
Distribution: S Chile.

Map 4. Distribution of *Hypopetalia* sp.

Large austropetaliids (78–86 mm), dark brown with 3 whitish spots on sides of pterothorax. Lateral ocelli and eyes separated (for a distance equal to width of ocellus, Fig. 75), occipital triangle with posterior margin transverse and bearing a dorsally directed spine (Fig. 76). Wings (Fig. 78) hyaline or tinged with brown, FW with 7 reddish spots along costal area; 2–3 celled subtriangles; brace vein proximal to pterostigma. Male auricles about as long (**A**) as wide (**B**) (Fig. 82); male cerci flat and shorter than epiproct (Fig. 79); contour of lateral projections of basal segment of vesica spermalis 'V-shaped' in posterior view (Fig. 80); female sternum S10 smooth (Fig. 73). **Unique characters**: Pterothorax with hyperde-veloped medio-dorsal carina and coarse tubercles on dorsum, and with 3 lateral whitish spots (Fig. 77); lateral margins of central depression of vesica spermalis basal segment (Fig. 81) moderately uplifted (projected portion (**A**) not higher than segment height (**B**) in lateral view), and not surpassing level of genital lobe in lateral view.

Status of classification: Good.

Potential for new species: Unlikely.

Habitat: Adults fly in dense mountain forest, following small streams close to water in darkest canyons, following one another (Peña, 1968; Jurzitza, 1989). Larva described by Schmidt (1941b) without indication of habitat.

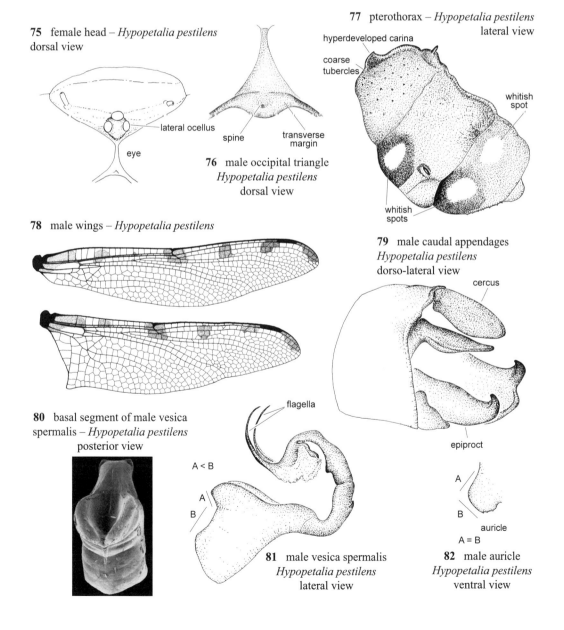

75 female head – *Hypopetalia pestilens* dorsal view

lateral ocellus

eye

76 male occipital triangle *Hypopetalia pestilens* dorsal view

spine

transverse margin

77 pterothorax – *Hypopetalia pestilens* lateral view

hyperdeveloped carina

coarse tubercles

whitish spot

whitish spots

78 male wings – *Hypopetalia pestilens*

79 male caudal appendages *Hypopetalia pestilens* dorso-lateral view

cercus

epiproct

80 basal segment of male vesica spermalis – *Hypopetalia pestilens* posterior view

flagella

A < B

A
B

81 male vesica spermalis *Hypopetalia pestilens* lateral view

A
B

auricle
A = B

82 male auricle *Hypopetalia pestilens* ventral view

Phyllopetalia Selys, 1858: 549.

[♂ pp. 19, couplet 1; ♀ pp. 20, couplet 1]

Type species: *Phyllopetalia stictica* Hagen *in* Selys, 1858 [by subsequent designation by Kirby, 1890: 82]

    syn *Rheopetalia* Carle, 1996: 236
        Type species: *Rheopetalia rex* Carle, 1996 [by original designation]
    syn *Odontopetalia* Carle, 1996: 236, 237
        Type species: *Odontopetalia apollo* Selys, 1878 [by original designation]
    syn *Eurypetalia* Carle, 1996: 237
        Type species: *Eurypetalia altarensis* Carle, 1996 [by original designation]
    syn *Ophiopetalia* Carle, 1996: 237
        Type species: *Ophiopetalia diana* Carle, 1996 [by original designation]
6 species:

altarensis Carle, 1996 [*Eurypetalia*]*
apicalis Selys, 1858*
    syn *decorata* McLachlan *in* Selys, 1878
    syn *rex* Carle, 1996 [*Rheopetalia*]
apollo Selys, 1878*
excrescens Carle, 1996 [*Eurypetalia*]*
pudu Dunkle, 1985*
    syn *araucana* Carle, 1996 [*Ophiopetalia*]
    syn *auregaster* Carle, 1996 [*Ophiopetalia*]
    syn *diana* Carle, 1996 [*Ophiopetalia*]
stictica Hagen *in* Selys, 1858* – **L?** [Schmidt, 1941b]

References: Schmidt, 1941b; Dunkle, 1985; Carle, 1996; von Ellenrieder, 2005.
Distribution: S Chile and SW Argentina.

Medium to large austropetaliids (57–78 mm), characterized by posterior margin of occipital triangle either transverse (Fig. 83) or anteriorly projected with (Figs. 85–86) or without (Fig. 84) a spine; pterothorax laterally and dorsally with yellow stripes (Fig. 88); FW (Figs. 89-90) with 5–6 reddish spots along costal area, bracevein at proximal end of pterostigma, and subtriangles with 1 cell. Lateral margins of

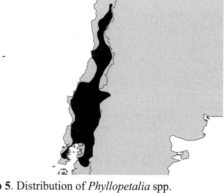

Map 5. Distribution of *Phyllopetalia* spp.

central depression of basal segment of vesica spermalis strongly uplifted (projected portion higher than segment height in lateral view), surpassing level of genital lobe in lateral view (Fig. 92); male auricles oblong, about 1.5–2 times as wide as long (Fig. 91); male cerci flat and foliaceus, with (Figs. 97–98) or without (Fig. 99–100) subbasal teeth; female sternum S10 with denticles (Fig. 101). **Unique characters**: Lateral ocelli and eyes almost contiguous (Fig. 87); contour of lateral projections of basal segment of vesica spermalis 'U-shaped' in posterior view (Fig. 93).

Status of classification: Confusion has characterized the taxonomic history of this small group of spectacular dragonflies (Carle, 1996; von Ellenrieder, 2005). The abundantly illustrated review by von Ellenrieder (2005) allows for species identification.
Potential for new species: Unlikely.
Habitat: Flight following narrow streams within forest and along forest paths and roads, close to water surface; feeding flight observed during late afternoon and dusk over pasture lands. According to Carle (1996) the larvae are semiterrestrial, and he found them at small seepage slopes, small springs, clinging to underside of damp rocks and sticks along tiny rivulets, and under rocks and logs, wet or submerged, in streams less than 50 cm wide. However, only one larva has been described so far (*P. stictica*, by supposition by Schmidt, 1941b).

83 occipital triangle
Phyllopetalia apicalis
dorsal view

84 occipital triangle
Phyllopetalia excrescens
dorsal view

margin anteriorly
projected

margin transverse

85 occipital triangle
Phyllopetalia pudu dorsal view

86 occipital triangle
Phyllopetalia stictica
dorsal view

spine

spine

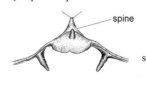

87 head – *Phyllopetalia altarensis* dorsal view

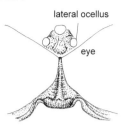

88 pterothorax – *Phyllopetalia apicalis* lateral view

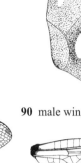

89 male wings – *Phyllopetalia apicalis*

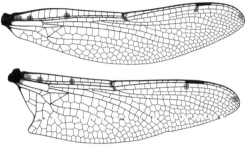

90 male wings – *Phyllopetalia pudu*

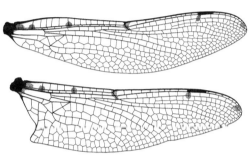

91 auricle
Phyllopetalia pudu
lateral view

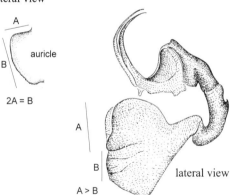

92 male vesica spermalis – *Phyllopetalia pudu*

93 basal segment of
male vesica spermalis
Phyllopetalia apicalis
posterior view

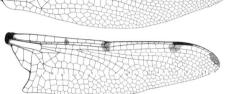

94 S7–8 – male *Phyllopetalia apicalis*
lateral view

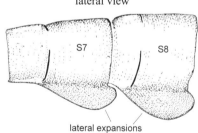

95 S7–8 – male *Phyllopetalia pudu*
lateral view

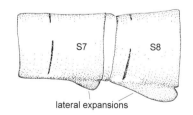

96 S7–8 – female *Phyllopetalia pudu*
lateral view

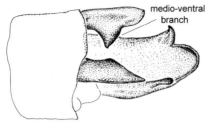

97   male caudal appendages – *Phyllopetalia apollo* dorso-lateral view

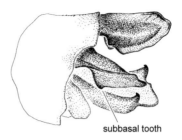

98   male caudal appendages – *Phyllopetalia pudu* dorso-lateral view

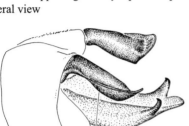

99   male caudal appendages – *Phyllopetalia apicalis* dorso-lateral view

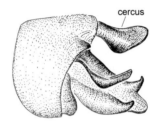

100   male caudal appendages *Phyllopetalia excrescens* dorso-lateral view

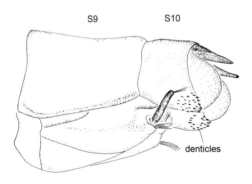

101   female S9–10 – *Phyllopetalia apollo* ventro-lateral view

5. Aeshnidae

Cosmopolitan: About 408 spp. in 50 genera.

New World: 152 spp. in 20 genera (14 endemic).

Diagnostic characters: Prementum cleft or entire; eyes in contact for some distance on top of head, forming an eye-seam (Fig. 115); vertex forming a prominent tubercle between ocelli (Figs. 129-130); FW and HW triangles similar in shape; terga of S2–3 with latero-longitudinal carinae; male anterior hamule with anteriorly directed process (Fig. 107); male distal segment of vesica spermalis with lateral lobes, and median lobes separated from each other dorsally throughout their length; bladelike gonapophyses forming a well developed endophytic ovipositor in females (Fig. 199); larvae with basal lamina of epiproct only in males.

Status of classification: The family was recently subject to phylogenetic analyses (Bechly, Nel, Martínez-Delclôs, Jarzembowski, Coram, Martill, Fleck, Escuillié, Wisshak, and Maisch, 2001; von Ellenrieder, 2002), but its classification still needs to be up-dated pending future studies. Sister group of austropetaliids (Bechly, 1996; 1999; Bechly, Nel, Martínez-Delclôs, Jarzembowski, Coram, Martill, Fleck, Escuillié, Wisshak, and Maisch, 2001; Carle, 1996; Misof, Rickert, Buckley, Fleck, and Sauer, 2001), with which they share the following derived characters: anterior lamina with elongate median cleft, anterior hamule directed medially, male genital ligula with anteroventral surface developed into a sharp-edged valve separator, abdomen bearing dorsolongitudinal carinae, and proventricular lobes of larval gizzard small and mound-like with eight or fewer clustered teeth.

[NOTE: Machet and Duquef (2004) reported the capture of the Old World species *Hemianax ephippiger* (Burmeister, 1839) in French Guiana, but unless the species is documented as resident there, it should not be considered part of the New World fauna.]

Key to males

[NOTE: The male of *Racenaeschna* is still unknown, therefore it is not included in the following key; check with genus account on page 57 for nongenital characters.]

1. IRP2 not forked (Fig. 102) .... **2**   1'. IRP2 forked (Figs. 103-104) **6**

102 FW distal half
Gomphaeschna antilope

103 FW distal half
Limnetron antarcticum

104 FW distal half
Gynacantha laticeps

2(1). Auricles with 5 to more than 20 minute denticles in a band of 2 or more rows (Fig. 105); anterior process of anterior hamule vertical or upturned, laminar and separated from base by a deep groove (Fig. 107); Mspl smoothly curved (parallel to MA) (Fig. 110); Rspl and IRP2 course nearly parallel (Figs. 102-103) **3**

2'. Auricles with 4-13 denticles in 1 row (Fig. 106); anterior process of anterior hamule approximately horizontal and not separated from base by a deep groove (Fig. 108); Mspl bent at distal portion (Fig. 109); Rspl and IRP2 course not parallel (Fig. 104) **5**

105 auricle – *Gomphaeschna furcillata*
ventro-lateral view

107 anterior hamules – *Gomphaeschna furcillata*
ventral view

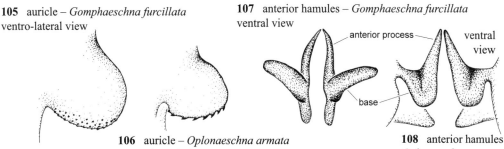

106 auricle – *Oplonaeschna armata*
ventro-lateral view

108 anterior hamules
Oplonaeschna armata

109 male HW basal half – *Oplonaeschna armata*   **110** male HW basal half – *Gomphaeschna antilope*

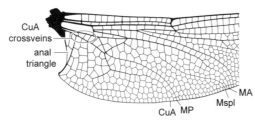

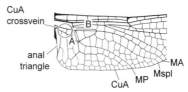

3(2). Median space usually free (Fig. 111); sides of thorax with pale stripes and/or with black spots (Fig. 113); RP2 undulated (convex and concave bends) (Fig. 102); wings usually without crossveins proximal to primary antenodal 1 (Fig. 111); HW triangle basal side (**A**) longer than 1/2 of costal side (**B**) (Fig. 110) **4**

3'. Median space crossed (Fig. 112); sides of thorax with 2 basal pale spots (Fig. 114); RP2 evenly curved (Fig. 103-104); wings with crossveins proximal to primary antenodal 1 (Fig. 112); HW triangle basal side (**A**) shorter than 1/2 of costal side (**B**) (Fig. 112) .... ***Boyeria*** (Page 44)

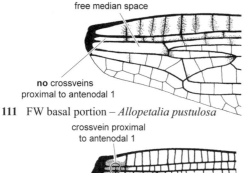

111 FW basal portion – *Allopetalia pustulosa*

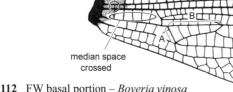

112 FW basal portion – *Boyeria vinosa*

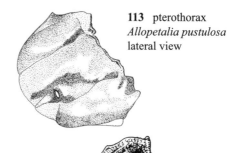

113 pterothorax
Allopetalia pustulosa
lateral view

114 pterothorax
Boyeria vinosa
lateral view

4(3). Cubito-anal space with 1 crossvein (Fig. 110); anal triangle with 1-2 cells (Fig. 110); Rspl not reaching distal border of wing (Fig. 102); occipital triangle in dorsal view approximately equilateral (as in Fig. 115); epiproct deeply forked (Fig. 117) .... ***Gomphaeschna*** (Page 49)

4'. Cubito-anal space with 2 or more crossveins (as in Figs. 109, 111); anal triangle with 3 or more cells (as in Fig. 109); Rspl reaching distal border of wing (as in Figs. 103-104); occipital triangle in dorsal view projected anteriorly for a distance greater than 1/2 of eye's suture (Fig. 116); epiproct roughly triangular or quadrangular (as in Fig. 118) .... ***Allopetalia*** (Page 39)

115 occipital triangle
Limnetron antarcticum
dorsal view

116 occipital triangle
Allopetalia pustulosa
dorsal view

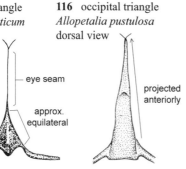

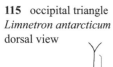

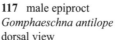

117 male epiproct
Gomphaeschna antilope
dorsal view

118 male epiproct
Boyeria vinosa
dorsal view

5(2). RP2 undulated (convex and concave bends) (as in Fig. 102); space between CuA and MP narrowing toward wing margin (Fig. 109); anal triangle with 3 or more cells (Fig.109); dorsum of abdominal S10 with spine higher than twice its width (Fig. 120); vesica spermalis distal segment with latero-basal lobes (as in Figs. 122a-b) .... *Oplonaeschna* (Page 56)

5'. RP2 evenly curved (as in Figs. 103-104); space between CuA and MP widening toward wing margin, or as wide at margin as at its base (as in Fig. 110); anal triangle with 1-2 cells (as in Fig. 102); dorsum of abdominal S10 smooth or with low crest or spine (as in Fig. 119); vesica spermalis distal segment lacking latero-basal lobes (Figs. 121a-b) .... *Basiaeschna* (Page 43)

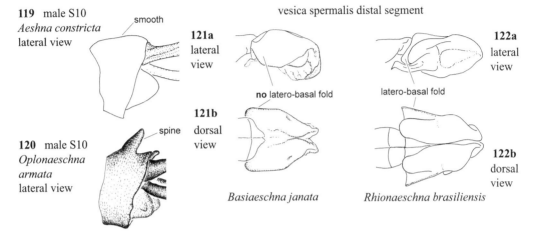

119 male S10
Aeshna constricta
lateral view
smooth

121a lateral view

122a lateral view

no latero-basal fold

latero-basal fold

vesica spermalis distal segment

120 male S10
Oplonaeschna armata
lateral view
spine

121b dorsal view

122b dorsal view

Basiaeschna janata

Rhionaeschna brasiliensis

6(1). Auricles approximately triangular or semicircular (wider at base than long) (Figs. 123-125) 7

6'. Auricles narrow and parallel-sided (narrower at base than long) (Fig. 126) .... *Andaeschna* (Page 42)

6''. Auricles absent .... *Anax* (Page 40)

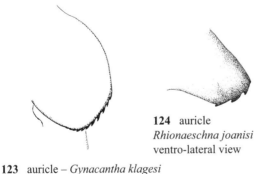

124 auricle
Rhionaeschna joanisi
ventro-lateral view

125 auricle
Rhionaeschna peralta
ventro-lateral view

126 auricle
Andaeschna andresi
ventro-lateral view

123 auricle – *Gynacantha klagesi*
ventro-lateral view

7(6). Subcosta extending up to nodus (Fig. 127) 8

7'. Subcosta prolonged beyond nodus (Fig. 128) **22**

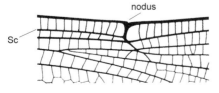

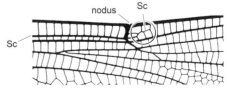

127 FW nodal area – *Gynacantha nervosa*

128 FW nodal area – *Neuraeschna calverti*

8(7). Frons in lateral view projected anterodorsally and flat (Fig. 129); anterior lamina lacking spines .... **Nasiaeschna** (Page 54)

8'. Frons in lateral view not projected anterodorsally and slightly convex (Fig. 130); anterior lamina with a pair of spines (these may be hidden or reduced) (Fig. 131) **9**

129 head *Nasiaeschna pentacantha* lateral view

130 head – *Rhionaeschna psilus* lateral view

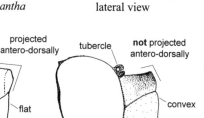

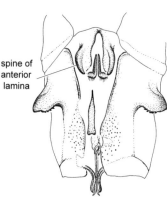

131 genital fossa – *Limnetron antarcticum* ventral view

9(8). Auricles with 5 to more than 20 minute denticles in a band of 2 or more rows (as in Fig. 105) .... **Epiaeschna** (Page 48)

9'. Auricles with 4-13 denticles in 1 row (Figs. 106, 123) **10**

9". Auricles with 2-3 strong denticles in 1 row (Figs. 124-125) **15**

10(9). Outer margin of anterior hamule in ventral view smoothly curved (Fig. 132) **11**

10'. Outer margin of anterior hamule in ventral view with a medial excision (Fig. 130) .... **Remartinia in part** (Page 58)

132 anterior hamules *Coryphaeschna perrensi* ventral view

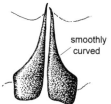

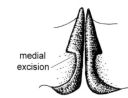

133 anterior hamules *Remartinia luteipennis* ventral view

11(10). Anterior process of anterior hamule vertical or upturned, laminar and separated from base by a deep groove (Fig. 134); membranule length (**A**) less than 1/4 of anal wing margin (**B**) (costal side longer than anal) (Fig. 136) **12**

11'. Anterior process of anterior hamule approximately horizontal and not separated from base by a deep groove (Fig. 135); membranule length (**A**) about as long as 1/3-1/2 of the anal wing margin (**B**) (Fig. 137) **13**

137 male HW anal region *Rhionaeschna planaltica*

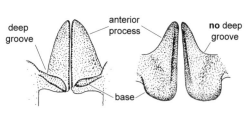

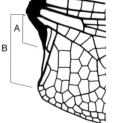

134 anterior hamules *Gynacantha klagesi* ventral view

135 anterior hamules *Triacanthagyna trifida* ventral view

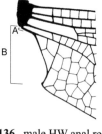

136 male HW anal region *Gynacantha tenuis*

12(11). Mspl smoothly curved (parallel to MA) (as in Fig. 110); Rspl and IRP2 course nearly parallel (Fig. 103); space between CuA and MP widening toward wing margin, or as wide at margin as at its base (as in Fig. 110); HW triangle basal side (**A**) longer than 1/2 of costal side (**B**) (as in Fig. 110); vesica spermalis distal segment lacking latero-basal lobes (Fig. 138) ....
.... ***Limnetron*** (Page 53)

12'. Mspl bent at distal portion (as in Fig. 109); Rspl and IRP2 course not parallel (Fig. 104); space between CuA and MP narrowing toward wing margin (as in Fig. 109); HW triangle basal side (**A**) shorter than 1/2 of costal side (**B**) (as in Fig. 112); vesica spermalis distal segment with latero-basal lobes (Fig. 139) ....
.... ***Gynacantha* in part** (Page 51)

138 vesica spermalis distal segment *Limnetron antarcticum* dorsal view

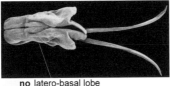

no latero-basal lobe

139 vesica spermalis distal segment *Gynacantha* sp. dorsal view

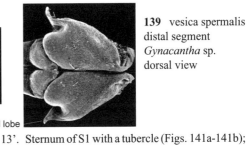

latero-basal lobe

13(11). Sternum of S1 approximately flat (Fig. 140); posteroventral tergal angles of S1 rounded (Fig. 140); venter of vesica spermalis distal segment lacking a flagellum (Fig. 142) ....
....… **14**

13'. Sternum of S1 with a tubercle (Figs. 141a-141b); posteroventral tergal angles of S1 produced into a pair of processes (Figs. 141a-141b); venter of vesica spermalis distal segment with a dorsally directed flagellum (Fig. 143) ....
.... ***Castoraeschna* in part** (Page 45)

140 male sternum S1 – *Coryphaeschna adnexa* ventral view

male sternum S1 – *Castoraeschna decurvata*

tergal angles rounded

141a ventral view

tubercle

tergal processes

141b lateral view

142 vesica spermalis distal segment *Coryphaeschna perrensi* lateral view

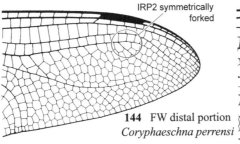

no latero-basal lobe   no flagellum

flagellum

143 vesica spermalis distal segment *Castoraeschna decurvata* lateral view

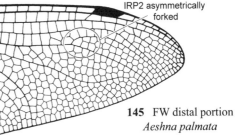

14(13). IRP2 forked symetrically (Fig. 144); HW triangle basal side (**A**) shorter than 1/2 of costal side (**B**) (as in Fig. 112); vesica spermalis distal segment lacking latero-basal lobes (Fig. 142) .... ***Coryphaeschna* in part** (Page 47)

14'. IRP2 forked asymmetrically (Fig. 145); HW triangle basal side (**A**) longer than 1/2 of costal side (**B**) (as in Fig. 110); vesica spermalis distal segment with latero-basal lobes (as in Figs. 122a-122b, 139) ..... ***Aeshna* in part** (Page 36)

IRP2 symmetrically forked

144 FW distal portion *Coryphaeschna perrensi*

IRP2 asymmetrically forked

145 FW distal portion *Aeshna palmata*

15(9). Outer margin of anterior hamule in ventral view smoothly curved (Fig. 132) **16**

15'. Outer margin of anterior hamule in ventral view with a medial excision (Fig. 133) .... *Remartinia* **in part** (Page 58)

16(15). Anterior process of anterior hamule vertical or upturned, laminar and separated from base by a deep groove (Fig. 134) .... *Gynacantha* **in part** (Page 51)

16'. Anterior process of anterior hamule approximately horizontal and not separated from base by a deep groove (Fig. 135) **17**

17(16). Sternum of S1 approximately flat (Figs. 140, 146, 148); venter of vesica spermalis distal segment lacking a flagellum (Fig. 142) **18**

17'. Sternum of S1 with a tubercle (Figs. 141a-b, 147, 149); venter of vesica spermalis distal segment with a dorsally directed flagellum (Fig. 143) **21**

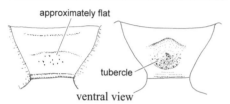
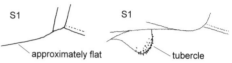

approximately flat

tubercle

ventral view

lateral view

S1   S1

approximately flat   tubercle

146 male sternum S1
'*Aeshna*' *williamsoniana*

147 male sternum S1
Rhionaeschna demarmelsi

148 male sternum S1
'*Aeshna*' *williamsoniana*

149 male sternum S1
Rhionaeschna demarmelsi

18(17). Cercus about as long as S9-10 (Fig. 150) .... '*Aeshna' williamsoniana* (Page 37)

18'. Cercus longer than S9-10 (Fig. 151) **19**

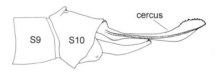

cercus

S9   S10

cercus

S9   S10

150 S9–10 – '*Aeshna' williamsoniana*
lateral view

151 S9–10 – *Aeshna eremita*
lateral view

19(18). IRP2 forked symetrically (Fig. 144); vesica spermalis distal segment lacking latero-basal lobes (Figs. 142, 151); HW triangle basal side (**A**) shorter than 1/2 of costal side (**B**) (as in Fig. 112) **20**

19'. IRP2 forked asymmetrically (Fig. 145); vesica spermalis distal segment with latero-basal lobes (as in Figs 122a-b); HW triangle basal side (**A**) longer than 1/2 of costal side (**B**) (as in Fig. 110) .... *Aeshna* **in part** (Page 36)

20(19). Anal triangle with 1-2 cells (as in Fig. 110); membranule length (**A**) about as long as 1/3-1/2 of the anal wing margin (**B**) (as in Fig. 137); vesica spermalis distal segment lacking ventral tubular structure (as in Fig. 152) .... *Coryphaeschna* **in part** (Page 47)

20'. Anal triangle with 3 or more cells (Fig. 109); membranule length (**A**) less than 1/4 of anal wing margin (**B**) (costal side longer than anal) (as in Fig. 136); vesica spermalis distal segment with ventral tubular structure (Fig. 153) .... *Triacanthagyna* (Page 64) .

tubular structure

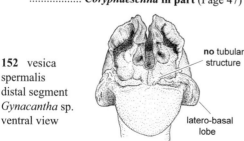
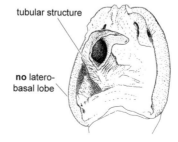

152 vesica
spermalis
distal segment
Gynacantha sp.
ventral view

no tubular
structure

no latero-
basal lobe

latero-basal
lobe

153 vesica
spermalis
distal segment
*Triacanthagyna
trifida*
ventral view

21(17). Anal triangle with 1-2 cells (as in Fig. 109); posteroventral tergal angles of S1 produced into a pair of processes (Figs. 141a-b); HW triangle basal side (**A**) shorter than 1/2 of costal side (**B**) (as in Fig. 112); vesica spermalis distal segment lacking latero-basal lobes (Fig. 143) .... ***Castoraeschna* in part** (Page 45)

21'. Anal triangle with 3 or more cells (as in Fig. 110); posteroventral tergal angles of S1 rounded (as in Fig. 140; Fig. 147); HW triangle basal side (**A**) longer than 1/2 of costal side (**B**) (as in Fig. 110); vesica spermalis distal segment with latero-basal lobes (Figs 122a-b) .... ***Rhionaeschna*** (Page 59)

22(7). Median space free (Fig. 154); sides of thorax without stripes or spots .... ***Staurophlebia*** (Page 62)

22'. Median space crossed (Fig. 155); sides of thorax with pale stripes and/or with black spots …................................ ***Neuraeschna*** (Page 55)

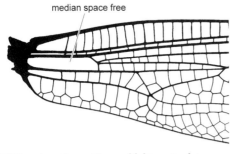

154 FW basal portion – *Staurophlebia reticulata*

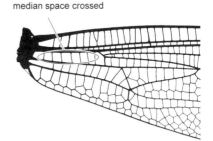

155 FW basal portion – *Neuraeschna calverti*

Key to females

1. IRP2 not forked (as in Fig. 156) **2**

1'. IRP2 forked (as in Figs. 157-158) **6**

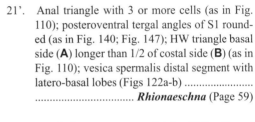

156 FW distal half
Gomphaeschna antilope

157 FW distal half
Limnetron antarcticum

158 FW distal half
Gynacantha laticeps

2(1). Cubito-anal space with 1 crossvein (as in Fig. 159); Rspl not reaching distal border of wing (as in Fig. 156) .... ***Gomphaeschna*** (Page 49)

2'. Cubito-anal space with 2 or more crossveins (as in Fig. 160); Rspl reaching distal border of wing (as in Figs. 157-158) **3**

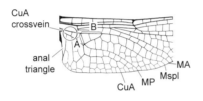

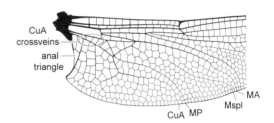

159 male HW basal half – *Gomphaeschna antilope*

160 male HW basal half – *Oplonaeschna armata*

3(2). Median space free (as in Fig. 161); sides of thorax with pale stripes and/or with black spots (as in Fig. 163); wings without crossveins proximal to primary antenodal 1 (as in Fig. 161); HW triangle basal side (**A**) longer than 1/2 of costal side (**B**) (as in Fig. 159) **4**

3'. Median space crossed (as in Fig. 162); sides of thorax with 2 basal pale spots (as in Fig. 164); wings with crossveins proximal to primary antenodal 1 (as in Fig. 162); HW triangle basal side (**A**) shorter than 1/2 of costal side (**B**) (as in Fig. 162) .... ***Boyeria*** (Page 44)

161 FW basal portion *Allopetalia pustulosa*

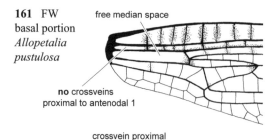

free median space

no crossveins proximal to antenodal 1

163 pterothorax *Allopetalia pustulosa* lateral view

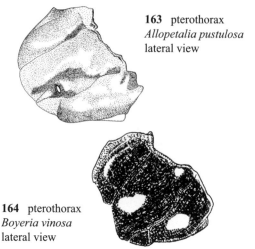

crossvein proximal to antenodal 1

162 FW basal portion *Boyeria vinosa*

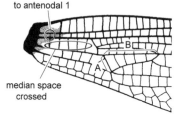

median space crossed

164 pterothorax *Boyeria vinosa* lateral view

4(3). Space between CuA and MP widening toward wing margin, or as wide at margin as at its base (as in Fig. 159) **5**

4'. Space between CuA and MP narrowing toward wing margin (as in Fig.160) .... ***Oplonaeschna*** (Page 56)

5(4). Mspl smoothly curved (parallel to MA) (as in Fig. 159); Rspl and IRP2 course nearly parallel (as in Figs. 156-157); cercus about length of S10 or shorter; tip of epiproct in dorsal view pointed or bearing a subapical spine (Fig. 165); distribution: South America (Map 8) .... ***Allopetalia*** (Page 39)

5'. Mspl bent at distal portion (as in Fig. 160); Rspl and IRP2 course not parallel (as in Fig. 158); cercus longer than S10 (CAUTION: make sure it is not broken off); tip of epiproct in dorsal view rounded (as in Fig. 166); distribution: North America (Map 11) ……..……............. ………………………...….... ***Basiaeschna*** (Page 43)

165 female cerci dorsal view

Allopetalia reticulosa

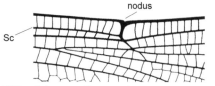

cercus

S10

epiproct pointed

166 female cerci

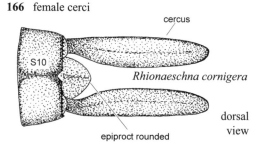

cercus

S10

Rhionaeschna cornigera

epiproct rounded

dorsal view

6(1). Subcosta extending up to nodus (as in Fig. 167) **7**

6'. Subcosta prolonged beyond nodus (as in Fig. 168) **20**

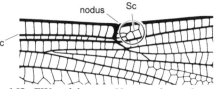

nodus

Sc

167 FW nodal area – *Gynacantha nervosa*

nodus   Sc

Sc

168 FW nodal area – *Neuraeschna calverti*

7(6). Sectors of arculus arising from its upper end (Fig. 169) **8**

7'. Sectors of arculus arising near its middle (Fig. 170) **9**

169 FW arculus – *Anax amazili*

170 FW arculus – *Staurophlebia reticulata*

8(7). Rspl not reaching distal border of wing (as in Fig. 171); sternum of S10 not produced posteriorly into a flat sclerotized process (as in Fig. 174); membranule length (**A**) greater than 1/2 of the anal wing margin (**B**) (as in Fig. 173) .... *Anax* (Page 40)

8'. Rspl reaching distal border of wing (as in Figs. 157-158); sternum of S10 produced posteriorly into a flat sclerotized process (Figs. 177a-b); membranule length (**A**) less than 1/4 of anal wing margin (**B**) (costal side longer than anal) (as in Fig. 172) .... *Gynacantha* **in part** (Page 51)

171 HW postnodal half – *Anax amazili*

172 male HW anal region *Gynacantha tenuis*

173 male HW anal region *Rhionaeschna planaltica*

9(7). HW triangle basal side (**A**) shorter than 1/2 of costal side (**B**) (as in Fig. 162) **10**

9'. HW triangle basal side (**A**) longer than 1/2 of costal side (**B**) (as in Fig. 159) **16**

10(9). Sternum of S10 produced posteriorly into a flat sclerotized process (Figs. 176-178) **11**

10'. Sternum of S10 not produced posteriorly into a flat sclerotized process (Figs. 174-175) **13**

174 female S9-10 *Rhionaeschna absoluta*

175 female S9-10 *Epiaeschna heros*

176a female S9-10 *Limnetron* sp.

176b female S10

177b female S10

177a female S9-10 *Gynacantha bifida*

178a female S9-10 *Triacanthagyna obscuripennis*

178b

11(10). Tip of epiproct in dorsal view rounded (as in Fig. 166) **12**

11'. Tip of epiproct in dorsal view pointed (as in Fig. 165) or bearing a subapical spine ....
.... ***Racenaeschna*** (Page 57)

12(11). Sternum of S10 projected posteriorly with 2 apical spines (Figs. 177a-b) ....
.... ***Gynacantha* in part** (Page 51)

12'.   Sternum of S10 projected posteriorly with 3 spines (Figs. 178a-b) ....
.... ***Triacanthagyna*** (Page 64)

13(10). Venter of S10 with sclerotized marginal arch bearing strong denticles (Fig. 179) **14**
13'. Venter of S10 with sclerotized posterior bar with denticles in a band of more or less uniform width (Fig. 180) **15**

13". Venter of S10 with sclerotized posterior bar with denticles in a band strongly constricted in the middle or in 2 separated lateral groups (Fig. 181) ....
.... ***Castoraeschna*** (Page 45)

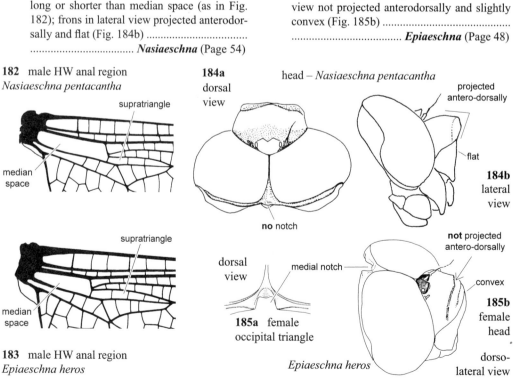

179   *Nasiaeschna pentacantha*   **180**   *Coryphaeschna adnexa*   **181**   *Castoraeschna longfieldae*

14(13). Mspl smoothly curved (parallel to MA) (as in Fig. 159); tip of epiproct in dorsal view pointed or bearing a subapical spine (as in Fig. 165); posterior margin of occiput straight or slightly concave (Figs. 184a-b); HW supratriangle as long or shorter than median space (as in Fig. 182); frons in lateral view projected anterodorsally and flat (Fig. 184b) ....
.... ***Nasiaeschna*** (Page 54)

14'. Mspl bent at distal portion (as in Fig. 160); tip of epiproct in dorsal view rounded (as in Fig 166); posterior margin of occiput medially notched (Figs. 185a-b); HW supratriangle longer than median space (as in Fig. 183); frons in lateral view not projected anterodorsally and slightly convex (Fig. 185b) ....
.... ***Epiaeschna*** (Page 48)

182 male HW anal region
Nasiaeschna pentacantha

184a dorsal view

head – *Nasiaeschna pentacantha*

184b lateral view

183 male HW anal region
Epiaeschna heros

185a female occipital triangle

Epiaeschna heros

185b female head

dorso-lateral view

15(13). Cercus about length of S10 or shorter (as in Fig. 181) .... *Remartinia* (Page 58)

15'. Cercus longer than S10 (CAUTION: make sure it is not broken off) (Fig. 180) ....
.... *Coryphaeschna* (Page 47)

16(9). IRP2 forked symmetrically (as in Fig. 186) **17**

16'. IRP2 forked asymmetrically (as in Fig. 187) **19**

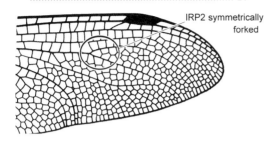

186 male HW distal region – *Rhionaeschna variegata*

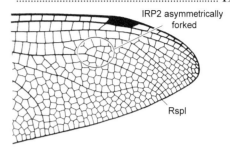

187 male HW distal region – *Aeshna palmata*

17(16). Sternum of S10 produced posteriorly into a flat sclerotized process (Figs. 176a-b); tip of epiproct in dorsal view pointed or bearing a subapical spine (as in Fig. 165; Figs. 176a-b); space between CuA and MP widening toward wing margin, or as wide at margin as at its base (as in Fig. 159); Mspl smoothly curved (parallel to MA) (as in Fig. 159); Rspl and IRP2 course nearly parallel (as in Figs. 156-157) ....
.... *Limnetron* (Page 53)

17'. Sternum of S10 not produced posteriorly into a flat sclerotized process (Fig. 174); tip of epiproct in dorsal view rounded (Figs. 166, 174); space between CuA and MP narrowing toward wing margin (as in Fig. 160); Mspl bent at distal portion (as in Fig. 160); Rspl and IRP2 course not parallel (as in Fig. 158) ....
.... **18**

18(17). Sternum S1 approximately flat (as in Figs. 188, 190) .... *'A.' williamsoniana* (Page 37)

18'. Sternum S1 with a tubercle (as in Figs.189, 191) .... *Rhionaeschna* (Page 59)

188 male sternum S1 '*Aeshna' williamsoniana* lateral view

189 male sternum *Rhionaeschna demarmelsi* lateral view

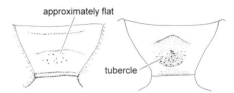

190 male sternum S1 '*Aeshna' williamsoniana* ventral view

191 male sternum S1 *Rhionaeschna demarmelsi* ventral view

19(16). Sides of thorax with pale stripes and/or with black spots; Rspl reaching distal border of wing (as in Fig. 187); cercus longer than S10 (CAUTION: make sure it is not broken off (as in Fig. 166); membranule length (**A**) about as long as 1/3-1/2 of the anal wing margin (**B**) (as in Fig. 173); distribution: North America ....
.... *Aeshna* (Page 36)

19'. Sides of thorax without stripes or spots; Rspl not reaching distal border of wing (as in Fig. 171); cercus about length of S10 or shorter (as in Fig. 166); membranule length (**A**) greater than 1/2 of the anal wing margin (**B**); distribution: South America ....
.... *Andaeschna* (Page 42)

20(6). Median space free (as in Figs. 182-183); sides of thorax without stripes or spots ....
.... *Staurophlebia* (Page 62)

20'. Median space crossed (as in Fig. 162); sides of thorax with pale stripes and/or with black spots .... *Neuraeschna* (Page 55)

192

female S9-10

Andaeschna rufipes lateral view

Aeshna Fabricius, 1775: 424.

[♂ pp. 29-30, couplets 14, 19; ♀ pp. 35, couplet 19]

Type species: *Aeshna grandis* Fabricius [Cowley, 1934c by subsequent designation]

[NOTE: *Aeschna* Illiger, 1801, is an unjustified emendation of *Aeschna* Fabricius, 1775 (Opinion 34, ICZN, 1999). Various authors in erecting new genera and subgenera with the suffix *Aeschna* subsequently used the revised name which accounts for the duality of spellings in aeshnid genera used today.]

     syn *Secundaeschna* Götz, 1923: 39
         Type species: *Libellula juncea* Linnaeus,
         1758 [by original designation]
40 species; 15 New World species:

canadensis Walker, 1908 [*Anax*]* – **L** [Walker, 1958]
     syn *maritimus* (*nomen oblitum*) (Provancher, 1895) [*Anax*]
clepsydra Say, 1840* – **L** [Walker, 1958]
constricta Say, 1840* – **L** [Walker, 1958]
eremita Scudder, 1866* – **L** [Musser, 1962]
     syn *hudsonica* Selys, 1875
interrupta interna Walker, 1908* – **L** [Musser, 1962]
interrupta interrupta Walker, 1908*
interrupta lineata Walker, 1908*
     syn *interrupta nevadensis* Walker, 1908
juncea (Linnaeus, 1758) [*Libellula*]* – **L** [Walker, 1912]
     syn *juncea americana* Bartenef, 1929
palmata Hagen, 1856* – **L** [Musser, 1962]
     syn *arida* Kennedy, 1918
persephone Donnelly, 1961*
septentrionalis Burmeister, 1839* – **L** [Walker, 1958]
sitchensis Hagen, 1861* – **L** [Walker, 1958; Musser, 1962]
subarctica subarctica Walker, 1908* – **L** [Walker, 1958]
tuberculifera Walker, 1908* – **L** [Walker, 1958]
umbrosa occidentalis Walker, 1912* – **L** [Musser, 1962]
umbrosa umbrosa Walker, 1908*
verticalis Hagen, 1861* – **L** [Walker, 1958]
     syn *propinqua* Scudder, 1866
walkeri Kennedy, 1917* – **L** [Kennedy, 1917]

References: Walker, 1912; Needham, Westfall, and May, 2000 (United States and Canada).
Distribution: Mainly holarctic; in the New World from Canada and United States to NW Mexico.

Medium to large aeshnids (55–79 mm); black T-spot on frons; pair of pale lateral thoracic stripes; mosaic pattern of pale spots on abdomen. Pale coloration of most males blue; some green or green and blue.

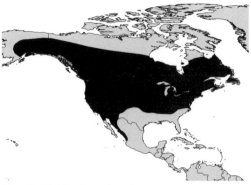

Map 6. Distribution of *Aeshna* spp.

Females of several species dichromatic with andromorph (male-like) and heteromorph (often green or yellow) color females. Wings (Figs. 194-195) with IRP2 asymmetrically forked, angled anal angle in males. Male auricles (Fig. 198) with 4–5 (occasionally 3) strong teeth. Distal segment of vesica spermalis with latero-basal folds (Fig. 193). Male cerci (Figs. 196a, 197a) longer than S9–10, female cerci longer than S10 (Fig. 199). Female sternum S10 with small denticles on posterior 1/2 (Figs. 200a-b). **Unique characters**: None known.

Status of classification: Systematics of this genus is yet to be worked out. In its present composition the genus is polyphyletic, including several species of uncertain position ('*A.' affinis* [Eurasia and northern Africa], '*A.' williamsoniana* [Central America], '*A.' ellioti* [central and eastern Africa], '*A.' mixta* [palearctic], and '*A.' isoceles* [Eurasia]), and some South African species ('*A.' moori*, '*A.' rileyi*, '*A.' subpupillata*, '*A.' yemenensis*, '*A.' minuscula*) belonging to a different genus (von Ellenrieder, 2003).

Potential for new species: Unlikely.

Habitat: Occurring at almost any body of water. In boreal America, *Aeshna* is the predominant dragonfly of bogs, marshes and rivers, where it can be very abundant. Information on its biology can be found in Walker (1912), Lincoln (1940), Cannings (1982a), and Peters (1998).

193
ventro-lateral view

latero-basal fold

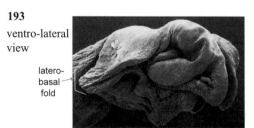

vesica spermalis distal segment – *Aeshna cyanea*

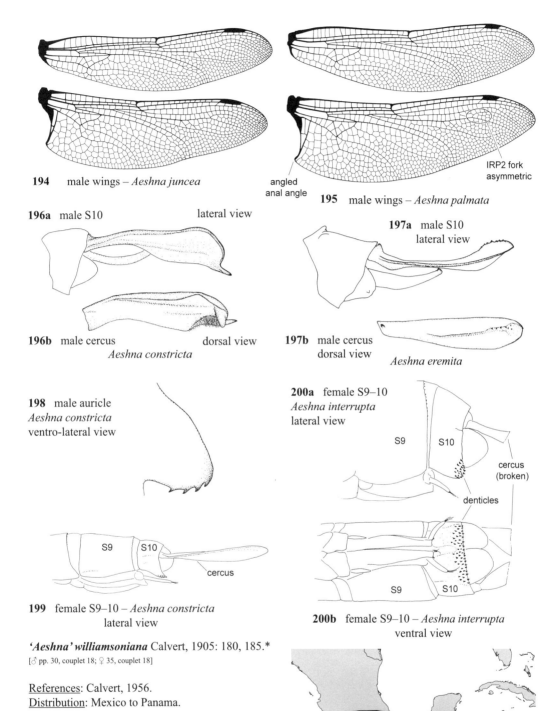

194 male wings – *Aeshna juncea*

195 male wings – *Aeshna palmata*

angled anal angle

IRP2 fork asymmetric

196a male S10   lateral view

197a male S10 lateral view

196b male cercus   dorsal view
Aeshna constricta

197b male cercus dorsal view   *Aeshna eremita*

198 male auricle *Aeshna constricta* ventro-lateral view

200a female S9–10 *Aeshna interrupta* lateral view

S9   S10

cercus (broken)

denticles

S9   S10   cercus

199 female S9–10 – *Aeshna constricta* lateral view

200b female S9–10 – *Aeshna interrupta* ventral view

'Aeshna' williamsoniana Calvert, 1905: 180, 185.*
[♂ pp. 30, couplet 18; ♀ 35, couplet 18]

References: Calvert, 1956.
Distribution: Mexico to Panama.

Black T-spot on frons (Fig. 201a); pair of pale lateral thoracic stripes (Fig. 202); mosaic pattern of pale spots on abdomen. Differs from all *Aeshna* species by symmetrical fork of IRP2 (Fig. 203), male cerci (Fig. 207a) shorter than S9–10, female cerci as long as S10, and from some species by pale green color, male anal triangle 3-celled and male cerci with sub-

Map 7. Distribution of *'Aeshna' williamsoniana*

basal tooth; differs from *Rhionaeschna* by flat abdominal sternum 1 (Fig. 204) and male cerci shorter than S9–10; from *Andaeschna* by symmetrical fork of IRP2, and male cerci shorter than S9–10, auricles approximately triangular, angled anal triangle and distal segment of vesica spermalis with latero-basal folds. **Unique characters**: Vesica spermalis distal segment with latero-basal fold limited to ventral margin and rounded in dorsal view (Figs. 205a-c).

<u>Status of classification</u>: A species of uncertain position; more closely related to *Rhionaeschna* and *Andaeschna* than to *Aeshna* (von Ellenrieder, 2003). Its generic identity needs to be analyzed taking into account the remaining *'Aeshna'* species from the Old World.

<u>Habitat</u>: Unknown, between 1150 - 1550 m.

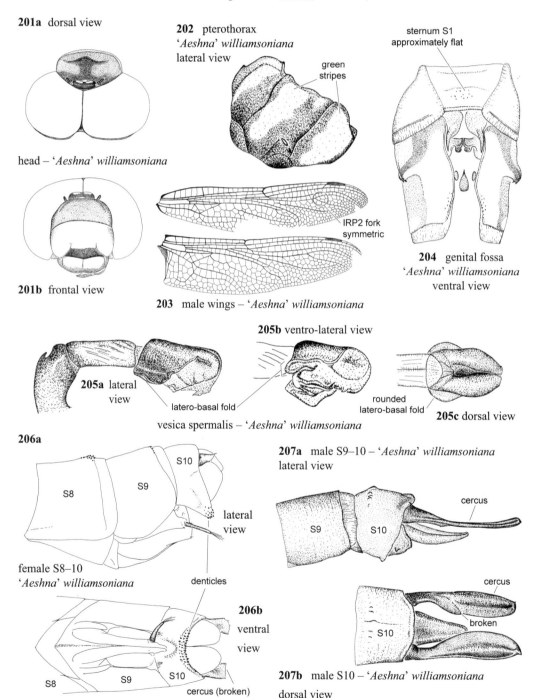

201a dorsal view

head – *'Aeshna' williamsoniana*

201b frontal view

202 pterothorax *'Aeshna' williamsoniana* lateral view

green stripes

203 male wings – *'Aeshna' williamsoniana*

IRP2 fork symmetric

sternum S1 approximately flat

204 genital fossa *'Aeshna' williamsoniana* ventral view

205b ventro-lateral view

205a lateral view

latero-basal fold

rounded latero-basal fold

205c dorsal view

vesica spermalis – *'Aeshna' williamsoniana*

206a

S8   S9   S10

lateral view

female S8–10 *'Aeshna' williamsoniana*

denticles

206b ventral view

S8   S9   S10

cercus (broken)

207a male S9–10 – *'Aeshna' williamsoniana* lateral view

S9   S10

cercus

cercus

broken

S10

207b male S10 – *'Aeshna' williamsoniana* dorsal view

Allopetalia Selys, 1873: 510 (65 reprint).
[♂ pp. 26, couplet 4; ♀ pp. 32, couplet 5]
Type species: *Allopetalia pustulosa* Selys, 1873
[Selys, 1883b by subsequent designation]
2 species:

pustulosa Selys, 1873* – **L** [De Marmels, 2000]
reticulosa Selys, 1873*
  syn *striatus* (Kirby, 1899) [*Anax*]

References: Ris, 1904; Martin, 1909.
Distribution: Andes from Venezuela to Peru (*A. pustulosa*) and Chile (*A. reticulosa*).

Map 8. Distribution of *Allopetalia* spp.

Large (71–79 mm), robust, brown aeshnids; dark T-spot on frons (Figs. 208a-b); pterothorax with small lateral pale and dark spots and stripes (Fig. 209). Wings (Fig. 211) with dark streaks or small spots along costal cross veins, arculus, and tip of triangles; unforked IRP2, Mspl and Rspl parallel to MA and IRP2, respectively, and undulated RP2. Male auricles (Fig. 210) with a marginal band of more than 20 small denticles; vesica spermalis distal segment (Fig. 214a) bearing two long flagella originated on its ventral base, with free portion shorter than length of segment and apices recurved and simple; basal segment (Fig. 214b) with elongate depression and moderately uplifted margins. Female sternum S10 (Fig. 212a-b) with small denticles on posterior 1/2. **Unique characters**: Occipital triangle in dorsal view projected anteriorly for a distance greater than 1/2 of eye's seam (Figs. 208a-b).
Status of classification: No comprehensive review exists for the genus. Likely related to *Boyeria* and *Limnetron* in the New World (von Ellenrieder, 2002) [see under *Limnetron*].
Potential for new species: Likely.
Habitat: Specimens rare in collections. Larvae of *A. pustulosa* described from small (1/2 m wide) streamlets and stony streams on steep mountain slopes within disturbed cloud forest areas of the Andes in W Venezuela (De Marmels, 2000).

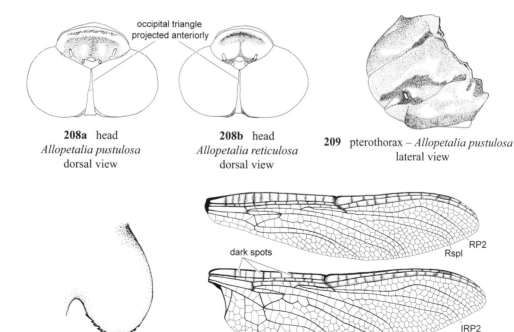

208a head
Allopetalia pustulosa
dorsal view

occipital triangle projected anteriorly

208b head
Allopetalia reticulosa
dorsal view

209 pterothorax – *Allopetalia pustulosa*
lateral view

dark spots

RP2
Rspl

IRP2

Mspl MA

210 male auricle
Allopetalia reticulosa
ventro-lateral view

211 male wings – *Allopetalia pustulosa*

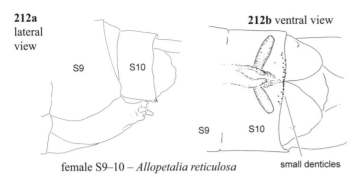

212a lateral view

212b ventral view

female S9–10 – *Allopetalia reticulosa*   small denticles

212c female S10
Allopetalia reticulosa
dorsal view

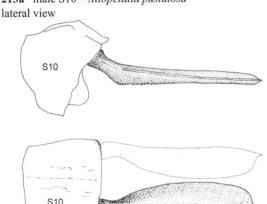

213a male S10 – *Allopetalia pustulosa*
lateral view

dorsal view

214a vesica spermalis distal segment
Allopetalia reticulosa

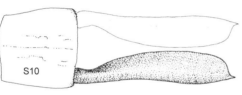

213b male S10 – *Allopetalia pustulosa*
dorsal view

dorso-lateral view

214b vesica spermalis basal segment
Allopetalia reticulosa

Anax Leach, 1815: 137.

[♂ pp. 27, couplet 6; ♀ pp. 33, couplet 8]

Type species: *Anax imperator* Leach, 1815 [by monotypy]

 syn *Cyrtosoma* Burmeister, 1839: 839

  Type species: *Aeschna ephippigera* Burmeister, 1839 [Kirby, 1890 by subsequent designation]

27 species; 5 New World species:

amazili (Burmeister, 1839) [*Aeschna*]* – **L** [Calvert, 1934; Rodrigues Capítulo, 1981]
 syn *maculatus* Rambur, 1842
concolor Brauer, 1865* – **L** [Geijskes, 1968]
junius (Drury, 1773) [*Libellula*]* – **L** [Musser, 1962]
 syn *spiniferus* Rambur, 1842
longipes Hagen, 1861* – **L** [Calvert, 1934; Geijskes, 1968]
walsinghami McLachlan, 1883* – **L** [Musser, 1962]
 syn *validus* (*nomen nudum*) Hagen, 1877

References: Geijskes, 1968; Dunkle, 2000; Needham, Westfall, and May, 2000.

Distribution: Cosmopolitan; in the New World from Canada to central Argentina.

Map 9. Distribution of *Anax* spp.

Large to very large aeshnids (67–110 mm); postfrons with black spot, except in *A. longipes* and *A.*

concolor; unmarked green pterothorax; abdomen red (*A. longipes*) or patterned with combinations of dark brown, green and blue. Wings with sectors of arculus arising from its upper end (Fig. 216b) (shared only with *Neuraeschna*); anal angle of male HW (Fig. 216a-b) rounded (shared only with *Andaeschna*). Distal segment of vesica spermalis (Fig. 215) with latero-basal folds. Female sternum S10 (Figs. 218a-b) with small denticles on posterior 1/2. **Unique characters**: RP2 with marked convex bend at distal end of pterostigma (Fig. 216b); posterior vein of male anal triangle (Fig. 216a) not thickened (anal triangle 'absent'); male auricles absent.

Status of classification: American species distinctive, Old World species require a revision. Related to *Andaeschna* and *Anaciaeschna* (von Ellenrieder, 2002).

Potential for new species: Unlikely in the New World.

Habitat: Almost any lentic water system; wide-ranging species often occurring at disturbed aquatic habitats (impounded water systems, stock ponds, etc.). Strong fliers; *A. junius* is known to reach England (Corbet, 2000). Some studies on their biology performed by May (1986), Córdoba-Aguilar (1995) and Young (1965).

215 vesica spermalis distal segment
– *Anax amazili*

latero-basal fold   lateral view

216b male wings – *Anax amazili*

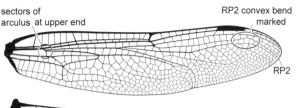

sectors of
arculus at upper end

RP2 convex bend
marked

RP2

216a anal region of male HW
Anax amazili

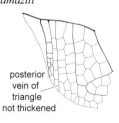

posterior
vein of
triangle
not thickened

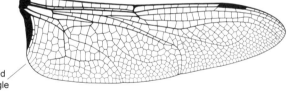

rounded
anal angle

218a lateral view

217a lateral view

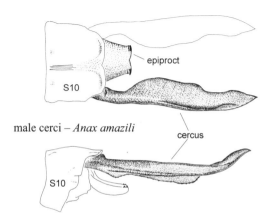

epiproct

S10

male cerci – *Anax amazili*

cercus

S10

217b dorsal view

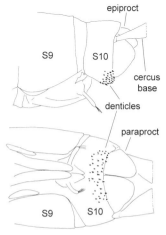

epiproct

S9   S10

cercus
base

denticles

paraproct

S9   S10

218b ventral view

female S9–10 – *Anax amazili*

Andaeschna De Marmels, 1994: 427.

[♂ pp. 27, couplet 6; ♀ pp. 35, couplet 19]

Type species: *Aeshna andresi* Rácenis, 1958 [De Marmels, 1994 by original designation]

4 species:

andresi (Rácenis, 1958) [*Aeshna*]* – **L** [De Marmels, 1992c]
rufipes (Ris, 1918) [*Aeshna*]* – **L** [De Marmels, 1982b; 1992b]
timotocuica De Marmels, 1994 – **L** [De Marmels, 1994]
unicolor [*Aeschna*] (Martin, 1908)

References: De Marmels, 1994.
Distribution: From Venezuela to N Argentina along the eastern slopes of the Andes.

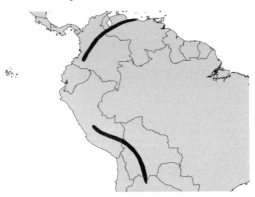

Map 10. Distribution of *Andaeschna* spp.

Medium to large aeshnids (66–78 mm); dull colored with concolorous frons, pterothorax and abdomen (largely brown or brick red). Wings (Fig. 222) with anal angle of male rounded (shared with *Anax*); distal segment of vesica spermalis (Fig. 223) lacking latero-basal folds (shared with Old World *'Aeshna' isosceles*); auricles (Fig. 219) narrow parallel sided (shared with Old World *Anaciaeschna* and *'Aeshna' isosceles*). Female sternum S10 (Fig. 221) with small denticles on posterior 1/2. **Unique characters**: Concolorous body; distal segment of vesica spermalis with a ventral fold as high as the segment height (Fig. 223c).

Status of classification: Good; apparently most closely related to *'Aeshna' isosceles* (De Marmels, 1994; von Ellenrieder, 2002).

Potential for new species: Likely.

Habitat: Mountain streams and rivulets; specimens, even of *A. rufipes*, the best known species, are scarce in collections.

220a dorsal view

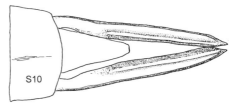

male cerci – *Andaeschna andresi*

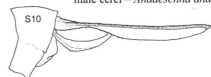

220b lateral view

219 male auricle
Andaeschna andresi

ventro-lateral view

narrow parallel-sided

221a
lateral view

S9   S10

221b
ventral view

denticles

S9   S10

rounded anal angle

female S9–10 – *Andaeschna rufipes*

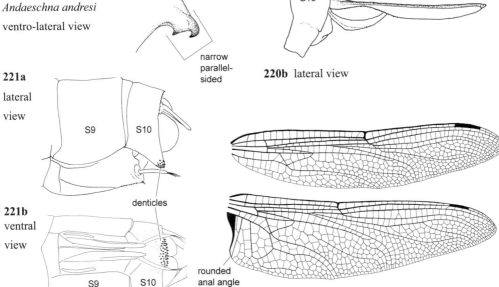

222 male wings – *Andaeschna andresi*

no latero-basal fold   vesica spermalis distal segment – *Andaeschna rufipes*

ventral fold

223a dorsal view   **223b** ventral view   **223c** lateral view

Basiaeschna Selys, 1883b: 735 (27 reprint).
[♂ pp. 27, couplet 5; ♀ pp. 32, couplet 5]
Type species: *Aeshna janata* Say, 1840 [Selys, 1883b by monotypy]
1 species:

janata (Say, 1840) [*Aeshna*]* – **L** [Walker, 1958; Louton, 1982b]
    syn *minor* (Rambur, 1842) [*Aeschna*]

References: Needham, Westfall, and May, 2000.
Distribution: SE Canada to E United States.

Small aeshnids (50–67 mm); concolorous frons; pterothorax with a pair of pale lateral thoracic stripes; intricate series of pale blue spots on abdomen. Similar to *Aeshna* or *Rhionaeschna* except for wings (Fig. 224) with unforked IRP2, small brown spot at base of each wing, and membranule shorter than 1/4 of wing anal margin length. **Unique characters**: None known.
Status of classification: According to von Ellenrieder (2002), this genus represents a basal Aeshninae.
Potential for new species: Unlikely.
Habitat: Forested rivers, streams and lakes with little shore vegetation, and oxygenated ponds (Dunkle, 2000).

Map 11. Distribution of *Basiaeschna* sp.

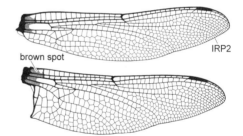

brown spot   IRP2

224 male wings – *Basiaeschna janata*

225a   dorsal view

male cerci – *Basiaeschna janata*

225b medio-
dorsal view

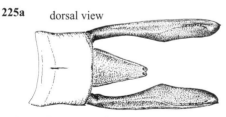

225c lateral view

226a   lateral view

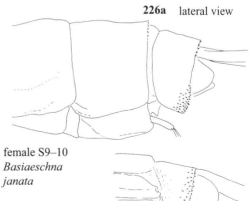

female S9–10
*Basiaeschna
janata*

226b ventral view

227 male auricle – *Basiaeschna janata*

ventro-lateral view

vesica spermalis distal segment – *Basiaeschna janata*

 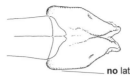 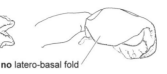 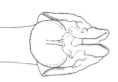

no latero-basal fold

228a dorsal view   **228b** lateral view   **228c** ventral view

Boyeria McLachlan, 1896: 424.

[♂ pp. 26, couplet 3; ♀ pp. 32, couplet 3]

Type species: *Aeschna irene* Fonscolombe, 1838 [Kirby, 1890 by subsequent designation]

   syn *Fonscolombia* Selys, 1883b: 736

      Type species: *Aeschna irene* Fonscolombe, 1838 [Kirby 1890 by subsequent designation]

[NOTE: *Boyeria* is a replacement name for *Fonscolombia* Selys, 1883, a junior primary homonym of *Fonscolombia* Lichtenstein, 1877 in Sternorrhyncha.]

5 species; 2 New World species:

grafiana Williamson, 1907* – **L** [Walker, 1913; Wright, 1949; Louton, 1982b]

vinosa (Say, 1840) [*Aeshna*]* – **L** [Needham, 1901; Needham and Hart, 1901; Wright, 1949; Louton, 1982b]

   syn *quadriguttata* (Burmeister, 1839) [*Aeschna*]

References: Needham, Westfall, and May, 2000 (North American species).

Distribution: Holarctic.

Medium aeshnids (60–71 mm); black T-spot on frons; pterothorax (Fig. 230) with a pair of pale yellow thoracic spots (in New World species); abdomen largely brown. Wing base with a brown spot; one crossvein proximal to first thickened antenodal (shared with some species of *Gynacantha* and *Neuraeschna*); median space with crossveins (shared with *Neuraeschna* in the New World); unforked IRP2, Mspl and Rspl parallel to MA and IRP2, respectively (Fig. 229). Male auricles (Fig. 232) with a marginal band of more than 20 small denticles; vesica spermalis distal segment (Fig. 231a) bearing two long flagella originated on its ventral base, with free portion as long as or shorter than length of segment and recurved and bifid apices; basal segment (Fig. 231b) with elongated depression and sides not uplifted. Female sternum 10 (Figs. 234a-b) with small denticles on posterior 1/2.

Unique characters: Bifid apices of vesica spermalis flagella (Fig. 231a).

Status of classification: No comprehensive review exists for the genus. Related to *Allopetalia* and *Limnetron* in the New World (von Ellenrieder, 2002) [see discussion under *Limnetron*].

Potential for new species: Unlikely in North America.

Habitat: Shady and cool forest streams, rivers and lakes (Dunkle, 2000), where the larvae cling to tangles of debris and root masses (Louton, 1982b).

Map 12. Distribution of *Boyeria* spp.

proximal antenodal crossvein

dark spots

IRP2
RP2

Mspl MA

229 male wings *Boyeria vinosa*

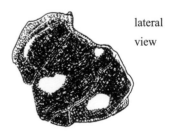

lateral view

230 pterothorax – *Boyeria vinosa*

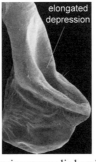
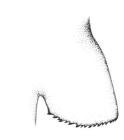

231a vesica spermalis distal segment – *Boyeria vinosa* lateral view

231b vesica spermalis basal segment – *Boyeria vinosa*

lateral view

232 male auricle – *Boyeria vinosa* ventro-lateral view

233a dorsal view

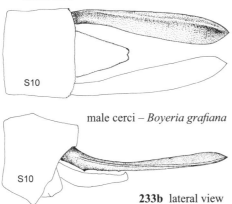

male cerci – *Boyeria grafiana*

233b lateral view

234a lateral view

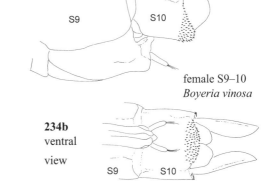

female S9–10 *Boyeria vinosa*

234b ventral view

Castoraeschna Calvert, 1952: 264.
[♂ pp. 29, 31, couplets 13, 21; ♀ p. 34, couplet 13]
Type species: *Aeschna castor* (Brauer, 1865) [Calvert, 1952 by original designation]
8 species:

castor (Brauer, 1865) [*Aeschna*]* – **L** [Santos, 1970b]
colorata (Martin, 1908) [*Aeschna*]
coronata (Ris, 1918) [*Coryphaeschna*]
decurvata Dunkle and Cook, 1984* – **L** [Rodrigues Capítulo and Jurzitza, 1989]
januaria (Hagen, 1867) [*Aeschna*]*
longfieldae (Kimmins, 1929) [*Coryphaeschna*]*
margarethae Jurzitza, 1979
tepuica De Marmels, 1989*

References: Calvert, 1956.
Distribution: Venezuela south to Peru and central Argentina.

Medium to large aeshnids (64–85 mm); yellow to pale green or blue face, with or without black spot on frons; pterothorax with pair of green lateral thoracic

Map 13. Distribution of *Castoraeschna* spp.

stripes; abdomen reddish brown to black with greenish spots. Wings (Fig. 235) with elongated triangles, basal side of HW triangle shorter than 1/2 of its costal side; Rspl reaching or not border of wing distally. In males, anal triangle 2 celled, and vesica spermalis distal segment (Figs. 239a-b) closed except for anterodorsal 'opening' (shared with *Coryphaeschna* and *Remartinia*), with patches of spines on base of flagellum. Female cerci (Figs. 236a-b) about as long as S10. **Unique characters**: S1 (Figs. 238a-b) with a sternal cylindrical tubercle (*Rhionaeschna* also has a tubercle, but lower and conical) and posteroventral tergal angles produced into a pair of processes; vesica spermalis distal segment with a long ventral flagellum directed dorsally (Figs. 239a-b); sternum of female S10 (Figs. 236a-b) with sclerotized posterior bar bearing denticles in a band, strongly constricted in the middle or, more often, separated into 2 lateral groups.

Status of classification: Poor. Several new species found since its only revision. It forms a monophyletic group together with *Coryphaeschna* and *Remartinia* (von Ellenrieder, 2002), constituting the sister taxon of these two genera (Carvalho, 1995).

Potential for new species: Very likely. Two new species under description by Peters (*pers. comm.*).

Habitat: Rivers and streams in lowland forested areas.

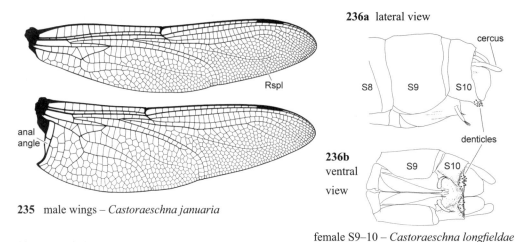

236a lateral view

cercus

S8   S9   S10

denticles

236b ventral view

S9   S10

female S9–10 – *Castoraeschna longfieldae*

235 male wings – *Castoraeschna januaria*

Rspl

anal angle

237a dorsal view

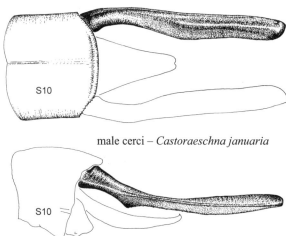

S10

male cerci – *Castoraeschna januaria*

S10

237b lateral view

238a ventral view

S1

cylindrical tubercle

postero-ventral tergal angles

S1   S2

postero-ventral tergal angles

238b lateral view

male S1 – *Castoraeschna decurvata*

239a lateral view

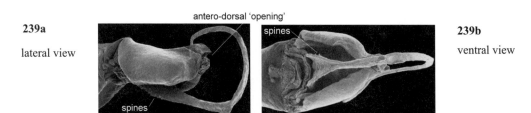

antero-dorsal 'opening'

spines

spines

spines

239b ventral view

vesica spermalis distal segment – *Castoraeschna decurvata*

Coryphaeschna Williamson, 1903: 2.

[♂ pp. 29-30, couplets 14, 20; ♀ pp. 35, couplet 15]

Type species: *Aeschna ingens* Rambur, 1842 [Williamson, 1903 by original designation]
8 species:

adnexa (Hagen, 1861) [*Aeschna*]* – **L** [Calvert, 1956; Santos, 1970c]
    syn *cyanifrons* (*nomen nudum*) (Hagen, 1861) [*Aeschna*]
    syn *guyanensis* Machet, 1991
    syn *macromia* (Brauer, 1865) [*Aeschna*]
amazonica De Marmels, 1989*
apeora Paulson, 1994*
diapyra Paulson, 1994*
huaorania Tennessen, 2001
ingens (Rambur, 1842) [*Aeschna*]* – **L** [Needham and Heywood, 1929; Klots, 1932]
    syn *abboti* (Hagen, 1863) [*Aeschna*]
perrensi (McLachlan, 1887) [*Aeschna*]* – **L** [Santos, 1969a; Carvalho, 1992b, 1993]
    syn *rufina* (*nomen nudum*) (Hagen, 1877) [*Aeschna*]
viriditas Calvert, 1952* – **L** [Geijskes, 1943; Calvert, 1956]
    syn *virens* (Rambur (*nec* Charpentier 1840), 1842) [*Aeschna*]

References: Calvert, 1956; Paulson, 1994; Tennessen, 2001.
Distribution: From SE United States to central Argentina.

Map 14. Distribution of *Coryphaeschna* spp.

Medium to large aeshnids (66.8–92.4 mm); pale green, blue or reddish face, with or without black spot on frons; pterothorax with pair of green lateral thoracic stripes to completely green or reddish; abdomen reddish brown to black with or without greenish spots. Wings (Fig. 242) with elongated triangles, basal side of HW triangle shorter than 1/2 of its costal side; Rspl reaching or not border of wing distally. In males, anal triangle 2 celled, and vesica sperma-

lis distal segment (Fig. 243) closed except for anterodorsal 'opening' (shared with *Castoraeschna* and *Remartinia*), and paired patches of spines on ventral process (absent in *Coryphaeschna adnexa*, which has a pair of long ventro-lateral flagella instead). Sternum of female S10 (Fig. 240a-b) with sclerotized posterior bar bearing denticles in one band; female cerci longer than S9–10. **Unique characters**: None.

Status of classification: Good. The phylogeny was recently analyzed in a thesis (Carvalho, 1995), where the genus was found to be monophyletic and constituting the sister taxon of *Remartinia*.
Potential for new species: Likely.
Habitat: Ponds, temporary pools and slow moving rivers in lowland forested areas.

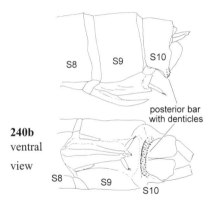

240a lateral view

240b ventral view

posterior bar with denticles

female S9–10 – *Coryphaeschna adnexa*

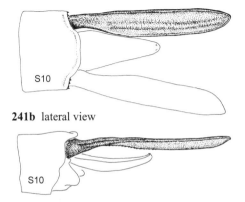

241a dorsal view

241b lateral view

male cerci – *Coryphaeschna perrensi*

242 male wings – *Coryphaeschna perrensi*

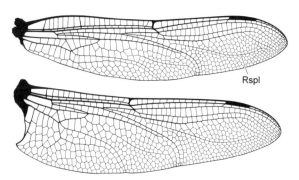

Rspl

243 vesica spermalis distal segment
Coryphaeschna perrensi

fronto-lateral
view

antero-dorsal 'opening'

patches of spines

ventral view

244 male anterior hamules
Coryphaeschna perrensi

Epiaeschna Hagen, 1877: 36.
[♂ pp. 28, couplet 9; ♀ pp. 34, couplet 14]
Type species: *Aeshna heros* Fabricius, 1798 [by monotypy]
1 species:
heros (Fabricius, 1798) [*Aeshna*]* – **L** [Needham, 1901; Walker, 1958]
      syn *multicincta* (Say, 1840) [*Aeshna*]

References: Dunkle, 2000; Needham, Westfall, and May, 2000.
Distribution: SE Canada to Mexico through E United States
[NOTE: Donnelly (1992a) records the capture of a male from Panama, but it is not known to breed there.]

Large aeshnids (82–91 mm); black spot on dark brown frons; pair of green lateral thoracic stripes; green rings on abdominal segments. Head (Fig. 246) with a lateral horn on each side of vertex (shared with *Nasiaeschna*). Wings (Fig. 247) with HW supratriangle as long as, or shorter than, median space, and Mspl slightly bent at distal portion (not parallel to MA). Constriction of S3 absent (shared with *Nasiaeschna*); dorsum of S10 of male (Fig. 249b) with a prominent spine (shared with *Oplonaeschna*, but shorter than twice its width at base). Male auricles (Fig. 248) with more than 20 small denticles on a band. Distal segment of vesica spermalis shaped like a boxing-glove in lateral view (Fig. 250a), suddenly narrowed at distal 1/3 in dorsal view (Fig. 250b, shared with *Nasiaeschna*). Female sternum S10 (Figs. 245a-b) with a sclerotized marginal arch bearing strong denticles (shared with *Nasiaeschna*).
Unique characters: Posterior margin of female occiput medially notched (Fig. 246).
Status of classification: Good. Related to *Nasiaeschna* in the New World (von Ellenrieder, 2002).
Potential for new species: Unlikely.
Habitat: Shady woodland ponds, swamps, temporary pools and slow streams (Dunkle, 2000).

Map 15. Distribution of *Epiaeschna* sp.

245a lateral view

245b ventral view

S9   S10

S9   S10

marginal arch

female S9–10 – *Epiaeschna heros*

246 female head – *Epiaeschna heros*

medial
notch

lateral horn

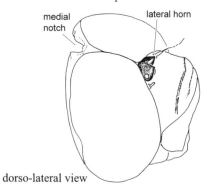

dorso-lateral view

247 male wings – *Epiaeschna heros*

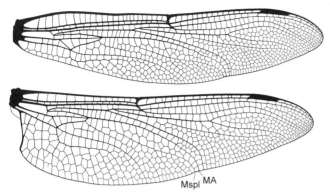

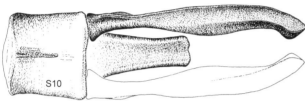

Mspl MA

248 male auricle – *Epiaeschna heros*
ventro-lateral view

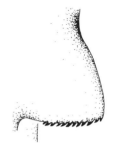

249a dorsal view

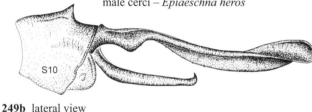

S10

male cerci – *Epiaeschna heros*

S10

249b lateral view

vesica spermalis distal segment
Epiaeschna heros

250a
dorsal
view

250b
lateral
view

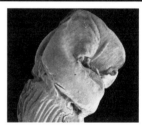

Gomphaeschna Selys, 1871b: 413.
[♂ pp. 26, couplet 4; ♀ pp. 31, couplet 2]
Type species: *Aeschna furcillata* Say, 1840 [by monotypy]
2 species:

antilope (Hagen, 1874) [*Aeschna*]* – **L** [Dunkle, 1977b]
furcillata (Say, 1840) [*Aeschna*]* – **L** [Dunkle, 1977b]
      syn *quadrifida* (Rambur, 1842) [*Gynacantha*]

References: Dunkle, 2000; Needham, Westfall, and May, 2000.
Distribution: SE Canada to SE United States.

Map 16. Distribution of *Gomphaeschna* spp.

Small aeshnids (52–60 mm); black T-spot on frons; sides of pterothorax greenish with blackish streaks ventrally; abdomen blackish with green spots in males and white or orange spots in females. Wings (Fig. 254) with 1 cubito-anal crossvein, RP2 undulated (with convex and concave bends) and IRP2 not

forked. Male auricles (Fig. 255) with 5 to more than 20 minute denticles on a band of 2 or more rows; genital ligula anterior surface with a mediolongitudinal furrow (Fig. 256a) and posterior surface smooth (Fig. 256b); central depression of vesica spermalis basal segment (Fig. 251c) shallow and wide, with greatly uplifted margins, and epiproct of male (Fig. 258) deeply forked (shared with Old World *Sarasaeschna*, *Oligoaeschna* and *Linaeschna*); 2 sclerotized flagella (Figs. 251a-b) originated on dorsal ends of distal segment of vesica spermalis (shared with *Sarasaeschna*). Female sternum S10 (Figs. 259a-b) with small denticles on posterior 1/2. **Unique characters**: Emargination of basal segment of vesica

spermalis 'U' shaped and projected margins narrow and arched posteriorly (Fig. 251c; emargination 'V' shaped and projections wide and not arched in *Sarasaeschna*).

Status of classification: Good. Related to the Old World *Sarasaeschna*, *Oligoaeschna* and *Linaeschna* (Karube and Yeh, 2001; von Ellenrieder, 2002).

Potential for new species: Unlikely.

Habitat: Swamps and bogs.

251a lateral view

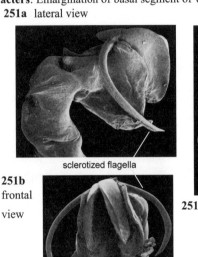

sclerotized flagella

251b frontal view

depression shallow and wide — greatly uplifted margin

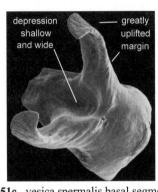

251c vesica spermalis basal segment *Gomphaeschna furcillata* dorso-lateral view

vesica spermalis distal segment *Gomphaeschna furcillata*

252 male anterior hamules *Gomphaeschna furcillata*

ventral view

posterior hamules — genital ligula

253 male genital fossa *Gomphaeschna furcillata* lateral view

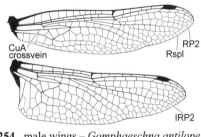

CuA crossvein

RP2 Rspl

IRP2

254 male wings – *Gomphaeschna antilope*

ventro-lateral view

255 male auricle *Gomphaeschna furcillata*

medio-longitudinal furrow

256a posterior view

256b anterior view

male genital ligula *Gomphaeschna furcillata*

257a dorsal view

S10

male cerci – *Gomphaeschna antilope*

S10

257b lateral view

258 male epiproct *Gomphaeschna antilope*

dorsal view

259a lateral view

S9 S10

denticles

259b ventral view S9 S10

female S9–10 *Gomphaeschna antilope*

Gynacantha Rambur, 1842: 209.

[♂ pp. 29-30, couplets 12,16; ♀ pp. 33-34, couplets 8, 12]

Type species: *Gynacantha nervosa* Rambur, 1842 [Calvert, 1905 by subsequent designation]

[NOTE: Rambur described *trifida* under *Gynacantha* together with another six species without designating a type species. Selys (1883b) included only *G. trifida* in his new genus *Triacanthagyna* which made it the type species. Kirby (1890), with no explanation, placed *trifida* back in *Gynacantha*, designated it as type species, and created the genus *Acanthagyna* for the previous *Gynacantha* species. The argument stems from Cowley (1934a) and has been recently re-opened by Hedge and Crouch (2000) who state that *Triacanthagyna* is therefore a junior objective synonym of *Gynacantha* Rambur, and that the names *Gynacantha sensu* Kirby, 1890 and *Acanthagyna* should be used instead. Although this is correct according to the nomenclature rules, the synonymy of *Gynacantha* and *Triacanthagyna* was unjustified, and this has been made a case to the ICZN (von Ellenrieder and Garrison, 2005b) in order to conserve the name usages for *Gynacantha sensu* Selys, 1883b and *Triacanthagyna*.]

> syn *Acanthagyna* Kirby, 1890: 94
>> Type species: *Gynacantha nervosa* Rambur, 1842 [Cowley, 1934a by subsequent designation]
> syn *Selysophlebia* Förster, 1905: 75
>> Type species: *Selysophlebia aratrix* Förster, 1905 [by original designation]
> syn *Subaeschna* Martin, 1908: 7
>> Type species: *Subaeschna francesca* Martin, 1909 [by monotypy]

86 species; 23 New World species:

adela Martin, 1909
> syn *martini* Navás, 1911
auricularis Martin, 1909*
bartai Paulson and von Ellenrieder, 2005*
bifida Rambur, 1842* – **L** [Carvalho, 1987]
> syn *robusta* Kolbe, 1888
caudata Karsch, 1891*
chelifera McLachlan, 1896
> syn *aratrix* (Förster, 1905) [*Selysiophlebia*]
convergens Förster, 1908
> syn *limai* Navás, 1916
croceipennis Martin, 1897*
ereagris Gundlach, 1888*
francesca Martin, 1909 [*Subaeschna*]*
gracilis (Burmeister, 1839) [*Aeschna*]* – **L** [Santos, 1973b]
helenga Williamson and Williamson, 1930*
interioris Williamson, 1923*
jessei Williamson, 1923
klagesi Williamson, 1923*
laticeps Williamson, 1923*
litoralis Williamson, 1923*
membranalis Karsch, 1891* – **L** [Santos, Costa, and Pujol-Luz, 1987]
> syn *jubilaris* Navás, 1915

mexicana Selys, 1868* – **L** [Carvalho and Ferreira, 1989]
nervosa Rambur, 1842* – **L** [Williams, 1937]
remartinia Navás, 1934
tenuis Martin, 1909*
tibiata Karsch, 1891* – **L** [Ramírez, 1994]

References: Williamson, 1923a.

Distribution: Tropics of the World; in the New World from S United States through N Argentina.

Map 17. Distribution of *Gynacantha* spp.

Small to large aeshnids (55–90 mm); black T-spot on frons; sides of pterothorax with pale or dark stripes or concolorous; reddish brown or pale brown abdomen with pale green, yellow or blue markings. Wings (Figs. 260-263) in some species with dark areas on basal or costal area; accessory crossveins proximal to first thickened antenodal (shared with *Neuraeschna*); IRP2 asymmetrically or symmetrically forked; basal side of HW triangle shorter than 1/2 of costal side; HW supratriangle as long as, longer or shorter than median space. Male auricles triangular to wide semicircular with 3–10 teeth (Fig. 268); anterior process of anterior hamule (Figs. 267a-b) separated from base by a groove (lacking groove in some Old World species); distal segment of vesica spermalis (Figs. 266a-b) with laterobasal folds. Male cerci (Figs. 264-265) longer than S9–10, female cerci longer than S10. Female sternum S10 (Figs. 269a-b) with sclerotized posterior process bearing 2 long apical spines.

Unique characters: None known.

Status of classification: Extremely variable genus; a revision including all the world species is needed in order to determine whether it constitutes a monophyletic group.

Potential for new species: Likely. One undescribed species from Peru known to the authors.

Habitat: Mud bottomed pools, temporary ponds and phytotelmata within tropical forests; habits often crepuscular (Louton, Garrison, and Flint, 1996; Bede, Piper, Peters, and Machado, 2000); often flies into lighted buildings in forest clearings at night.

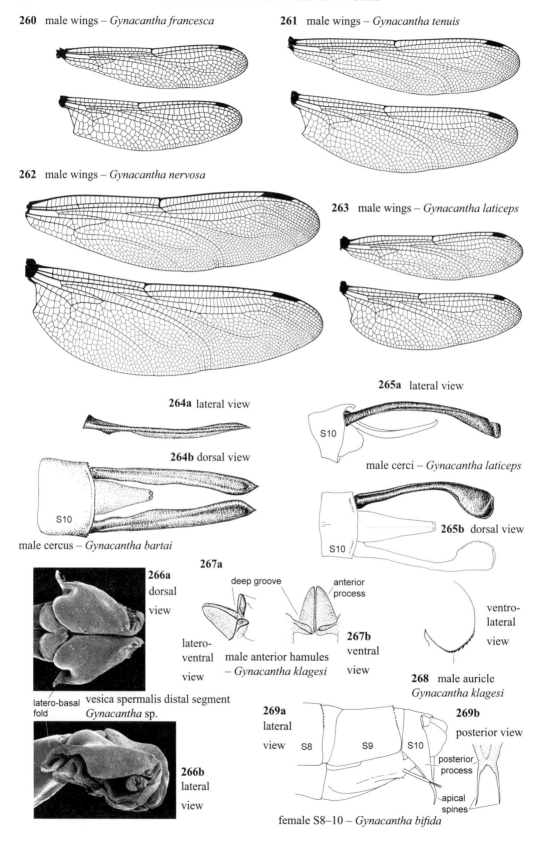

260 male wings – *Gynacantha francesca*

261 male wings – *Gynacantha tenuis*

262 male wings – *Gynacantha nervosa*

263 male wings – *Gynacantha laticeps*

264a lateral view

264b dorsal view

male cercus – *Gynacantha bartai*

S10

265a lateral view

S10

male cerci – *Gynacantha laticeps*

265b dorsal view

S10

266a dorsal view

latero-basal fold

vesica spermalis distal segment *Gynacantha* sp.

266b lateral view

267a

deep groove

anterior process

latero-ventral view

male anterior hamules – *Gynacantha klagesi*

267b ventral view

ventro-lateral view

268 male auricle *Gynacantha klagesi*

269a lateral view

S8 S9 S10

posterior process

apical spines

269b posterior view

female S8–10 – *Gynacantha bifida*

Limnetron Förster, 1907: 163.

[♂ pp. 29, couplet 12; ♀ pp. 35, couplet 17]

Type species: *Limnetron antarcticum* Förster, 1907 [by original designation]
2 species:

antarcticum Förster, 1907*
  ?syn *viridivittatum* (Fraser, 1947) [*Aeshna*]
debile (Karsch, 1891) [*Epiaeschna*]* – **L** [Santos, 1970a; Assis, Carvalho, and Dorvillé, 2000]

References: Karsch, 1891; Förster, 1907.
Distribution: SE Brazil, Paraguay, N Argentina, and Peru.

Map 18. Distribution of *Limnetron* spp.

Medium to large (63–72 mm) aeshnids; face yellow to pale green or pale blue, frons with or without dark T-spot (Fig. 270); pterothorax reddish brown with four green lateral stripes; abdomen reddish brown to black with green rings. Wings (Fig. 271) with symmetrically forked IRP2; Mspl parallel to MA and Rspl parallel to IRP2. Male auricles (Fig. 273) sub-quadrate, with 10–12 marginal teeth in a row; vesica spermalis distal segment (Fig. 275a) bearing two long flagella originated on its ventral base (shared with *Allopetalia* and *Boyeria*), with free portion longer than length of segment and apices simple; basal segment (Fig. 275b) with elongated depression and moderately uplifted margins. Female cerci longer than S10; female sternum S10 (Figs. 272a-c) with sclerotized posterior process bearing 4–5 apical spines (shared with *Racenaeschna*, the Australian *Telephlebia* and the oriental *Tetracanthagyna*); epiproct with a sub-apical spine (shared with *Allopetalia*, *Racenaeschna*, Mediterranean *Caliaeschna*, and Australian *Notoaeschna* and *Spinaeschna*). **Unique characters**: Free portion of flagella of vesica spermalis distal segment longer than length of segment (Fig. 275a).

Status of classification: No comprehensive review exists for the genus, and the species need to be diagnosed. De Marmels (2000) considers it to be a Gynacanthini related to *Racenaeschna*, because of the sclerotized process of female sternum S10, crepuscular habits and large eyes they share. According to von Ellenrieder (2002), it is not a Gynacanthini because of its primitive wing venation (MA and Mspl parallel), and is instead related to *Allopetalia* and *Boyeria* in the New World, with which it shares vesica spermalis morphology (similitude of female sternum S10 would be a convergence, also shared by *Telephlebia* and *Tetracanthagyna*, as well as the crepuscular habits and large eyes, shared by several dragonflies).

Potential for new species: Likely.

Habitat: Streams within forests.

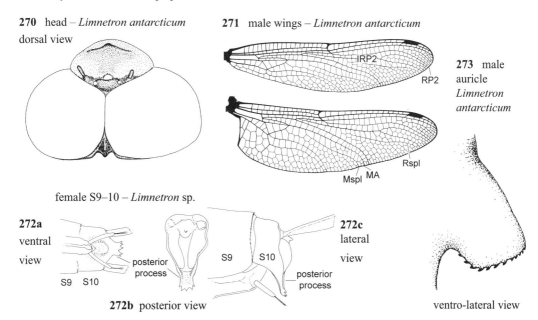

270 head – *Limnetron antarcticum* dorsal view

271 male wings – *Limnetron antarcticum*

IRP2
RP2
Rspl
Mspl MA

273 male auricle *Limnetron antarcticum*

female S9–10 – *Limnetron* sp.

272a ventral view
S9 S10
posterior process

272b posterior view

S9 S10
posterior process

272c lateral view

ventro-lateral view

274a dorsal view

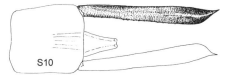

male cerci – *Limnetron antarcticum*

274b lateral view

275a vesica spermalis distal segment
Limnetron antarcticum   dorsal view

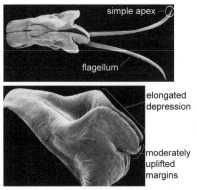

275b vesica spermalis basal segment
Limnetron antarcticum   ventro-lateral view

Nasiaeschna Selys *in* Förster, 1900: 93.
[♂ pp. 28, couplet 8; ♀ pp. 34, couplet 14]
Type species: *Aeschna pentacantha* Rambur, 1842
[by monotypy]
1 species:

pentacantha (Rambur, 1842) [*Aeschna*]* – **L** [Need-ham and Hart, 1901; Walker, 1958]

References: Dunkle, 2000; Needham, Westfall, and May, 2000.
Distribution: SE Canada to SE United States.

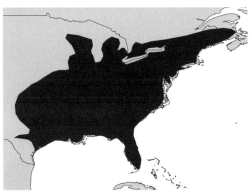

Map 19. Distribution of *Nasiaeschna* sp.

Medium (62–73 mm), dark brown aeshnids; frons con-colorous; pair of green lateral pterothoracic stripes; blue-green stripes on abdomen. Head (Fig. 276) with a lateral tubercle on each side of vertex (shared with *Epiaeschna*). Wings (Fig. 277) with HW supratriangle longer than median space, and Mspl parallel to MA. Constriction of S3 absent (shared with *Epiaeschna*); male abdomen tapering to tip. Male auricles (Fig. 280) with 4–10 teeth. Distal segment of vesica sper-malis shaped like a boxing-glove in lateral view (Fig. 279b), suddenly narrowed at distal 1/3 in dorsal view (Fig. 279a; shared with *Epiaeschna*). Female sternum S10 (Figs. 281a-b) with a sclerotized marginal arch bearing strong denticles (shared with *Epiaeschna*).
Unique characters: In lateral view frons projected anterodorsally and flat (Fig. 276); anterior lamina of male lacking spines (condition shared only by one *Rhionaeschna* species, *R. decessus*, from Brazil).
Status of classification: Good. Related to *Epiaeschna* in the New World (von Ellenrieder, 2002).
Potential for new species: Unlikely.
Habitat: Lakes, ponds and swampy streams (Dunkle, 2000).

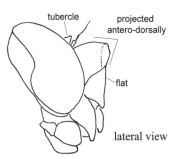

276 head – *Nasiaeschna pentacantha*

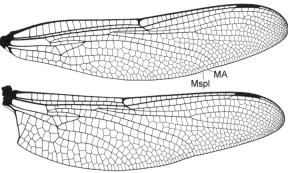

277 male wings – *Nasiaeschna pentacantha*

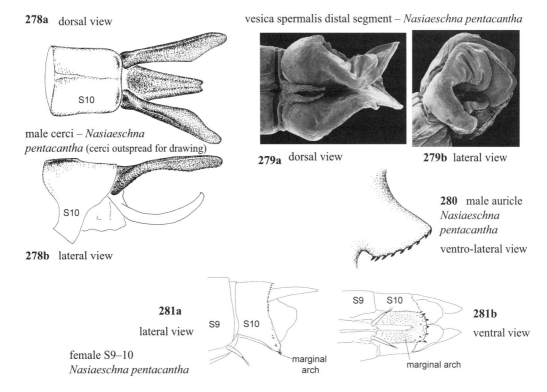

278a dorsal view

vesica spermalis distal segment – *Nasiaeschna pentacantha*

S10

male cerci – *Nasiaeschna pentacantha* (cerci outspread for drawing)

279a dorsal view   **279b** lateral view

S10

278b lateral view

280 male auricle *Nasiaeschna pentacantha* ventro-lateral view

281a lateral view

female S9–10 *Nasiaeschna pentacantha*

S9   S10

marginal arch

S9   S10

281b ventral view

marginal arch

Neuraeschna Hagen, 1867: 54.

[♂ pp. 31, couplet 22; ♀ pp. 35, couplet 20]

Type species: *Aeschna costalis* Burmeister, 1839 [by monotypy]

15 species:

calverti Kimmins, 1951*
capillata Machet, 1990
claviforcipata Martin, 1909
clavulata Machet, 1990
cornuta Belle, 1989
costalis (Burmeister, 1839) [*Aeschna*]* – **L** [Carvalho, 1989]
   syn *ferox* (Erichson, 1848) [*Gynacantha*]
   syn var. *hyalinata* Belle, 1989
   syn var. *marginata* Belle 1989
dentigera Martin, 1909*
   syn *inarmata* Kimmins, 1951
harpya Martin, 1909* – **L** [Belle, 1989]
maxima Belle, 1989
maya Belle, 1989*
mayoruna Belle, 1989
mina Williamson and Williamson, 1930
producta Kimmins, 1933
tapajonica Machado, 2002
titania Belle, 1989*

References: Belle, 1989; Machet, 1990.
Distribution: Honduras to Peru and Brazil.

Map 20. Distribution of *Neuraeschna* spp.

Medium to very large aeshnids (64–100 mm); face pale green to brown with or without black T-spot on frons; pterothorax brown with two lateral green stripes; abdomen reddish brown to black with green to yellow spots. Wings (Fig. 282) in some species with dark areas on basal or costal area; subcosta prolonged beyond nodus (shared with *Staurophlebia*); accessory crossveins proximal to first thickened antenodal in costal (some species) and subcostal (all species) interspaces (shared with some *Gynacantha*); median space crossed (shared with *Boyeria* in the New World); sectors of arculus arising from its upper end (shared with *Anax*); HW triangle basal side shorter than 1/2 of costal side; HW supratriangle longer than median space. Distal segment of vesica spermalis (Fig. 284) with a ventral mediolongitudinal laminar fold bearing one or two flap-like projections at its anterior end (shared with *Staurophlebia*). Female cerci longer than S10; sclerotized posterior process on venter of S10 (Figs. 283a-b) bearing 2 apical spines and a variable number of smaller spines.

Unique characters: Small apical fold anterior to medio-longitudinal fold on venter of vesica spermalis distal segment (Fig. 284).

Status of classification: Good. Sister genus of *Staurophlebia* (von Ellenrieder, 2002).

Potential for new species: Likely.

Habitat: Swampy areas, creeks and rivers in woodlands; crepuscular habits.

282   male wings – *Neuraeschna calverti*

median space   subcosta

accesory cross veins   arculus

283a   female S9–10 – *Neuraeschna costalis*

lateral view

S9   S10

posterior process

283b female posterior process of S10 – *Neuraeschna costalis* posterior view

284   vesica spermalis distal segment – *Neuraeschna calverti*

lateral view

apical fold

medio-longitudinal fold

dorsal view

lateral view

S10

S10

285a   male cerci – *Neuraeschna calverti*

285b   male cerci – *Neuraeschna calverti*

Oplonaeschna Selys, 1883b: 735 (27 reprint).

[♂ pp. 27, couplet 5; ♀ pp. 32, couplet 4]

Type species: *Aeschna armata* Hagen, 1861 [by monotypy]

    syn *Hoplonaeschna* Karsch, 1891: 283

       Type species: *Aeschna armata* Hagen, 1861 [incorrect replacement name for *Oplonaeschna* Selys, 1883]

2 species:

armata (Hagen, 1861) [*Aeschna*]* – **L** [Musser, 1962; González-Soriano and Novelo-Gutiérrez, 1998]

    syn var. *polernista* Navás, 1917

magna González and Novelo, 1998* – **L** [González-Soriano and Novelo-Gutiérrez, 1998]

References: González-Soriano and Novelo-Gutiérrez, 1998; Needham, Westfall, and May, 2000.

Distribution: SW United States to Guatemala.

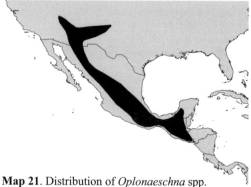

Map 21. Distribution of *Oplonaeschna* spp.

Medium to large aeshnids (66–80 mm); black T-spot on frons; face brown-orange, postfrons pale green; pair of pale lateral pterothoracic stripes; intricate series of pale spots on abdomen. Similar to an *Aeshna* or *Rhionaeschna* except for unforked IRP2, RP2 undulated (with convex and concave bends), and membranule shorter than 1/4 of wing anal margin length (Fig. 286). Male auricles (Fig. 289) with 4–10 teeth. Distal segment of vesica spermalis (Fig. 288) with laterobasal folds. **Unique characters**: Dorsum of S10 of male (Fig. 290b) with a spine higher than twice its width at base.

Status of classification: Good.

Potential for new species: Unlikely.

Habitat: Rocky streams in mountain oak and pine woodland and subtropical lower elevations (Dunkle, 2000). Larvae lay on the streambeds, cling to the underside of rocks and hide among vegetation; adults patrol over streams and females oviposit endophytically in dead leaves, without male guarding (Johnson, 1968).

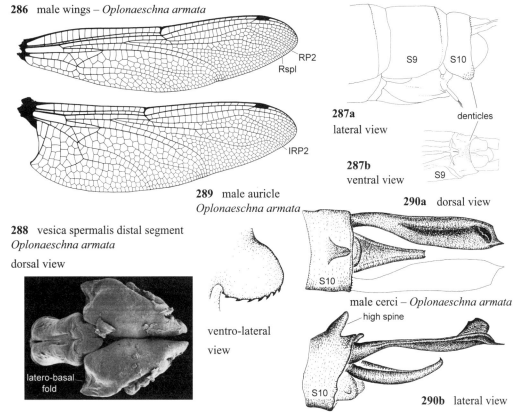

286   male wings – *Oplonaeschna armata*

RP2
Rspl
IRP2

female S9–10 – *Oplonaeschna armata*

S9   S10

287a
lateral view

denticles

287b
ventral view   S9

289   male auricle
Oplonaeschna armata

288   vesica spermalis distal segment
Oplonaeschna armata

dorsal view

latero-basal fold

ventro-lateral
view

290a   dorsal view

S10

male cerci – *Oplonaeschna armata*

high spine

S10

290b   lateral view

Racenaeschna Calvert, 1958: 227.
[♀ pp. 34, couplet 11]
Type species: *Racenaeschna angustistrigis* Calvert, 1958 [by original designation]
1 species:

angustistrigis Calvert, 1958** – **L** [De Marmels, 1990a]

References: Calvert, 1958.
Distribution: Venezuelan tepuis (highlands).

Male unknown. Large aeshnids (68–75 mm); face brown-orange, pale green lacking black spot; pterothorax brown with two lateral green stripes. Wings (Fig. 291) with HW triangle basal side shorter than 1/2

Map 22. Distribution of *Racenaeschna* sp.

of costal side; HW supratriangle longer than median space; Mspl bent at distal portion; Rspl and IRP2 not parallel to each other; IRP2 asymmetrically forked. Female cerci longer than S9–10 or as long as S8–10; female sternum S10 (Figs. 292a-b) with sclerotized posterior process bearing 4–5 apical spines (shared with *Limnetron*, the Australian *Telephlebia* and the oriental *Tetracanthagyna*); epiproct with a subapical spine (shared with *Allopetalia*, *Limnetron*, Mediterranean *Caliaeschna*, and Australian *Notoaeschna* and *Spinaeschna*). **Unique characters**: None.

Status of classification: Because male is still un-

known, its phylogenetic position is still uncertain. It is considered to be a Gynacanthini (De Marmels, 2000; von Ellenrieder, 2002), and De Marmels (2000) considers it to be related to *Limnetron* [see discussion under *Limnetron*].

Potential for new species: Likely. Extremely rare in collections.

Habitat: Pools in small, narrow mountain streams enclosed in primary forest, with shallow (ca. 50 cm) slow flowing, tannin rich water (De Marmels, *pers. comm.*).

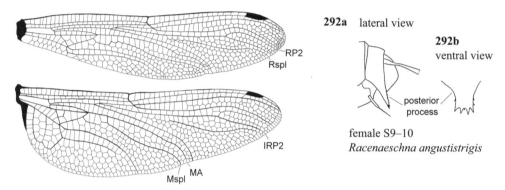

292a   lateral view

292b
ventral view

posterior process

female S9–10
Racenaeschna angustistrigis

291   female wings – *Racenaeschna angustistrigis*

Remartinia Navás, 1911: 479.

[♂ pp. 28, 30, couplets 10, 15; ♀ pp. 35, couplet 15]

Type species: *Aeschna luteipennis* Burmeister, 1839 [Carvalho, 1992a by subsequent designation]

4 species:

luteipennis florida (Hagen, 1861) [*Aeschna*]* – **L** [Calvert, 1956]
luteipennis luteipennis (Burmeister, 1839) [*Aeschna*]*
  syn *barbiellina* Navás, 1911
  syn *excisa* (Brauer, 1865) [*Aeschna*]
luteipennis peninsularis (Calvert, 1941) [*Aeshna (Coryphaeschna)*]*
restricta Carvalho, 1992
rufipennis (Kennedy, 1941) [*Aeshna*]*
secreta (Calvert, 1952) [*Coryphaeschna*]* – **L** [Novelo-Gutiérrez, 1998]

References: Calvert, 1956; Carvalho, 1992a.

Distribution: Neotropical, from S United States to N Argentina.

Large aeshnids (71–85 mm); pale green or blue face, black T- spot on frons; pterothorax with pair of green lateral thoracic stripes; abdomen reddish brown to black with fine green lines. Wings (Fig. 293) with elongated triangles, basal side of HW triangle shorter

Map 23. Distribution of *Remartinia* spp.

than 1/2 of its costal side; Rspl not reaching border of wing distally. In males, anal triangle 2 celled, and vesica spermalis distal segment (Fig. 294a) closed except for anterodorsal 'opening' (shared with *Castoraeschna* and *Coryphaeschna*), and paired patches of spines on base of ventral process (Fig. 294b). Sternum S10 of female (Figs. 295a-b) with sclerotized posterior bar bearing denticles in one band; female cerci about as long as S10. **Unique characters**: Male anterior hamule (Fig. 296) with a medial excision on its outer margin; vesica spermalis distal segment subquadrate, with maximum width at its distal portion

(Figs. 294a-b).
Status of classification: Good. Most closely related to *Coryphaeschna* (Carvalho, 1995).

Potential for new species: Likely.
Habitat: Ponds and temporary pools with abundant aquatic vegetation in forests.

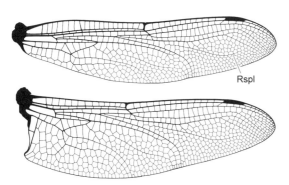

Rspl

293 male wings – *Remartinia luteipennis*

vesica spermalis distal segment

antero-dorsal 'opening'

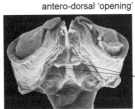

patches
of spines

294a
dorsal
view

Remartinia
luteipennis

294b
ventral
view

295a lateral view

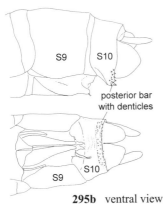

S9   S10

posterior bar
with denticles

S9   S10

295b ventral view

female S9–10 – *Remartinia luteipennis*

medial
excision

296 male anterior hamules – *Remartinia luteipennis*

ventral view

S10

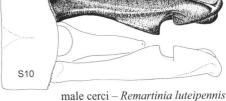

297a dorsal view

S10

male cerci – *Remartinia luteipennis*

S10

297b lateral view

Rhionaeschna Förster, 1909: 220.
[♂ pp. 31, couplet 21; ♀ pp. 35, couplet 18]
Type species: *Rhionaeschna maita* Förster, 1909 [by original designation]
   syn *Neureclipa* Navás, 1911: 476, 478
      Type species: *Aeschna bonariensis* Rambur, 1842 [Calvert, 1952 by subsequent designation]
   syn *Aeshna* (*Hesperaeshna*) Cockerell, 1913: 581
      Type species: *Aeshna californica* Calvert, 1895 [by monotypy]
   syn *Aeshna* (*Marmaraeschna*) Calvert, 1952: 256
      Type species: *Aeshna intricata* Martin, 1908 [by original designation]
   syn *Aeshna* (*Schizuraeschna*) Calvert, 1952: 256
      Type species: *Aeschna multicolor* Hagen, 1861 [by original designation]

41 species:

absoluta (Calvert, 1952) [*Aeshna*]* – **L** [von Ellenrieder, 2001]
biliosa (Kennedy, 1938) [*Aeshna*]*
bonariensis (Rambur, 1842) [*Aeshna*]* – **L** [Rodrigues Capítulo, 1980; von Ellenrieder, 2001]
   syn *litigatrix* Navás, 1911 [*Aeschna*]
   syn var. *lutea* Navás, 1920 [*Aeshna*]
brasiliensis (von Ellenrieder and Costa, 2002) [*Aeshna*]* – **L** [Santos, 1966b as *A.peralta*; von Ellenrieder and Costa, 2002]
brevicercia (Muzón and von Ellenrieder, 2001) [*Aeshna*]* – **L** [De Marmels, 2001b as *A. vigintipunctata*]
brevifrons (Hagen, 1861) [*Aeshna*]*

syn *maita* Foerster, 1909

californica (Calvert, 1895) [*Aeschna*]* – L [Needham and Hart, 1901; Musser, 1962]

condor (De Marmels, 2001) [*Aeshna*]* – L [De Marmels, 2001b]

confusa (Rambur, 1842) [*Aeschna*]* – L [von Ellenrieder, 2001]

cornigera (Brauer, 1865) [*Aeschna*]* – L [De Marmels, 1982b]

decessus (Calvert, 1953) [*Aeshna*]

demarmelsi von Ellenrieder, 2003* – L [De Marmels, 2001b as *A. joannisi*]

diffinis (Rambur, 1842) [*Aeschna*]* – L [Calvert, 1956; von Ellenrieder, 2001]

draco (Rácenis, 1958) [*Aeshna*]* – L [De Marmels, 1990a]

dugesi (Calvert, 1905) [*Aeschna*]* – L [Novelo-Gutiérrez and González-Soriano, 1991]

eduardoi (Machado, 1984) [*Aeschna*]*

elsia (Calvert, 1952) [*Aeschna*]*

fissifrons (Muzón and von Ellenrieder, 2001) [*Aeshna*]*

galapagoensis (Currie, 1901) [*Aeschna*]* – L [Needham, 1904, based on half-grown larvae]

haarupi (Ris, 1908) [*Aeschna*]*

intricata (Martin, 1908) [*Aeschna*]*

itatiaia (Carvalho and Salgado, 2004) [*Aeshna*]

jalapensis (Williamson, 1908) [*Aeschna*]* – L [Calvert, 1956]

joannisi (Martin, 1897) [*Aeschna*]*

manni (Williamson and Williamson, 1930) [*Aeshna*]*

marchali (Rambur, 1842) [*Aeschna*]* – L [Limongi, 1983]

multicolor (Hagen, 1861) [*Aeschna*]* – L [Walker, 1912; Musser, 1962]

syn *furcifera* (Karsch, 1891) [*Aeschna*]

mutata (Hagen, 1861) [*Aeschna*]* – L [Walker, 1958]*

nubigena (De Marmels, 1989) [*Aeschna*]*

obscura (Muzón and von Ellenrieder, 2001) [*Aeshna*]*

pallipes (Fraser, 1947) [*Aeshna*]* – L [von Ellenrieder and Muzón, 2003]

peralta (Ris, 1918) [*Aeschna*]*

pauloi (Machado 1994) [*Aeshna*]*

planaltica (Calvert, 1952) [*Aeschna*]* – L [De Marmels, 1992d; von Ellenrieder, 1999]

psilus (Calvert, 1947) [*Aeshna*]* – L [Needham and Westfall, 1955; Calvert, 1956]

syn ?*dominicana* (*nomen nudum*) (Hagen, 1861) [*Aeschna*]

punctata (Martin, 1908) [*Aeschna*]* – L [Santos, 1966a]

syn *depravata* (*nomen nudum*) (Hagen, 1861) [*Aeschna*]

syn *lobata* (*nomen nudum*) (Hagen, 1861) [*Aeschna*]

serrana (Carvalho and Salgado, 2004) [*Aeshna*]

tinti (von Ellenrieder, 2000) [*Aeshna*]*

variegata (Fabricius, 1775) [*Aeschna*]* – L [Muzón and von Ellenrieder, 1996; von Ellenrieder, 2001]

syn *diffinis* var. *risi* (Enderlein, 1912) [*Aeschna*]

vazquezae (González, 1986) [*Aeschna*]*

vigintipunctata (Ris, 1918) [*Aeschna*]*

References: Calvert, 1956; Muzón and von Ellenrieder, 2001; von Ellenrieder, 2003.

Distribution: Mainly neotropical, with only three nearctic species (*R. mutata, R. multicolor* and *R. californica*); from S Canada to S Argentina, where *R. variegata* represents the southernmost odonate species, with breeding populations in Tierra del Fuego (54° S). Absent from the Amazonian basin, the genus shows its highest species richness in the Andes from Venezuela to Bolivia.

Map 24. Distribution of *Rhionaeschna* spp.

Small to large aeshnids (50–80 mm); black T-spot on frons (Figs. 298a-300a); pterothorax (Figs. 305-307) with a pair of pale lateral thoracic stripes, or with pale and dark spots, or concolorous; mosaic of pale spots on abdomen. Pale coloration of males blue, green or yellow. Females of several species dichromatic with andromorph (male-like) and heteromorph (often green or yellow) color females. Wings (Figs. 301-304) with IRP2 symmetrically forked, angled anal angle in males. Male auricles (Figs. 308-309) with 2–3 strong teeth. Distal segment of vesica spermalis (Fig. 314) with laterobasal folds. Male cerci (Figs. 317-319) longer than S9–10. **Unique characters**: Sternum S1 of both males and females (Figs. 310-312) with a conical tubercle (*Castoraeschna* has also a tubercle, but it is higher and cylindrical).

Status of classification: Fairly good. Two species recently described (*R. itatiaia* and *R. serrana*, Carvalho and Salgado, 2004) most likely represent variation of *R. punctata*; their synonymy awaits further study. According to von Ellenrieder (2003), the genus could be related to some African species of *'Aeshna'* (*'A.' rileyi* Calvert, *'A.' subpupillata* McLachlan, *'A.' moori* Pinhey, *'A.' yemenensis* Waterston, *'A.' minuscula* McLachlan), and *Rhionaeschna* plus the African clade would constitute the sister group of *Adversaeschna*, *Andaeschna*, *Anaciaeschna*, *Anax* and several species of *'Aeshna'* of uncertain affinities (*i.e. 'A.' affinis* vander Linden, *'A.' ellioti* Kirby, *'A.' mixta* Latreille, *'A.' isoceles* Müller, and *'A.' williamsoniana*).

Phylogenetic relationships within this complex are not yet resolved.

Potential for new species: Likely.

Habitat: Occurring at almost any body of water, from lowlands to high elevations in the Andes (i.e. *R. fissifrons* at 3800 m), and from dry steppes to rainforests. In austral South America, *Rhionaeschna* is the predominant dragonfly of bogs, marshes and rivers, where it can be abundant.

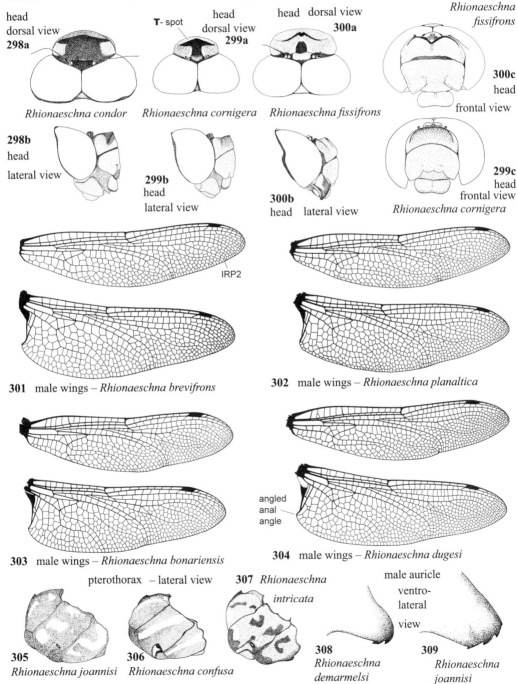

head dorsal view **298a**

Rhionaeschna condor

298b head lateral view

T- spot head dorsal view **299a**

Rhionaeschna cornigera

299b head lateral view

head dorsal view **300a**

Rhionaeschna fissifrons

300b head lateral view

Rhionaeschna fissifrons

300c head frontal view

299c head frontal view

Rhionaeschna cornigera

IRP2

301 male wings – *Rhionaeschna brevifrons*

302 male wings – *Rhionaeschna planaltica*

303 male wings – *Rhionaeschna bonariensis*

angled anal angle

304 male wings – *Rhionaeschna dugesi*

pterothorax – lateral view

307 *Rhionaeschna intricata*

male auricle ventro-lateral view

305 *Rhionaeschna joannisi*

306 *Rhionaeschna confusa*

308 *Rhionaeschna demarmelsi*

309 *Rhionaeschna joannisi*

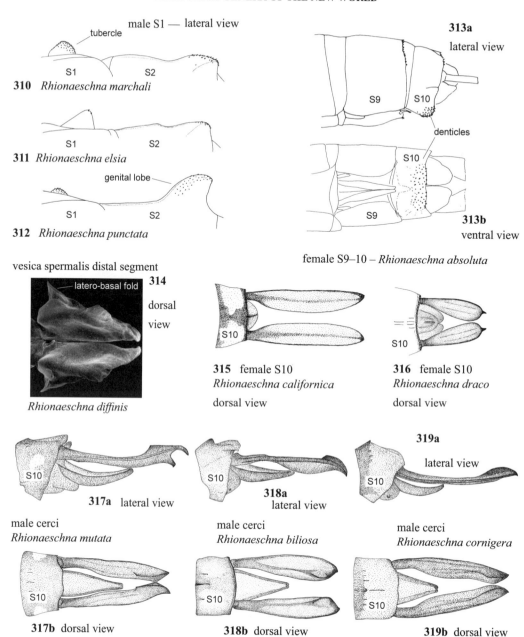

male S1 — lateral view

tubercle

S1 S2

310 *Rhionaeschna marchali*

S1 S2

311 *Rhionaeschna elsia*

genital lobe

S1 S2

312 *Rhionaeschna punctata*

313a lateral view

S9 S10

denticles

S10

S9

313b ventral view

vesica spermalis distal segment

latero-basal fold **314** dorsal view

Rhionaeschna diffinis

female S9–10 – *Rhionaeschna absoluta*

S10

S10

315 female S10 *Rhionaeschna californica* dorsal view

316 female S10 *Rhionaeschna draco* dorsal view

S10

317a lateral view

male cerci *Rhionaeschna mutata*

S10

318a lateral view

male cerci *Rhionaeschna biliosa*

319a lateral view

S10

male cerci *Rhionaeschna cornigera*

S10

317b dorsal view

S10

318b dorsal view

S10

319b dorsal view

Staurophlebia Brauer, 1865: 907.

[♂ pp. 31, couplet 22; ♀ pp. 35, couplet 20]

Type species: *Staurophlebia magnifica* Brauer, 1865 [by monotypy]

5 species:

auca Kennedy, 1937
bosqi Navás, 1927 – **L** [Bachmann, 1963]
gigantula Martin, 1909
reticulata guatemalteca Walker, 1915* – **L** [Needham, 1904]

reticulata obscura Walker, 1915
reticulata reticulata (Burmeister, 1839) [*Aeschna*]*
– **L** [Geijskes, 1959]
    syn ?*platyura* Navás, 1920
    syn *gigas* (Rambur, 1842) [*Aeschna*]
    syn *magnifica* Brauer, 1865
wayana Geijskes, 1959*

References: Walker, 1915; Geijskes, 1959, 1964a.
Distribution: From Belize to E Argentina.

Map 25. Distribution of *Staurophlebia* spp.

Large to very large aeshnids (75–96 mm); face pale green to pale brown with black T-spot on frons; pterothorax concolorous, green to reddish brown; abdomen reddish brown or yellow-orange to black with green spots. Wings (Fig. 320) with subcosta prolonged beyond nodus (shared with *Neuraeschna*); free median space; HW triangle basal side shorter than 1/2 of costal side; HW supratriangle longer than median space. Distal segment of vesica spermalis (Figs. 323-324) with a ventral medio-longitudinal laminar fold bearing one or two flap-like projections at its anterior end (shared with *Neuraeschna*). Female cerci longer than S10; sclerotized posterior process on sternum of S10 (Figs. 321a-c) bearing two apical spines. **Unique characters**: No apical fold anterior to medio-longitudinal fold on venter of vesica spermalis distal segment (Figs. 323a, 324a).

<u>Status of classification</u>: Genus well defined, most closely related to *Neuraeschna* in the New World (von Ellenrieder, 2002). In spite of Geijskes (1959) revision, a review of the genus is needed to clarify the status of Navás' names.

<u>Potential for new species</u>: Likely.

<u>Habitat</u>: Creeks and narrow streams within forests.

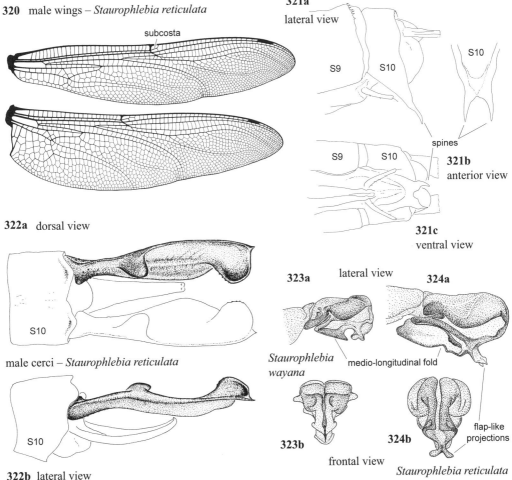

320 male wings – *Staurophlebia reticulata*

subcosta

female S9–10 –*Staurophlebia reticulata*

321a
lateral view

S9   S10

S10

S9   S10

spines

321b
anterior view

321c
ventral view

322a dorsal view

S10

male cerci – *Staurophlebia reticulata*

S10

322b lateral view

323a   lateral view   324a

Staurophlebia wayana   medio-longitudinal fold

323b   324b

frontal view

flap-like projections

Staurophlebia reticulata

vesica spermalis distal segment

Triacanthagyna Selys, 1883b: 745 (37 reprint).
[♂ pp. 30, couplet 20; ♀ pp. 34, couplet 12]

Type species: *Gynacantha trifida* Rambur, 1842 [by monotypy]

9 species:

caribbea Williamson, 1923* – **L** [Santos, 1973a]
dentata (Geijskes, 1943) [*Coryphaeschna*]* – **L** [De Marmels, 1992d]
ditzleri Williamson, 1923*
nympha (Navás, 1933) [*Gynacantha*]* – **L** [Carvalho, 1988 as *T. ditzleri*]
obscuripennis (Blanchard, 1845) [*Aeschna*]*
satyrus (Martin, 1909) [*Gynacantha*]*
septima (Selys *in* Sagra, 1857) [*Gynacantha*]* – **L** [Calil and Carvalho, 1999]
trifida (Rambur, 1842) [*Gynacantha*]* – **L** [Needham and Westfall, 1955]
　　syn *needhami* Martin, 1909
williamsoni von Ellenrieder and Garrison, 2003*

References: Williamson, 1923a; von Ellenrieder and Garrison, 2003.
Distribution: From SE United States to central Argentina.

Small to medium aeshnids (about 55–77 mm), with black spot on frons, pterothorax with pair of green lateral thoracic stripes or concolorous pale brown or pale green, and pale to dark brown abdomen with green or bluish spots. Wings (Fig. 325) with HW su-

Map 26. Distribution of *Triacanthagyna* spp.

pratriangle longer than median space. Male auricles with 2–3 strong teeth; anterior process of anterior hamules not separated from base by a groove (Fig. 328). Female sclerotized posterior process on sternum of S10 (Figs. 326a-b) bearing three apical spines (unique among New World aeshnids, shared with the oriental *Agyrtacantha*). **Unique characters**: Tubular structure on ventral surface of vesica spermalis distal segment (Fig. 327).
Status of classification: Good. Related to *Staurophlebia*, *Neuraeschna*, *Racenaeschna* and *Gynacantha* in the New World (von Ellenrieder, 2002).
Potential for new species: Likely.
Habitat: Crepuscular habits in forests, where they breed in temporary pools and phytotelmata.

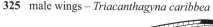

325　male wings – *Triacanthagyna caribbea*

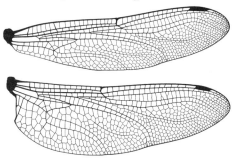

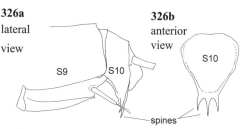

326a lateral view　　　　**326b** anterior view

S9　　S10　　S10

spines

female S9–10 – *Triacanthagyna obscuripennis*

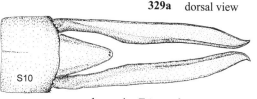

329a　dorsal view

S10

male cerci – *Triacanthagyna septima*

328　male anterior hamules
Triacanthagyna trifida

anterior process

base

ventral view

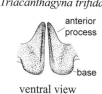

tubular structure

ventral view

327　vesica spermalis distal segment
Triacanthagyna nympha

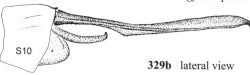

S10

329b　lateral view

6. Gomphidae

Cosmopolitan: About 941 spp. in 92 genera.

New World: 355 spp. in 34 genera (31 endemic).

Diagnostic characters: Prementum of adult entire (Carle, 1995); eyes separated on top of head by space equal to distance between lateral ocelli (Fig. 346); trapezoidal vertex composed, in most genera, of a raised transverse postocellar ridge variously modified (Figs. 530-531); trapezoidal occiput well developed, rounded (Fig. 341) to ridged posteriorly (Fig. 342); some genera/ species with basal subcostal crossvein (Fig. 331); midbasal space free (Figs. 330-331); numerous bridge crossveins (Figs. 330-331); anal triangle present in males of most genera (Figs. 330-331), and when absent, tornal area of HW angled; males with auricles on S2 (Figs. 332-333), these structures present in females of several genera in a rudimentary state (Fig. 469); anterior hamules small and often inconspicuous, posterior hamules prominent and well-developed (Figs. 337-338); female vulvar lamina present as a bifid arcuate flap (Figs. 438-442).

Status of classification: Gomphidae is a well-studied group comprising the largest number of genera and species next to the Libellulidae. Several workers (Tillyard, 1917; Williamson, 1920, Tillyard and Fraser, 1940; Carle, 1986; Bechly, 1996; Belle, 1996) have offered reclassifications based primarily on wing venation. Bechly (1996) raised the family to a new taxonomic level (Gomphata), including in it Progomphidae, Zonophoridae, Lindeniidae, Epigomphidae, and Hageniidae, and allocated numerous subordinate infrafamilial names according to previous classifications primarily based on wing venation. Carle (1986) proposed a world-wide "phylogenetic" classification without supported illustrations, cladogram, or indication of material examined; his classification was emended by Belle (1996), but it referred only to the South American fauna. Many new neotropical taxa were described in a series of over 60 papers by the late Jean Belle from 1963 to 1996. However, he (Belle, 1996) provided references to wing illustrations for all South American species, thus allowing reference to the scattered literature

Key to males

[NOTE: The males of *Anomalophlebia, Brasiliogomphus* and *Praeviogomphus* are still unknown, therefore they are not included in the following key; check with genera accounts on pages 89, 94, and 124 for nongenital characters.]

1. Crossveins in HW space between sectors of arculus and point of branching of RP in number of 1 (Fig. 330) **2**

1'. Crossveins in HW space between sectors of arculus and point of branching of RP in number of 2 or more (Fig. 331) **14**

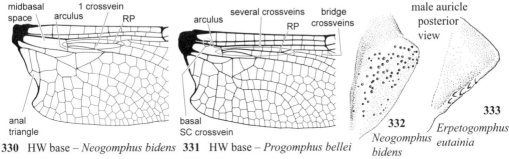

330 HW base – *Neogomphus bidens*   **331** HW base – *Progomphus bellei*

332 *Neogomphus bidens*   **333** *Erpetogomphus eutainia*

2(1). Auricle in postero-lateral view with small denticles in a cluster (Fig. 332) **3**

2'. Auricle in postero-lateral view with small denticles in a row (Fig. 333) **13**

3(2). Cercus unbranched (Figs. 334-335) **4**

3'. Cercus branched (Fig. 336) **12**

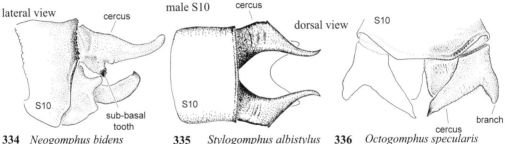

334 *Neogomphus bidens*   **335** *Stylogomphus albistylus*   **336** *Octogomphus specularis*

4(3). Anterior hamule not digitiform (Fig. 337) .... 4'. Anterior hamule digitiform (Fig. 338) ....
.... **5** .... *Stylurus* (Page 128)

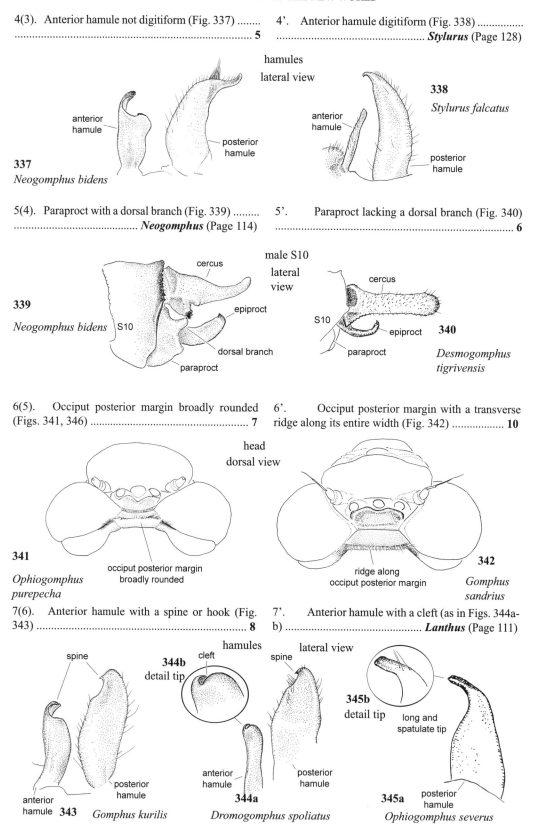

hamules
lateral view

338
Stylurus falcatus

anterior
hamule

anterior
hamule

posterior
hamule

337
Neogomphus bidens

posterior
hamule

5(4). Paraproct with a dorsal branch (Fig. 339) .... 5'. Paraproct lacking a dorsal branch (Fig. 340)
.... *Neogomphus* (Page 114) **6**

male S10
lateral view

cercus

339

Neogomphus bidens S10

epiproct

cercus

S10

epiproct

340

paraproct

dorsal branch

paraproct

*Desmogomphus
tigrivensis*

6(5). Occiput posterior margin broadly rounded 6'. Occiput posterior margin with a transverse
(Figs. 341, 346) .... **7** ridge along its entire width (Fig. 342) **10**

head
dorsal view

341

*Ophiogomphus
purepecha*

occiput posterior margin
broadly rounded

342

ridge along
occiput posterior margin

*Gomphus
sandrius*

7(6). Anterior hamule with a spine or hook (Fig. 7'. Anterior hamule with a cleft (as in Figs. 344a-
343) .... **8** b) .... *Lanthus* (Page 111)

hamules
lateral view

spine

344b
detail tip

cleft

spine

345b
detail tip

long and
spatulate tip

posterior
hamule

anterior
hamule

anterior
hamule **343** *Gomphus kurilis*

posterior
hamule

anterior
hamule

344a

Dromogomphus spoliatus

345a

posterior
hamule

Ophiogomphus severus

8(7). Posterior hamule with long and spatulate posterior tip lacking teeth (Figs. 345a-b) .... ***Ophiogomphus* in part** (Page 117)

8'.   Posterior hamule lacking long and spatulate posterior tip, ending in a single tooth (Figs. 343-344a) **9**

9(8). Occiput largely bare or with scattered hairs (as in Fig. 346) .... ***Stylogomphus*** (Page 127)

9'.   Occiput with a fringe of hairs along its summit (as in Figs. 341-342) .. ***Gomphus* in part** (Page 106)

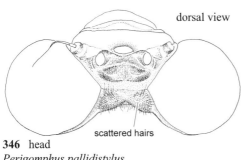

dorsal view

HW base

346 head
Perigomphus pallidistylus

scattered hairs

no anal loop

347
Gomphus australis

anal loop

348
Ophiogomphus arizonicus

10(6). Anterior hamule with a spine or hook (Fig. 343) **11**

10'.   Anterior hamule with a cleft (Figs. 344a-b) .... ***Dromogomphus*** (Page 99)

11(10). Anal loop not defined (Fig. 347); posterior hamule lacking long and spatulate posterior tip, ending in a single tooth (Fig. 343) .... ***Arigomphus* in part / *Gomphus* in part ***

11'.   Anal loop defined (Fig. 348); posterior hamule with long and spatulate posterior tip lacking teeth (Figs. 345a-b) ... ***Ophiogomphus* in part** (Page 117)

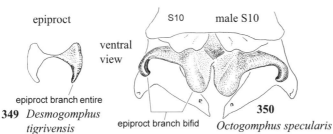

epiproct

S10   male S10

hamules lateral view

entire

ventral view

posterior hamule

epiproct branch entire

349 *Desmogomphus tigrivensis*

epiproct branch bifid

350
Octogomphus specularis

anterior hamule

351
Octogomphus specularis

12(3).   Epiproct branches entire (as in Fig. 349); anal loop not defined (Fig. 347); anterior hamule with a spine or hook (Fig. 343) .... ***Arigomphus* in part / *Gomphus* in part ***

12'. Epiproct branches bifid (Fig. 350); anal loop defined (as in Fig. 348); anterior hamule lacking cleft, hooks or spines (Fig. 351)................................ .... ***Octogomphus*** (Page 116)

* *Arigomphus* (Page 92): Middorsal thoracic dark stripe absent, indistinct or incomplete; when present, not extending anteriorly to collar; lateral thoracic stripes reduced or absent (as in Figs. 352-353)

* *Gomphus* (Page 106): Middorsal thoracic dark stripe present, usually extending anteriorly to collar; lateral thoracic stripes present (Fig. 354)

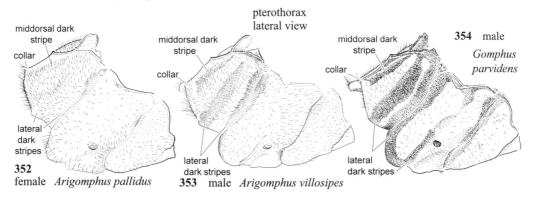

pterothorax lateral view

middorsal dark stripe

collar

lateral dark stripes

352
female *Arigomphus pallidus*

middorsal dark stripe

collar

lateral dark stripes

353 male *Arigomphus villosipes*

middorsal dark stripe

collar

lateral dark stripes

354 male

Gomphus parvidens

13(2).　Anal loop not defined (Fig. 355); posterior hamule lacking long and spatulate posterior tip (Fig. 357) .... ***Erpetogomphus*** (Page 103)

13'.　Anal loop defined (Fig. 356); posterior hamule with long and spatulate posterior tip (Figs. 358a-b) .... ***Ophiogomphus*** **in part** (Page 117)

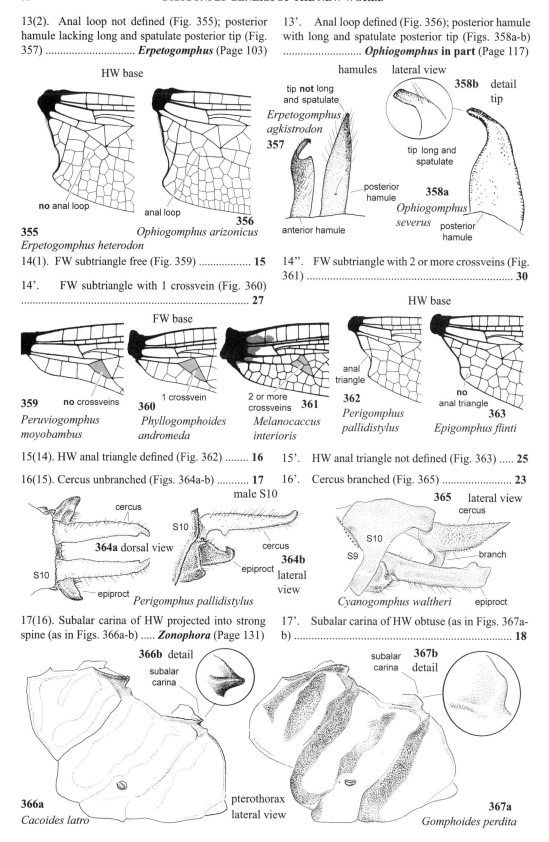

HW base

no anal loop

anal loop

355
Erpetogomphus heterodon

356
Ophiogomphus arizonicus

hamules　　lateral view

tip **not** long and spatulate

Erpetogomphus agkistrodon

357

anterior hamule

posterior hamule

358b detail tip

tip long and spatulate

358a
Ophiogomphus severus posterior hamule

14(1).　FW subtriangle free (Fig. 359) **15**

14'.　FW subtriangle with 1 crossvein (Fig. 360) **27**

14".　FW subtriangle with 2 or more crossveins (Fig. 361) **30**

FW base

HW base

359 **no** crossveins
Peruviogomphus moyobambus

360 1 crossvein
Phyllogomphoides andromeda

361 2 or more crossveins
Melanocaccus interioris

anal triangle

362
Perigomphus pallidistylus

no anal triangle

363
Epigomphus flinti

15(14).　HW anal triangle defined (Fig. 362) **16**

15'.　HW anal triangle not defined (Fig. 363) **25**

16(15).　Cercus unbranched (Figs. 364a-b) **17**

16'.　Cercus branched (Fig. 365) **23**

male S10

cercus

S10

364a dorsal view

S10

epiproct

cercus

364b lateral view

epiproct
Perigomphus pallidistylus

365 lateral view
cercus

S10

S9

branch

epiproct

Cyanogomphus waltheri

17(16).　Subalar carina of HW projected into strong spine (as in Figs. 366a-b) ..... ***Zonophora*** (Page 131)

17'.　Subalar carina of HW obtuse (as in Figs. 367a-b) **18**

366b detail

subalar carina

subalar carina

367b detail

366a
Cacoides latro

pterothorax lateral view

367a
Gomphoides perdita

368 FW – *Perigomphus pallidistylus*

369 FW – *Hagenius brevistylus*

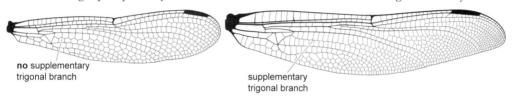

no supplementary
trigonal branch

supplementary
trigonal branch

18(17). FW with supplementary trigonal branch (Fig. 369); vesica spermalis basal segment entire (Fig. 370) .... ***Hagenius*** (Page 109)

18'.　FW lacking supplementary trigonal branch (Fig. 368); vesica spermalis basal segment bifid (Figs. 371a-b) **19**

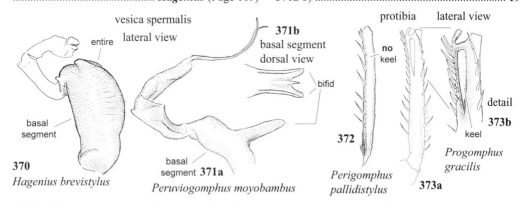

vesica spermalis
lateral view

entire

371b
basal segment
dorsal view

bifid

protibia　　lateral view

no
keel

detail
373b

keel

basal
segment

370
Hagenius brevistylus

basal
segment **371a**
Peruviogomphus moyobambus

372

*Perigomphus
pallidistylus*

373a

*Progomphus
gracilis*

19(18). HW anal triangle reaching anal angle of wing (Fig. 374a); inner margin of anal triangle with a series of closely spaced comb-like denticles (Fig. 374b) .... ***Peruviogomphus* in part** (Page 119)

19'　HW anal triangle not reaching anal angle of wing (Fig. 375); inner margin of anal triangle with irregularly placed denticles or lacking denticles **20**

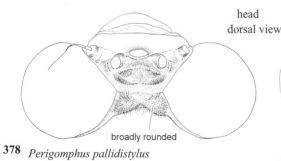

detail anal triangle
374b

*Peruviogomphus
moyobambus*

*Perigomphus
pallidistylus*

comb-like
denticles

anal
angle **374a**

anal
triangle

anal
angle
375

male HW base

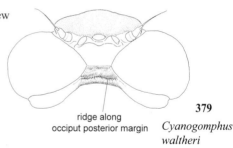

cercus

male S10
lateral
view

S10

cercus

S10

paraproct
dorsal branch

epiproct

377

376
*Perigomphus
pallidistylus*

paraproct

*Desmogomphus
tigrivensis*

20(19). Occiput posterior margin broadly rounded (Fig. 378); paraproct with a dorsal branch (Fig. 376); tibial keel in fore leg absent (Fig. 372) .... ***Perigomphus*** (Page 118)

20'.　Occiput posterior margin with a transverse ridge along its entire width (as in Fig. 379); paraproct lacking a dorsal branch (Fig. 377); tibial keel in fore leg present (Figs. 373a-b) **21**

head
dorsal view

broadly rounded

378 *Perigomphus pallidistylus*

ridge along
occiput posterior margin

379
*Cyanogomphus
waltheri*

21(20). Posterior hamule with no denticles on anterior edge nor a chitinous ridge at base (Figs. 380-381) **22**

21'. Posterior hamule with a row of denticles on anterior edge or a chitinous ridge at base (Fig. 382) ... . ***Progomphus*** **in part** (Page 125)

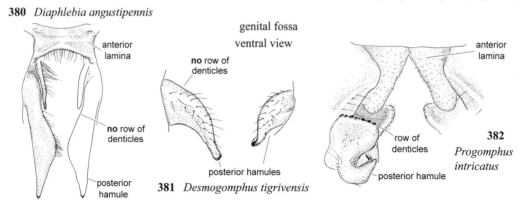

380 *Diaphlebia angustipennis*

anterior lamina

no row of denticles

posterior hamule

genital fossa
ventral view

no row of denticles

posterior hamules

381 *Desmogomphus tigrivensis*

anterior lamina

row of denticles

posterior hamule

382 *Progomphus intricatus*

22(21). Postocellar ridge lacking projections (as in Fig. 383); IR1 separated from RP1 at distal end of pterostigma by 1 cell row (Fig. 385) .... ***Desmogomphus*** (Page 97)

22'. Postocellar ridge with projections (Fig. 384); IR1 separated from RP1 at distal end of pterostigma by 2 or more cell rows (Fig. 386) .... ***Diaphlebia*** (Page 98)

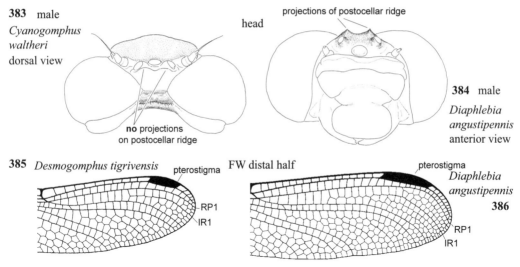

383 male
Cyanogomphus waltheri
dorsal view

head

no projections on postocellar ridge

projections of postocellar ridge

384 male
Diaphlebia angustipennis
anterior view

385 *Desmogomphus tigrivensis*
pterostigma

FW distal half

RP1
IR1

pterostigma
Diaphlebia angustipennis
386

RP1
IR1

23(16). Cercus with branch not directed ventrally (Figs. 387a-b); HW anal triangle reaching anal angle of wing (Fig. 389a); inner margin of anal triangle with a series of closely spaced comb-like denticles (Fig. 389b); supratriangle with crossveins (Fig. 389a) .... ***Peruviogomphus*** **in part** (Page 119)

23'. Cercus with branch directed ventrally (Fig. 388); HW (Fig. 390) anal triangle not reaching anal angle of wing; inner margin of anal triangle with irregularly placed denticles or lacking denticles; supratriangle free (Fig. 390) **24**

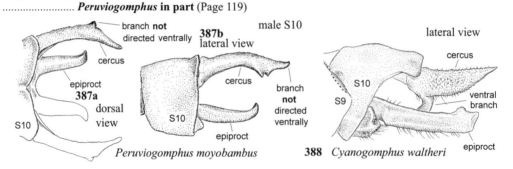

branch **not** directed ventrally

cercus

epiproct
387a
dorsal view

S10

387b
lateral view

male S10

cercus

S10

branch **not** directed ventrally

epiproct

Peruviogomphus moyobambus

lateral view

cercus

S10

S9

ventral branch

epiproct

388 *Cyanogomphus waltheri*

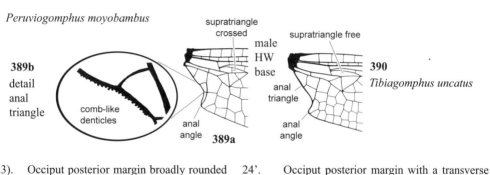

Peruviogomphus moyobambus

389b detail anal triangle

comb-like denticles

supratriangle crossed

male HW base

anal triangle

anal angle **389a**

supratriangle free

390
Tibiagomphus uncatus

anal triangle

anal angle

24(23). Occiput posterior margin broadly rounded (Fig. 391) or with a transverse ridge medially interrupted .... *Ebegomphus* (Page 101)

24'. Occiput posterior margin with a transverse ridge along its entire width (Fig. 392) ....
.... *Cyanogomphus* / *Tibiagomphus*

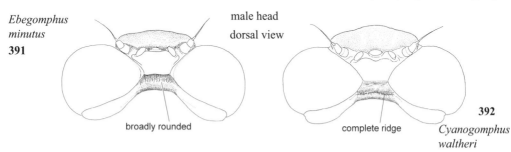

Ebegomphus minutus
391

male head dorsal view

broadly rounded

392
Cyanogomphus waltheri

complete ridge

Cyanogomphus (Page 96): Hind tibiae spines gradually tapering distally (Fig. 393a), or when swollen basally, projected at distal edge for a long distance (Fig. 393b)

Tibiagomphus (Page 129): Hind tibiae spines approximately quadrangular (Fig. 394a), projected at distal edge for a short distance (Fig. 394b)

detail spines
393b

hind tibia lateral view

393a

394a

detail spines
394b

Cyanogomphus waltheri

Tibiagomphus uncatus

25(15). Cercus unbranched (Fig. 398); pterostigma bracevein absent (Figs. 395-396) **26**

[NOTE: A bracevein may appear to be present sometimes in *Archaeogomphus*, since there is a vein occasionally aligned with the proximal angle of the pterostigma (Fig. 396); however, it is not considered a bracevein because it is not thicker than neighboring crossveins and it is usually parallel to them.]

25'. Cercus branched (Fig. 399); pterostigma bracevein usually present (Fig. 397) ....
.... *Agriogomphus* (Page 88)

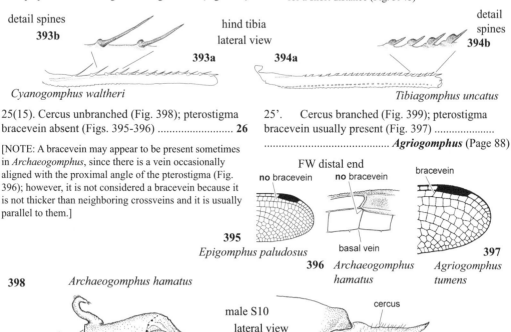

FW distal end

no bracevein

no bracevein

bracevein

395
Epigomphus paludosus

basal vein

396 *Archaeogomphus hamatus*

397
Agriogomphus tumens

398 *Archaeogomphus hamatus*

male S10 lateral view

cercus

S10

cercus

cercus branch

epiproct

399
Agriogomphus tumens

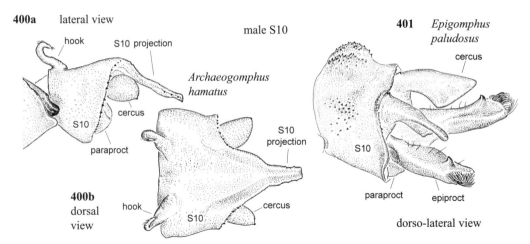

400a lateral view

male S10

hook   S10 projection

*Archaeogomphus
hamatus*

cercus

S10

paraproct

S10
projection

400b
dorsal
view   hook   cercus

S10

401 *Epigomphus
paludosus*

cercus

paraproct   epiproct

dorso-lateral view

26(25). Epiproct rudimentary, not visible in lateral view (Fig. 400a); S10 anterior margin projected into a pair of latero-dorsal hooks, and posterior margin projected medio-dorsally into a triangular, flat structure (Fig. 400b); hind tibial spines each gradually tapering distally (as in Fig. 402); anterior hamule digitiform (Fig. 405) .... ***Archaeogomphus*** (Page 91)

26'. Epiproct well developed (Fig. 401); S10 entire (Fig. 401); hind tibial spines approximately quadrangular (Fig. 403), sometimes projected at distal edge (as in Fig. 404); anterior hamule not digitiform (Fig. 406) .... ***Epigomphus*** (Page 102)

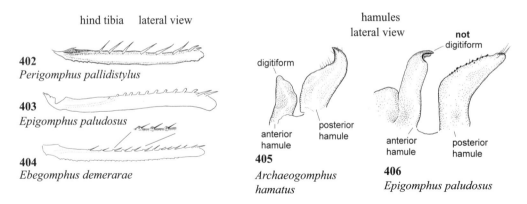

hind tibia   lateral view

hamules
lateral view   **not**
digitiform

402
Perigomphus pallidistylus

digitiform

403
Epigomphus paludosus

posterior
hamule

anterior
hamule

404
Ebegomphus demerarae

anterior
hamule

posterior
hamule

405
*Archaeogomphus
hamatus*

anterior
hamule

posterior
hamule

406
Epigomphus paludosus

27(14). Postero-dorsal rim of S10 distinguished from anterior part by polished surface, less dense setation, different color, and/or basal groove or ridge (Figs. 407a-b) **28**

27'. Postero-dorsal rim of S10 not differentiated from remainder of segment (as in Fig. 408) **29**

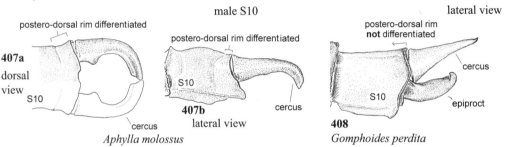

male S10

lateral view

postero-dorsal rim differentiated

postero-dorsal rim differentiated

postero-dorsal rim
not differentiated

407a
dorsal
view   S10

S10

cercus

S10

cercus

epiproct

cercus

407b lateral view

Aphylla molossus

408
Gomphoides perdita

28(27). Vesica spermalis distal segment with 2 long (Fig. 409a) serrulated (Fig. 409b) flagella or cornua .... ***Phyllocycla*** (Page 120)

28'. Vesica spermalis distal segment with 2 short (Fig. 410a-b) smooth flagella or cornua .... ***Aphylla* in part** (Page 90)

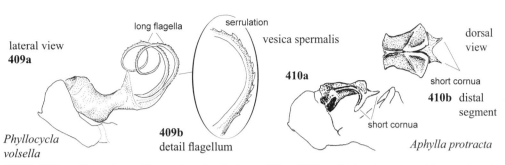

lateral view
409a

long flagella

serrulation

vesica spermalis

410a

dorsal view

short cornua

410b distal segment

short cornua

Phyllocycla volsella

409b detail flagellum

Aphylla protracta

29(27). HW anal triangle reaching anal angle of wing (Fig. 414); anterior lamina entire (as in Fig. 411); posterior hamule with no denticles on anterior edge nor a chitinous ridge at base (as in Figs. 411-412) .... ***Phyllogomphoides*** in part (Page 122)

29'. HW anal triangle not reaching anal angle of wing (Fig. 415); anterior lamina v-shaped (Fig. 413); posterior hamule with a row of denticles on anterior edge or a chitinous ridge at base (Fig. 413) .... ***Progomphus*** in part (Page 125)

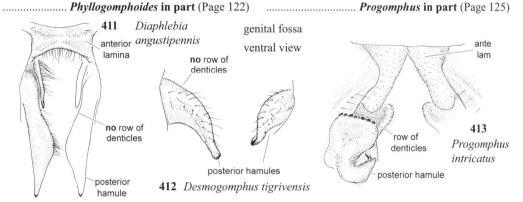

411 *Diaphlebia angustipennis*

anterior lamina

no row of denticles

no row of denticles

posterior hamule

genital fossa
ventral view

no row of denticles

posterior hamules

412 *Desmogomphus tigrivensis*

ante lam

row of denticles

413 *Progomphus intricatus*

posterior hamule

30(14). Subalar carina of HW projected into strong spine (Figs. 419a-b); cubito-anal crossveins in HW numbering 2 or more (Fig. 416); HW anal triangle not reaching anal angle of wing (as in Figs. 415-416); vesica spermalis basal segment with an entire antero-transverse ridge (Figs. 417a-b) **31**

30'. Subalar carina of HW obtuse (Figs. 420a-b); cubito-anal crossveins in HW numbering 1 (as in Fig. 415); HW anal triangle reaching anal angle of wing (as in Fig. 414); vesica spermalis basal segment bifid (as in Figs. 418a-b) **33**

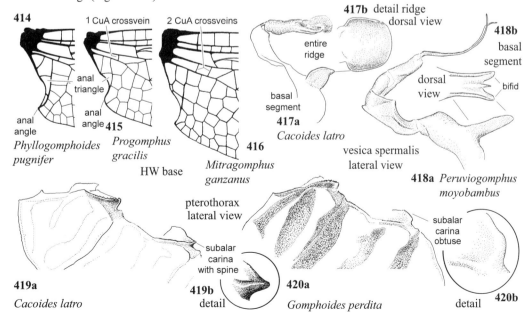

414

1 CuA crossvein

2 CuA crossveins

anal triangle

anal angle

anal angle **415**

anal angle

Phyllogomphoides pugnifer

Progomphus gracilis

416

Mitragomphus ganzanus

HW base

417b detail ridge
dorsal view

entire ridge

basal segment

417a
Cacoides latro

418b basal segment

dorsal view

bifid

vesica spermalis
lateral view

418a *Peruviogomphus moyobambus*

pterothorax
lateral view

subalar carina with spine

subalar carina obtuse

419a
Cacoides latro

419b detail

420a
Gomphoides perdita

detail **420b**

31(30). Pterostigma bracevein present; FW with supplementary trigonal branch; anal loop defined (Fig. 421) **32**

31'. Pterostigma bracevein absent; FW lacking supplementary trigonal branch; anal loop not defined (Fig. 422) .... ***Mitragomphus*** (Page 113)

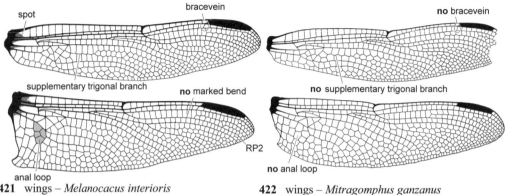

421 wings – *Melanocacus interioris*

422 wings – *Mitragomphus ganzanus*

32(31). RP2 without marked bend proximal to pterostigma (Fig. 421); wing bases colored (Fig. 421); spines of distal 1/2 of hind femur as long as 1/4 of femur width (Fig. 423) ....
.... ***Melanocacus*** (Page 112)

32'. RP2 with marked bend proximal to pterostigma (Fig. 425); wing bases hyaline (Fig. 425); spines of distal 1/2 of hind femur as long as 1/2 of femur width (Fig. 424) .... ***Cacoides*** (Page 95)

423 *Melanocacus mungo*

425 HW – *Cacoides latro*

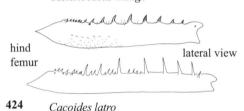

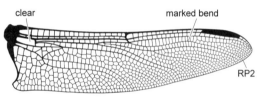

424 *Cacoides latro*

33(30). Cubito-anal crossveins in FW numbering 1 (Fig. 426); HW anal triangle with 4 or less cells (Fig. 426) **34**

33'. Cubito-anal crossveins in FW numbering 2 or more (Fig. 427); HW anal triangle with 5-6 cells (Fig. 427) .... ***Idiogomphoides*** (Page 110)

wing bases

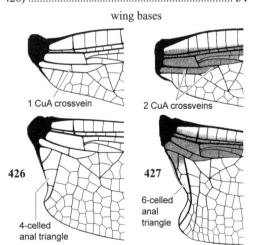

Gomphoides perdita *Idiogomphoides demoulini*

male S8–10
lateral view

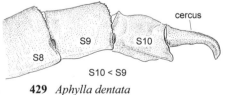

428 *Gomphoides perdita*

429 *Aphylla dentata*

34(33). S10 longer than S9 (Fig. 428); vesica spermalis distal segment with 2 long flagella or cornua (Figs. 430a-b) .... ***Gomphoides*** (Page 105)

34'. S10 shorter than S9 (Fig. 429); vesica spermalis distal segment with 2 short flagella or cornua (Figs. 431a-b) **35**

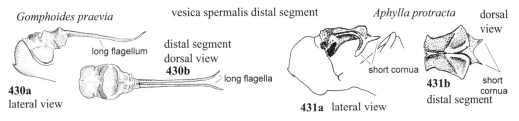

Gomphoides praevia

vesica spermalis distal segment

distal segment
dorsal view
430b
long flagella

long flagellum

430a
lateral view

Aphylla protracta   dorsal view

short cornua

431b

short cornua
distal segment

431a lateral view

35(34). Postero-dorsal rim of S10 distinguished from anterior part by polished surface, less dense setation, different color, and/or basal groove or ridge (Figs. 432a-b) .... *Aphylla* in part (Page 90)

35'.   Postero-dorsal rim of S10 not differentiated from remainder of segment (Fig. 433) ....
.... ***Phyllogomphoides*** in part (Page 122)

male S10

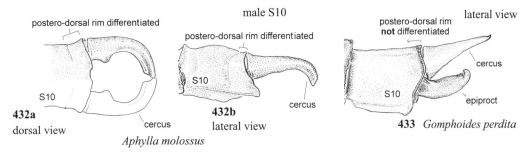

postero-dorsal rim differentiated

S10

432a
dorsal view

cercus

postero-dorsal rim differentiated

S10

432b
lateral view

cercus

Aphylla molossus

postero-dorsal rim
not differentiated   lateral view

cercus

S10

epiproct

433 *Gomphoides perdita*

Key to females

[NOTE: The female of *Mitragomphus* is still unknown, therefore it is not included in the following key; check with genus account on page 113 for nongenital characters.]

1. Crossveins in HW space between sectors of arculus and point of branching of RP in number of 1 (as in Fig. 434) **2**

1'.   Crossveins in HW space between sectors of arculus and point of branching of RP in number of 2 or more (as in Fig. 435) **19**

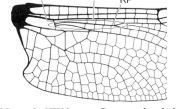

1 crossvein
arculus   RP

434 male HW base – *Neogomphus bidens*

several crossveins
arculus   RP

435 male HW base – *Progomphus bellei*

2(1).   Rear of head with 2 lateral depressions (Fig. 436) ………………........... **Octogomphus** (Page 116)

2'.   Rear of head lacking lateral depressions (as in Fig. 437) **3**

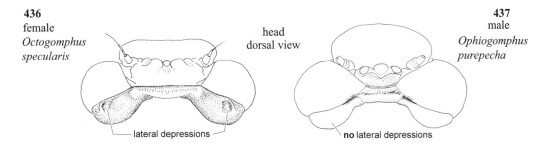

436
female
Octogomphus
specularis

head
dorsal view

437
male
Ophiogomphus
purepecha

lateral depressions

no lateral depressions

3(2). Vulvar lamina reaching 1/2 of S9 or shorter (Figs. 438-441) **4**

3'. Vulvar lamina surpassing 1/2 of S9 (Fig. 442) **14**

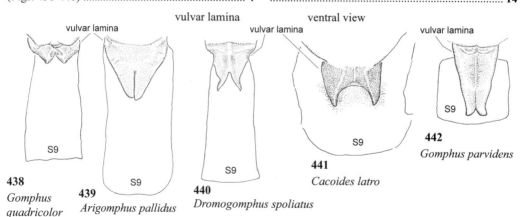

438 *Gomphus quadricolor*

439 *Arigomphus pallidus*

440 *Dromogomphus spoliatus*

441 *Cacoides latro*

442 *Gomphus parvidens*

4(3). Vulvar lamina extending to about a 1/4 of S9 (Fig. 438) **5**

4'. Vulvar lamina extending to about 1/2 of S9 (as in Fig. 441) **6**

4". Vulvar lamina extending to about a 1/3 of S9 (Figs. 439-440) **9**

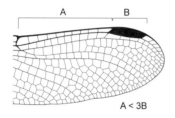

443 FW distal half – *Gomphus australis*

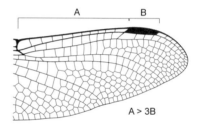

444 FW distal half – *Erpetogomphus eutainia*

5(4). Distance from nodus to pterostigma (**A**) less than 3 times the costal length of pterostigma (**B**) (Fig. 443) .... *Gomphus* **in part** / *Stylurus* **in part**

5'. Distance from nodus to pterostigma (**A**) more than 3 times the costal length of pterostigma (**B**) (Fig. 444) ...... *Erpetogomphus* **in part** / *Stylurus* **in part**

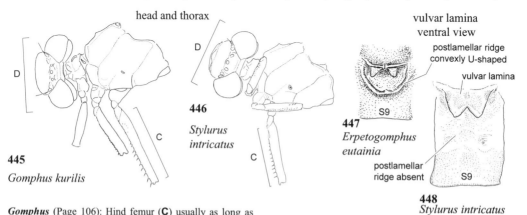

445 *Gomphus kurilis*

446 *Stylurus intricatus*

447 *Erpetogomphus eutainia*

448 *Stylurus intricatus*

Gomphus (Page 106): Hind femur (**C**) usually as long as or longer than width of head (**D**) (Fig. 445) (does not work for some species of subgenus *Hylogomphus*, e.g. *G. abbreviatus*, *G. adelphus*, and *G. geminatus*)

Stylurus (Page 128): Hind femur (**C**) usually shorter than width of head (**D**) (Fig. 446)

Erpetogomphus (Page 103): Postlamellar ridge on S9 convexly u-shaped or v-shaped (Fig. 447)

Stylurus (Page 128): Postlamellar ridge on S9 concavely u-shaped or absent (Fig. 448)

6(4). Anal loop not defined (Fig. 449) .... **7**   6'.   Anal loop defined (Figs. 450-451) **8**

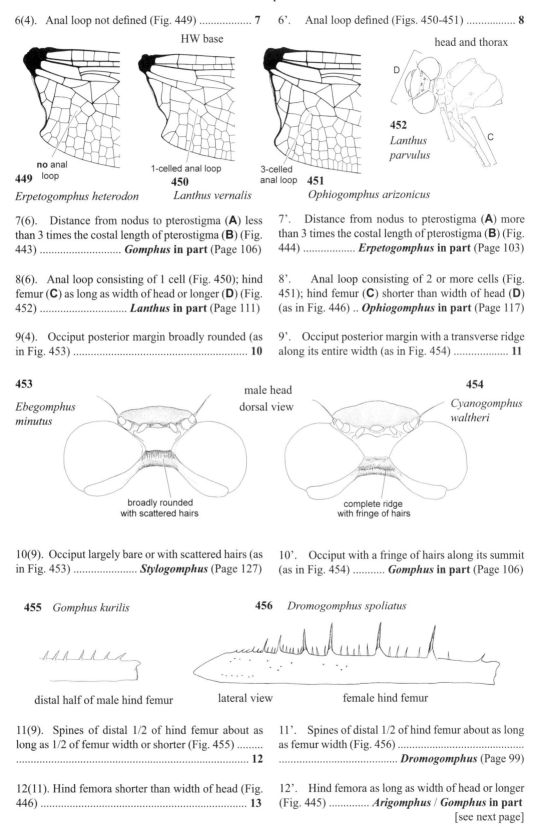

HW base

449 no anal loop

Erpetogomphus heterodon

1-celled anal loop
450
Lanthus vernalis

3-celled anal loop **451**
Ophiogomphus arizonicus

452
Lanthus parvulus

head and thorax

7(6).   Distance from nodus to pterostigma (**A**) less than 3 times the costal length of pterostigma (**B**) (Fig. 443) .... ***Gomphus*** **in part** (Page 106)

7'.   Distance from nodus to pterostigma (**A**) more than 3 times the costal length of pterostigma (**B**) (Fig. 444) .... ***Erpetogomphus*** **in part** (Page 103)

8(6). Anal loop consisting of 1 cell (Fig. 450); hind femur (**C**) as long as width of head or longer (**D**) (Fig. 452) .... ***Lanthus*** **in part** (Page 111)

8'.   Anal loop consisting of 2 or more cells (Fig. 451); hind femur (**C**) shorter than width of head (**D**) (as in Fig. 446) .. ***Ophiogomphus*** **in part** (Page 117)

9(4). Occiput posterior margin broadly rounded (as in Fig. 453) **10**

9'.   Occiput posterior margin with a transverse ridge along its entire width (as in Fig. 454) **11**

453

Ebegomphus minutus

male head
dorsal view

454

Cyanogomphus waltheri

broadly rounded
with scattered hairs

complete ridge
with fringe of hairs

10(9). Occiput largely bare or with scattered hairs (as in Fig. 453) .... ***Stylogomphus*** (Page 127)

10'.   Occiput with a fringe of hairs along its summit (as in Fig. 454) .... ***Gomphus*** **in part** (Page 106)

455 *Gomphus kurilis*

456 *Dromogomphus spoliatus*

distal half of male hind femur

lateral view   female hind femur

11(9). Spines of distal 1/2 of hind femur about as long as 1/2 of femur width or shorter (Fig. 455) **12**

11'.   Spines of distal 1/2 of hind femur about as long as femur width (Fig. 456) .... ***Dromogomphus*** (Page 99)

12(11). Hind femora shorter than width of head (Fig. 446) **13**

12'.   Hind femora as long as width of head or longer (Fig. 445) .... ***Arigomphus*** / ***Gomphus*** **in part** [see next page]

Arigomphus (Page 92): Vulvar lamina about 1/3 or less as long as S9 *and* divided into 2 triangular or rounded lobes (Fig. 457); middorsal thoracic dark stripe absent, indistinct or incomplete; when present, not extending anteriorly to collar; lateral thoracic stripes reduced or absent (as in Figs. 460, 461)

Gomphus (Page 106): Vulvar lamina about 1/2 or more the length of S9 (Fig. 458) and of varied outline (subgenera *Gomphurus*, *Hylogomphus* and *Gomphus* (*G.*) *borealis* and *G.* (*G.*) *diminutus*) *or* less than 1/4 as long as S9 (Fig. 459) and divided into 2 triangular or rounded lobes (in subgenus *Gomphus*); middorsal thoracic dark stripe present, usually extending anteriorly to collar; lateral thoracic stripes present (as in Fig. 462)

vulvar lamina
ventral view

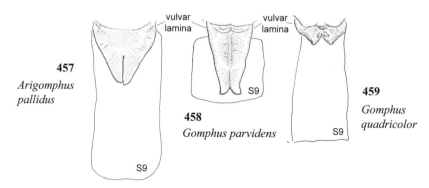

457 *Arigomphus pallidus*

458 *Gomphus parvidens*

459 *Gomphus quadricolor*

pterothorax
lateral view

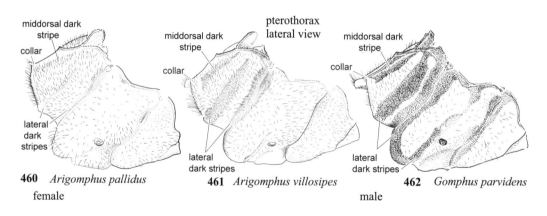

460 *Arigomphus pallidus* female

461 *Arigomphus villosipes*

462 *Gomphus parvidens* male

13(12). Distance from nodus to pterostigma (**A**) less than 3 times the costal length of pterostigma (**B**) (Fig. 463) .... **Gomphus** in part / **Stylurus** in part

13'. Distance from nodus to pterostigma (**A**) more than 3 times the costal length of pterostigma (**B**) (Fig. 464) .... **Erpetogomphus** in part / **Stylurus** in part

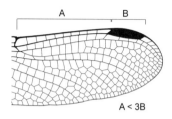

A < 3B

463 FW distal half – *Gomphus australis*

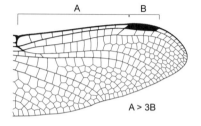

A > 3B

464 FW distal half – *Erpetogomphus eutainia*

Gomphus (Page 106): Hind femur (**C**) usually as long as or longer than width of head (**D**) (Fig. 465) (does not work for some species of subgenus *Hylogomphus*, e.g. *G. abbreviatus*, *G. adelphus*, and *G. geminatus*)

Stylurus (Page 128): Hind femur (**C**) usually shorter than width of head (**D**) (Fig. 466)

head and thorax

vulvar lamina
ventral view

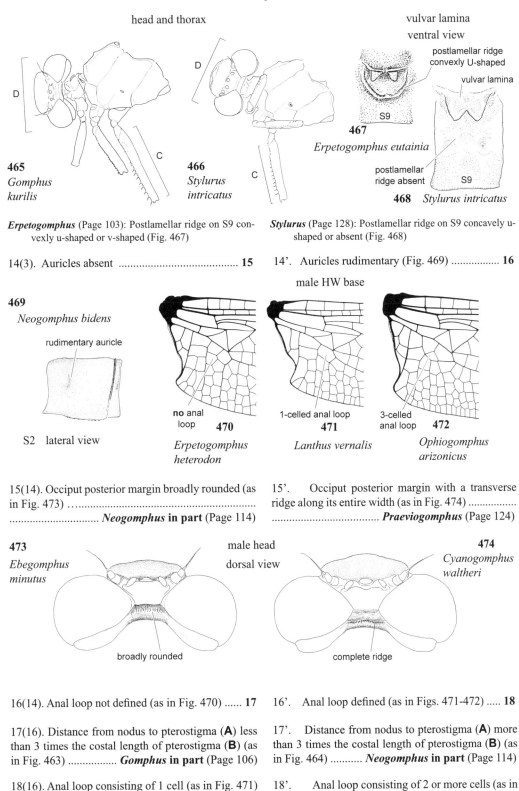

465
*Gomphus
kurilis*

466
*Stylurus
intricatus*

postlamellar ridge
convexly U-shaped

vulvar lamina

S9

467
Erpetogomphus eutainia

postlamellar
ridge absent   S9

468   *Stylurus intricatus*

Erpetogomphus (Page 103): Postlamellar ridge on S9 convexly u-shaped or v-shaped (Fig. 467)

Stylurus (Page 128): Postlamellar ridge on S9 concavely u-shaped or absent (Fig. 468)

14(3). Auricles absent **15**

14'. Auricles rudimentary (Fig. 469) **16**

male HW base

469
Neogomphus bidens

rudimentary auricle

S2   lateral view

no anal
loop   **470**
*Erpetogomphus
heterodon*

1-celled anal loop
471
Lanthus vernalis

3-celled
anal loop   **472**
*Ophiogomphus
arizonicus*

15(14). Occiput posterior margin broadly rounded (as in Fig. 473) ….. .... ***Neogomphus*** **in part** (Page 114)

15'.   Occiput posterior margin with a transverse ridge along its entire width (as in Fig. 474) .... ***Praeviogomphus*** (Page 124)

473
*Ebegomphus
minutus*

male head
dorsal view

474
*Cyanogomphus
waltheri*

broadly rounded

complete ridge

16(14). Anal loop not defined (as in Fig. 470) **17**

16'.   Anal loop defined (as in Figs. 471-472) **18**

17(16). Distance from nodus to pterostigma (**A**) less than 3 times the costal length of pterostigma (**B**) (as in Fig. 463) .... ***Gomphus*** **in part** (Page 106)

17'.   Distance from nodus to pterostigma (**A**) more than 3 times the costal length of pterostigma (**B**) (as in Fig. 464) .... ***Neogomphus*** **in part** (Page 114)

18(16). Anal loop consisting of 1 cell (as in Fig. 471) .... ***Neogomphus*** **in part** / ***Lanthus*** **in part**
[see next page]

18'.   Anal loop consisting of 2 or more cells (as in Fig. 472) .... ***Ophiogomphus*** **in part** (Page 117)

Neogomphus (Page 114): Vulvar lamina about as long as S9, divided at base into semicircular, widely separated, digit-like lobes converging at tips (Fig. 475)

Lanthus (Page 111): Vulvar lamina about 2/3 of S9 length, with a shallow and broad u-shaped cleft (Fig. 476)

vulvar lamina   ventral view

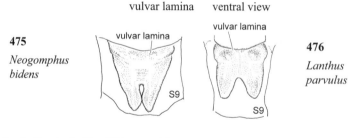

475 *Neogomphus bidens*

476 *Lanthus parvulus*

19(1). FW subtriangle free (as in Fig. 477) **20**

19'.   FW subtriangle with 1 crossvein (as in Fig. 478) **37**

19".   FW subtriangle with 2 or more crossveins (Fig. 479) **39**

male FW base

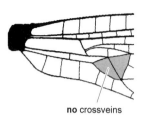

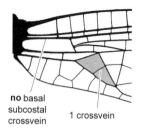

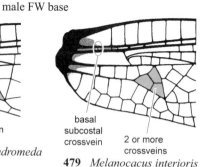

477 *Peruviogomphus moyobambus*

478 *Phyllogomphoides andromeda*

479 *Melanocacus interioris*

20(19). Epiproct clearly visible in dorsal view (Fig. 480) **21**

20'.   Epiproct rudimentary, barely or not visible in dorsal view (Fig. 481) **36**

S10 dorsal view

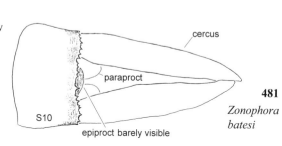

480 *Desmogomphus tigrivensis*

481 *Zonophora batesi*

21(20). Basal subcostal crossvein in FW present (as in Fig. 479) **22**

21'.   Basal subcostal crossvein in FW absent (as in Fig. 478) **30**

22(21). S10 longer than S9 (as in Fig. 482) .... *Peruviogomphus* (Page 119)

22'.   S10 shorter than S9 (as in Fig. 483) **23**

male S8–10 lateral view

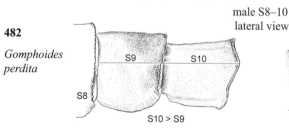

482 *Gomphoides perdita*

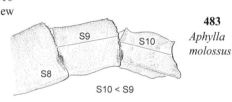

483 *Aphylla molossus*

23(22). Vulvar lamina reaching 1/2 of S9 or shorter (Figs. 484-487) **24**

23'.   Vulvar lamina surpassing 1/2 of S9 (Fig. 487) **29**

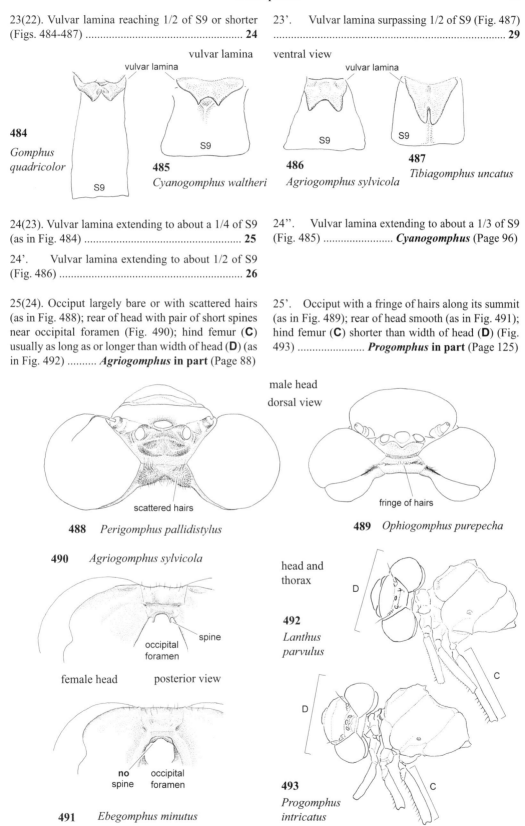

vulvar lamina

vulvar lamina

ventral view

vulvar lamina

484

Gomphus quadricolor

S9

485

Cyanogomphus waltheri

S9

486

Agriogomphus sylvicola

S9

487

Tibiagomphus uncatus

S9

24(23). Vulvar lamina extending to about a 1/4 of S9 (as in Fig. 484) **25**

24'.   Vulvar lamina extending to about 1/2 of S9 (Fig. 486) **26**

24".   Vulvar lamina extending to about a 1/3 of S9 (Fig. 485) .... *Cyanogomphus* (Page 96)

25(24). Occiput largely bare or with scattered hairs (as in Fig. 488); rear of head with pair of short spines near occipital foramen (Fig. 490); hind femur (**C**) usually as long as or longer than width of head (**D**) (as in Fig. 492) .... *Agriogomphus* **in part** (Page 88)

25'.   Occiput with a fringe of hairs along its summit (as in Fig. 489); rear of head smooth (as in Fig. 491); hind femur (**C**) shorter than width of head (**D**) (Fig. 493) .... *Progomphus* **in part** (Page 125)

male head dorsal view

scattered hairs

488   *Perigomphus pallidistylus*

fringe of hairs

489   *Ophiogomphus purepecha*

490   *Agriogomphus sylvicola*

spine

occipital foramen

head and thorax

492

Lanthus parvulus

D

C

female head   posterior view

no spine   occipital foramen

D

C

491   *Ebegomphus minutus*

493

Progomphus intricatus

26(24). Dorsal surface of pterothorax with dark areas to the side of the carina (as in Fig. 494) **27**

26'.   Dorsal surface of pterothorax pale to the sides of the carina (as in Fig. 495) ....
.... ***Brasiliogomphus*** (Page 94)

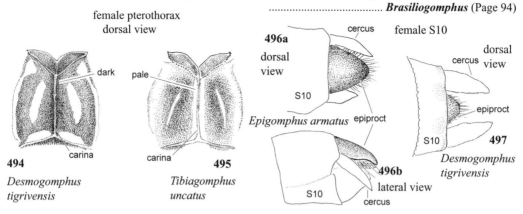

female pterothorax
dorsal view

dark | pale

carina | carina

494   **495**

Desmogomphus tigrivensis   *Tibiagomphus uncatus*

496a dorsal view
S10
cercus   female S10
dorsal view
cercus

Epigomphus armatus   epiproct

epiproct

S10   **497**

496b
S10   lateral view
cercus

Desmogomphus tigrivensis

27(26). Epiproct polished and about as long as cerci (Figs. 496a-b); pterostigma bracevein absent (Fig. 498); auricles rudimentary (as in Fig. 500); hind femur (**C**) shorter than width of head (**D**) (as in Fig. 502) .... ***Epigomphus* in part** (Page 102)

27'.   Epiproct not polished, covered with hairs, and usually not as long as cerci (as in Fig. 497); pterostigma bracevein present (Fig. 499); auricles absent; hind femur (**C**) as long as width of head or longer (**D**) (as in Fig. 501) **28**

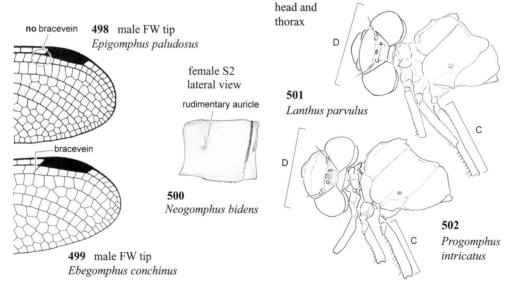

no bracevein   **498**   male FW tip
Epigomphus paludosus

bracevein

499   male FW tip
Ebegomphus conchinus

female S2
lateral view

rudimentary auricle

500
Neogomphus bidens

head and thorax

D

501
Lanthus parvulus

C

D

502
Progomphus intricatus

C

28(27). Rear of head smooth (Fig. 503) ....
.... ***Ebegomphus* in part** (Page 101)

28'.   Rear of head with pair of short spines near occipital foramen (Fig. 504) ....
.... ***Agriogomphus* in part** (Page 88)

female head
posterior view

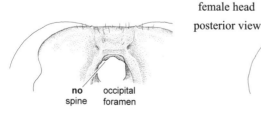

no occipital
spine foramen

503   *Ebegomphus minutus*

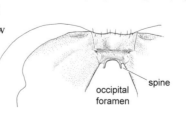

occipital foramen
spine

504   *Agriogomphus sylvicola*

29(23). Occiput largely bare or with scattered hairs (as in Fig. 505) ..... **Ebegomphus in part** (Page 101)

29'.   Occiput with a fringe of hairs along its summit (as in Fig. 506) .... **Tibiagomphus** (Page 129)

505
Perigomphus pallidistylus

male head dorsal view

506
Ophiogomphus purepecha

scattered hairs

fringe of hairs

30(21). FW with supplementary trigonal branch (as in Fig. 507) .... **Hagenius** (Page 109)

30'.   FW lacking supplementary trigonal branch (as in Fig. 508) **31**

supplementary trigonal branch

no supplementary trigonal branch

507   male FW basal half – *Hagenius brevistylus*

508   male FW basal half – *Progomphus gracilis*

31(30). Vulvar lamina bifid (as in Figs. 509-512) **32**

31'.   Vulvar lamina split into 2 parralel-sided lobes (Fig. 513) .... **Archaeogomphus** (Page 91)

vulvar lamina ventral view

vulvar lamina bifid

vulvar lamina bifid

vulvar lamina split

S9

S9

S9

S9

510
Cyanogomphus waltheri

511
Agriogomphus sylvicola

512
Tibiagomphus uncatus

513
Archaeogomphus furcatus

S9

509   *Gomphus quadricolor*

32(31). Epiproct polished and about as long as cerci (Figs. 496a-b) .... **Epigomphus in part** (Page 102)

32'.   Epiproct not polished, covered with hairs, and usually not as long as cerci (Fig. 497) **33**

33(32). Vulvar lamina reaching 1/2 of S9 or shorter (as in Figs. 509-511) **34**

33'.   Vulvar lamina surpassing 1/2 of S9 (as in Fig. 512) **35**

34(33). Vulvar lamina extending to about a 1/4 of S9 (as in Fig. 509) ...... **Progomphus in part** (Page 125)

34'.   Vulvar lamina extending to about a 1/3 of S9 (as in Fig. 510) ....
.... **Desmogomphus in part** (Page 97)

34''.   Vulvar lamina extending to about 1/2 of S9 (as in Fig. 511) ....
.... **Anomalophlebia / Desmogomphus in part**
[see next page]

Anomalophlebia (Page 89): Dorsal margin of occipital crest with a medial prominence (Fig. 514); primary antenodal 2 in HW closer to nodus than to primary antenodal 1 (Fig. 516)

Desmogomphus (Page 97): Dorsal margin of occipital crest linear (Fig. 515); primary antenodal 2 in HW slightly closer to primary antenodal 1 than to nodus (Fig. 517)

female occipital crest   frontal view

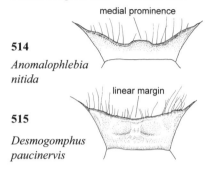

medial prominence

514

Anomalophlebia nitida

linear margin

515

Desmogomphus paucinervis

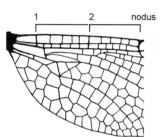

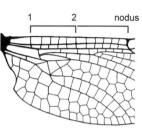

516 female HW basal half
Anomalophlebia nitida

517 male HW basal half
Desmogomphus tigrivensis

35(33). Occiput posterior margin broadly rounded (as in Fig. 518); spines of distal 1/2 of hind femur about as long as femur width (Fig. 520); hind femur (**C**) as long as width of head or longer (**D**) (as in Fig. 524) .... ***Perigomphus*** (Page 118)

35'.   Occiput posterior margin with a transverse ridge along its entire width (as in Fig. 519); spines of distal 1/2 of hind femur about as long as 1/2 of femur width or shorter (as in Fig. 521); hind femur (**C**) shorter than width of head (**D**) (as in Fig. 525) .... ***Desmogomphus*** **in part** (Page 97)

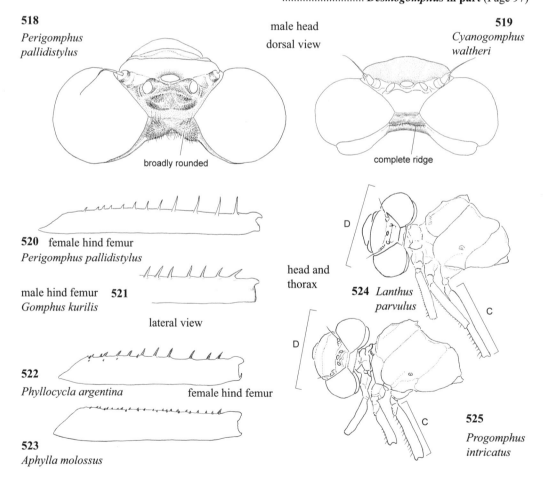

518

Perigomphus pallidistylus

male head
dorsal view

519

Cyanogomphus waltheri

broadly rounded

complete ridge

520 female hind femur
Perigomphus pallidistylus

male hind femur **521**
Gomphus kurilis

lateral view

head and thorax

524 *Lanthus parvulus*

522

Phyllocycla argentina   female hind femur

525

Progomphus intricatus

523

Aphylla molossus

36(20). Subalar carina of HW projected into strong spine (as in Figs. 526a-b); epiproct lacking an upturned hook (Fig. 528); postocellar ridge complete (Fig. 530); postfrons about 4 times as wide as long or less (Fig. 530) .... **Zonophora** (Page 131)

36'. Subalar carina of HW obtuse (as in Figs. 527a-b); epiproct with an upturned hook (Figs. 529a-b); postocellar ridge absent or incomplete (Fig. 531); postfrons more than 4 times as wide as long (Fig. 531) .... **Diaphlebia** (Page 98)

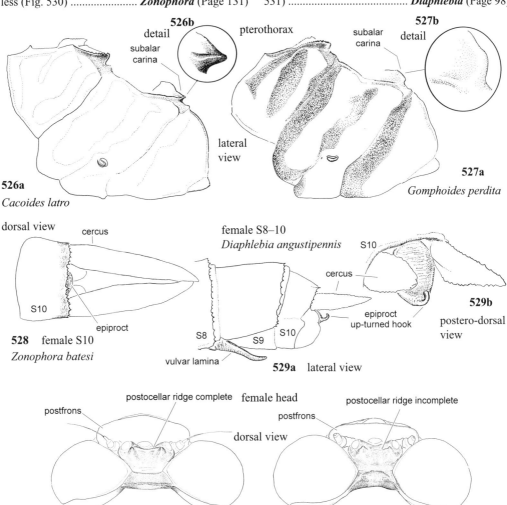

526b detail subalar carina

pterothorax

subalar carina

527b detail

526a
Cacoides latro

lateral view

527a
Gomphoides perdita

dorsal view

cercus

S10

epiproct

528 female S10
Zonophora batesi

female S8–10
Diaphlebia angustipennis

S8   S9   S10

vulvar lamina

S10

cercus

epiproct up-turned hook

529a lateral view

529b
postero-dorsal view

postocellar ridge complete

female head

postocellar ridge incomplete

postfrons

dorsal view

postfrons

530 *Zonophora calippus klugi*

531 *Diaphlebia angustipennis*

37(19). Postero-dorsal rim of S10 distinguished from anterior part by polished surface, less dense setation, different color, and/or basal groove or ridge (as in Figs. 532a-b) .... **Phyllocycla / Aphylla in part**

37'. Postero-dorsal rim of S10 not differentiated from remainder of segment (as in Fig. 533) **38**

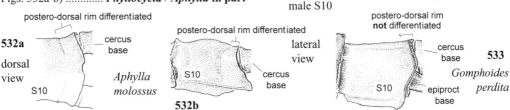

male S10

postero-dorsal rim differentiated

532a
dorsal view

cercus base

S10

Aphylla molossus

postero-dorsal rim differentiated

lateral view

S10

cercus base

532b

postero-dorsal rim **not** differentiated

cercus base

S10

epiproct base

533
Gomphoides perdita

Phyllocycla (Page 120): Outer distal spines of hind femur about 1/4 the diameter of femur (Fig. 522); venter of S9-10 usually brown

Aphylla (Page 90): Outer distal spines of hind femur about 1/6 the diameter of femur (Fig. 523); venter of S9-10 usually reddish

38(37). Epiproct visible in dorsal view (as in Fig. 534) .... ***Progomphus* in part** (Page 125)

38'.   Epiproct rudimentary, not visible in dorsal view (Fig. 535) .... ***Phyllogomphoides* in part** (Page 122)

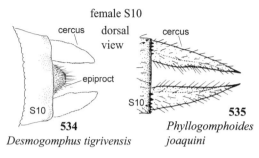

female S10

cercus   dorsal view   cercus

epiproct

S10

534
Desmogomphus tigrivensis

S10

535
Phyllogomphoides joaquini

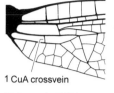

1 CuA crossvein

536   male FW base
Gomphoides perdita

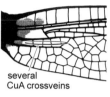

several CuA crossveins

537   female FW base
Melanocacus interioris

39(19). Cubito-anal crossveins in FW numbering 1 (as in Fig. 536) **40**

39'.   Cubito-anal crossveins in FW numbering 2 or more (Fig. 537) **42**

40(39). S10 longer than S9 (as in Fig. 538) .... ***Gomphoides*** (Page 105)

40'.   S10 shorter than S9 (as in Fig. 539) **41**

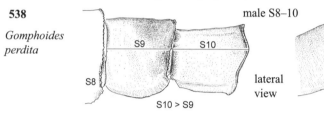

538

Gomphoides perdita

male S8–10

S9   S10

S8

S10 > S9

lateral view

S9   S10

S8

S10 < S9

539

Aphylla molossus

41(40). Postero-dorsal rim of S10 distinguished from anterior part by polished surface, less dense setation, different color, and/or basal groove or ridge (as in Figs. 540a-b) .... ***Aphylla* in part** (Page 90)

41'.   Postero-dorsal rim of S10 not differentiated from remainder of segment (as in Fig. 541) .... ***Phyllogomphoides* in part** (Page 122)

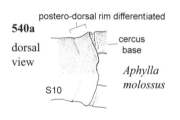

postero-dorsal rim differentiated

540a
dorsal view

cercus base

Aphylla molossus

S10

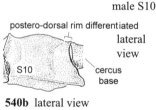

male S10

postero-dorsal rim differentiated

lateral view

S10

cercus base

540b   lateral view

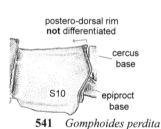

postero-dorsal rim **not** differentiated

cercus base

S10

epiproct base

541   *Gomphoides perdita*

42(39). Subalar carina of HW projected into strong spine (as in Figs. 544a-b); epiproct visible in dorsal view (as in Fig. 534); cubito-anal crossveins in HW numbering 2 or more (Fig. 548); FW with supplementary trigonal branch (Fig. 542) **43**

42'.   Subalar carina of HW obtuse (as in Figs. 545a-b); epiproct rudimentary, not visible in dorsal view (as in Fig. 535); cubito-anal crossveins in HW numbering 1 (Fig. 549); FW lacking supplementary trigonal branch (as in Fig. 543) .... ***Idiogomphoides*** (Page 110)

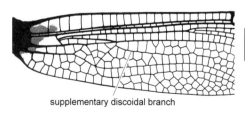

supplementary discoidal branch

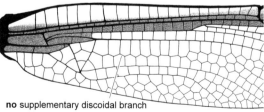

no supplementary discoidal branch

542   female FW basal half – *Melanocacus interioris*   **543**   male FW basal half – *Idiogomphoides demoulini*

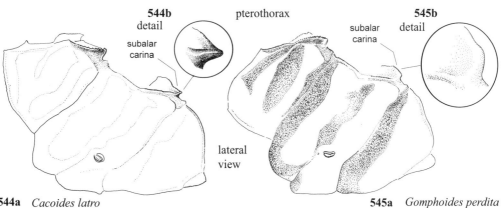

544b
detail

subalar
carina

pterothorax

subalar
carina

545b
detail

lateral
view

544a *Cacoides latro*

545a *Gomphoides perdita*

43(42). Vulvar lamina reaching 1/2 of S9 or shorter (Fig. 546); RP2 with marked bend proximal to pterostigma (as in Fig. 550); wing bases hyaline (as in Fig. 550) .... **Cacoides** (Page 95)

43'. Vulvar lamina surpassing 1/2 of S9 (Fig. 547); RP2 without marked bend proximal to pterostigma (Fig. 551); wing bases colored (Fig. 551) .... **Melanocacus** (Page 112)

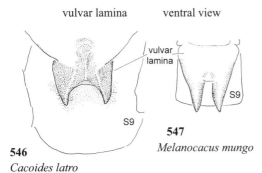

vulvar lamina

ventral view

vulvar
lamina

S9

S9

547
Melanocacus mungo

546
Cacoides latro

1 CuA
crossvein

several
CuA crossveins

548 female HW base
Melanocacus interioris

549 male HW base
Idiogomphoides demoulini

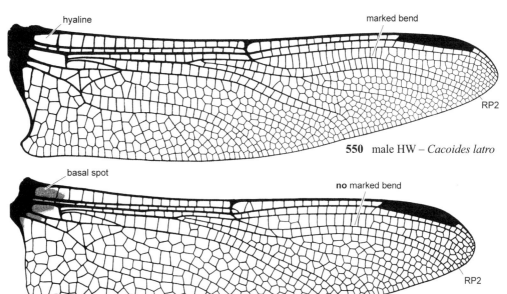

hyaline

marked bend

RP2

550 male HW – *Cacoides latro*

basal spot

no marked bend

RP2

551 female HW – *Melanocacus interioris*

Agriogomphus Selys, 1869a: 189 (26 reprint).

[♂ pp. 71, couplet 27; ♀ pp. 81-82, couplets 25, 28]

Type species: *Agriogomphus sylvicola* Selys, 1869 [by monotypy]

   syn *Ischnogomphus* Williamson, 1918: 6.

   Type species: *Ischnogomphus jessei* Williamson, 1918 [by original designation]

4 species:

ericae (Belle, 1966)* [*Ischnogomphus*] – **L** [Belle, 1966b]
jessei (Williamson, 1918) [*Ischnogomphus*] – **L**? [Arango and Roldán, 1983]
sylvicola Selys, 1869* – **L** [Needham, 1944; Belle, 1966b]
   syn *aquicola* Fraser, 1943
tumens (Calvert, 1905) [*Cyanogomphus* (?)]* – **L** [Novelo Gutiérrez, 1989]

References: Belle, 1966b; 1972a; 1975.
Distribution: Central Mexico south to Peru and Brazil.

Map 27. Distribution of *Agriogomphus* spp.

Small gomphids (40–41 mm); postfrons narrow; postocellar ridge complete; occiput broadly rounded, sometimes with transverse ridge interrupted medially, and with a pair of short spines near occipital foramen in the female (Fig. 552); pterothorax largely greenish gray to gray with dark brown stripes; abdomen banded dark brown and gray. Wings (Fig. 553) with basal subcostal crossveins; 2 or more crossveins in HW space between arculus branches and point of branching of RP; subtriangles and supratriangles free; male anal triangle not defined (shared with *Archaeogomphus* and *Epigomphus*); 1 cubito-anal crossvein; CuA and MP in HW divergent. Male cercus (Fig. 554) with ventrally directed branch; epiproct bifid; anterior hamule (Fig. 555) fused to anterior lamina; vesica spermalis distal segment with 2 long flagella. Female lacking auricles; vulvar lamina bifid, extending to about middle of S9 or less (Fig. 556). **Unique characters**: None known.

Status of classification: Species resolution seems to be good.

Potential for new species: Very likely. Most species are rare in collections allowing for potential discovery of new species in poorly explored areas. Belle (1996) established the tribe Agriogomphini under the Epigomphinae to receive *Agriogomphus*, *Brasiliogomphus* and *Ebegomphus*.

Habitat: Adults occur at streams within barely penetrable jungle although *A. tumens* is frequently found in disturbed areas bordering streams in eastern Mexico. Larvae found under decayed leaves on mud banks with slow water flow (Novelo Gutiérrez, 1989), and leaf packs in small streams (Louton, Garrison, and Flin, 1996). Belle (1966b) described the male of *A. ericae* as flying a short distance to land on leaves of surrounding understory or squatting close to water's edge in partially shaded areas.

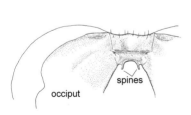

552 female head
Agriogomphus sylvicola

   posterior view

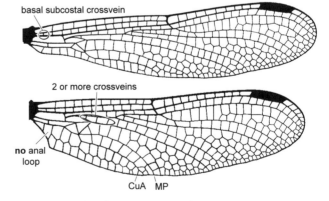

553 male wings – *Agriogomphus tumens*

554   male S10 – *Agriogomphus tumens*

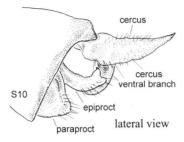

cercus

cercus
ventral branch

S10

epiproct

paraproct   lateral view

555   hamules – *Agriogomphus tumens*

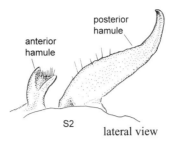

posterior
hamule

anterior
hamule

S2   lateral view

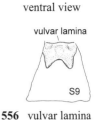

ventral view

vulvar lamina

S9

556   vulvar lamina
Agriogomphus sylvicola

Anomalophlebia Belle, 1995: 20.
[♀ pp. 83-84, couplet 34]
Type species: *Anomalophlebia nitida* Belle, 1995 [by original designation]
1 species:

nitida Belle, 1995: 20

References: Belle, 1995; 1996.
Distribution: Known only from the holotype female from Bolivar State, Venezuela.

Map 28. Distribution of *Anomalophlebia* sp.

Small gomphids (38 mm); occiput with well developed transverse ridge and scattered hairs; dorsal margin of occipital crest with a medial prominence (Fig. 558); pterothorax yellow with wide brown stripes; abdomen brown, with pale green or yellow areas or stripes laterally. Wings (Fig. 557) lacking basal subcostal crossveins; 2 or more crossveins in HW space between arculus branches and point of branching of RP; subtriangles and supratriangles free; 1 cubitoanal crossvein; CuA and MP in HW parallel. Female lacking auricles; vulvar lamina bifid, extending to about midlength of S9 (Fig. 559). **Unique characters**: None known.

Status of classification: Known only from female holotype. Belle (1996) erected the tribe Anomalophlebiini under the Octogomphinae to receive this genus. Until the male is discovered, we believe the affinities of *Anomalophlebia* will remain doubtful.

Potential for new species: Unknown.

Habitat: Type locality is within a bauxite mine. Current preservation of locality is unknown. Holotype was collected on a small rivulet in a very damp forest where she was laying eggs. Another small wary gomphid, perhaps the male, was flying around as well as settling on fallen trunks that crossed the creek (De Marmels, *pers. comm.*). Larva unknown.

557   female wings – *Anomalophlebia nitida*

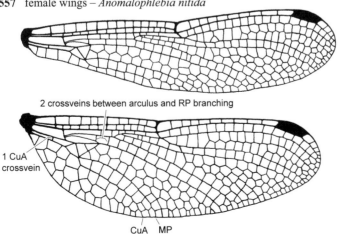

2 crossveins between arculus and RP branching

1 CuA
crossvein

CuA   MP

558   female head
Anomalophlebia nitida

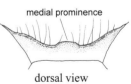

medial prominence

dorsal view

ventral
view

559   vulvar lamina
Anomalophlebia nitida

Aphylla Selys, 1854: 78 (59 reprint).

[♂ pp. 72, 75, couplets 30, 37; ♀ pp. 85-86, couplets 37, 41]

Type species: *Aphylla brevipes* Selys, 1854 [Kirby, 1889 by subsequent designation]

24 species:

alia Calvert, 1948
angustifolia Garrison, 1986* – **L**? [Needham, 1940 as *protracta*; Needham, Westfall, and May, 2000]
barbata Belle, 1994
boliviana Belle, 1972*
brasiliensis Belle, 1970
brevipes Selys, 1854 – **L**? [Belle, 1970 as *albinensis*]
    syn *albinensis* Belle, 1970
caraiba Selys, 1854* – **L** [Needham, 1940; Needham, Westfall, and May, 2000]
    syn *cubana* (Navás, 1917) [*Cyclophylla*]
caudalis Belle, 1987
dentata Selys, 1859* – **L**? [Belle, 1964 as *simulata*]
    syn *simulata* Belle, 1964
distinguenda (Campion, 1920) [*Gomphoides*]
edentata Selys, 1869*
exilis Belle, 1994
janirae Belle, 1994
linea Belle, 1994
molossus Selys, 1869*
producta Selys, 1854* – **L**? [Needham, 1944; Belle, 1964]
    syn *?curvata* (Navás, 1933) [*Gomphoides*]
protracta* (Hagen *in* Selys, 1859) [*Cyclophylla*]
    syn *ambigua* (Selys, 1873) [*Gomphoides*]
robusta Belle, 1976
scapula Belle, 1992*
silvatica Belle, 1992
spinula Belle, 1992
tenuis Selys, 1859*
    syn *elegans* Belle, 1970
    syn *obscura* (Kirby, 1899) [*Cyclophylla*]
theodorina (Navás, 1933) [*Gomphoides*]* – **L**? [Belle, 1992a]
williamsoni (Gloyd, 1936) [*Gomphoides*]* – **L**? [Needham, 1940; Needham, Westfall, and May, 2000]

References: Needham, Westfall, and May, 2000 (North America); Garrison, 1986b (Middle America), Belle, 1970; 1992b; 1994b (South America).

Distribution: SE United States south through N Argentina and Uruguay.

Medium to large gomphids (51–76 mm); occiput flat, with well developed transverse ridge along posterior margin bearing a fringe of hairs; pterothorax green or greenish yellow with brown stripes; abdomen brown becoming paler distally (orange to pale brown), with pale green lateral markings. Wings (Figs. 561-562)

Map 29. Distribution of *Aphylla* spp.

with or without basal subcostal crossveins in HW; 2 or more crossveins in HW space between arculus branches and point of branching of RP; subtriangle of FW with 1 crossvein, subtriangle of HW free; supratriangles crossed; male HW anal triangle reaching anal angle of wing; 1 cubito-anal crossvein; CuA and MP in HW divergent. Male cercus arcuate, semicircular, unbranched (Fig. 565b); epiproct rudimentary, not visible in lateral view (Fig. 565a). Vesica spermalis distal segment with 2 short flagella (Figs. 564a-b). Female lacking auricles; vulvar lamina bifid, extending to about a fourth of S9 or less; epiproct rudimentary and barely visible in dorsal view. Postero-dorsal rim of S10 in both males and females distinguished from anterior part by polished surface, less dense setation, different color and basal groove or ridge (Figs. 565a-b, shared with *Phyllocycla*). **Unique characters**: None known.

Status of classification: Good. The genus is most speciose in South America and the keys for adults by Belle (1992b; 1994b) allow for recognition of most species.

Potential for new species: Very likely. Fifteen of the 24-known species have been described since 1970 mostly by Belle.

Habitat: Adults found along jungle trails or at margins of streams or stagnant water areas where they alight on ground or snags. Some species found in cultivated fields bordering trees next to streams. Larvae, with S10 developed as a tube as long as previous 4 or 5 abdominal segments, are borrowers in soft muddy bottom zones in slowly flowing streams or still water areas (Belle, 1992b). Paulson (1999) provided detailed information on the biology of *A. williamsoni*.

lateral view

560 female hind femur – *Aphylla molossus*

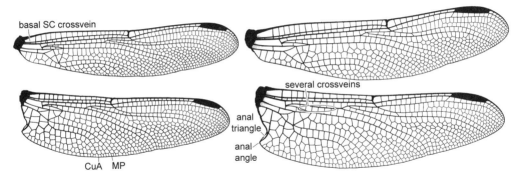

561 male wings – *Aphylla molossus*

562 male wings – *Aphylla protracta*

563 hamules – *Aphylla protracta*

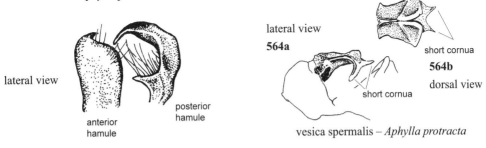

vesica spermalis – *Aphylla protracta*

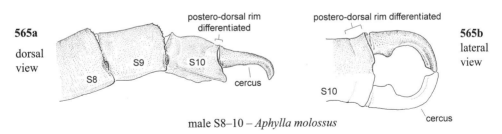

male S8–10 – *Aphylla molossus*

Archaeogomphus Williamson, 1919: 2.
[♂ pp. 72, couplet 28; ♀ pp. 83, couplet 31]
Type species: *Agriogomphus hamatus* Williamson, 1918 [by original designation]
7 species:
    syn *Austroarchaeogomphus* Carle, 1986: 305
        Type species: *Agriogomphus infans* Ris, 1913c: 72 (18 separate) [by original designation]
densus Belle, 1982
furcatus Williamson, 1923* – **L** [Belle, 1992a]
globulus Belle, 1994
hamatus (Williamson, 1918) [*Agriogomphus*]*
infans (Ris, 1913) [*Agriogomphus*]
nanus Needham, 1944 – **L** [Belle, 1970]
vanbrinkae Machado, 1994

References: Belle, 1982.

Map 30. Distribution of *Archaeogomphus* spp.

Distribution: NE Mexico south through NE Argentina.

Small gomphids (27–35 mm); occiput broadly round-ed with scattered hairs; rear of head with 2 short spines near occipital foramen in some species; ptero-thorax pale green or yellow to gray with dark brown stripes; abdomen brown with yellow markings. Wings (Fig. 566) lacking basal subcostal crossveins; 2 or more crossveins in HW space between arculus branches and point of branching of RP; subtriangles and supratriangles free; male anal triangle not defined (shared with *Agriogomphus* and *Epigomphus*); 1 cu-bito-anal crossvein; CuA and MP in HW convergent. Male cercus unbranched, short and rounded; epiproct rudimentary, not visible in lateral view (Fig. 569a). Anterior hamule (Fig. 567) fused to anterior lamina, digitiform, lacking teeth or spines; vesica spermalis distal segment with 2 long flagella. Female lacking auricles; vulvar lamina divided along entire length into 2 long narrow branches, apressed to each other and extending beyond posterior margin of S9 (Fig. 568). **Unique characters**: Male S10 with anterior margin projected into 2 latero-dorsal hooks, and pos-terior margin projected medio-dorsally into a triangu-lar, flat structure (Figs. 569a-b).

Status of classification: Good. Belle's (1982) illus-trated revision should allow for easy species recog-nition. *Austroarchaeogomphus* was proposed as a subgenus of *Archaeogomphus* by Carle (1986) in a key based on characters of wing venation, expansion of lateral margin of S8, and presence of submedial spines on dorsal edge of the occipital foramen, all of which we consider to represent no more than specific differences. *Austroarchaeogomphus* has not gener-ally been recognized by odonatologists and we treat the name here as a synonym.

Potential for new species: Good. These small gom-phids are rarely encountered in the field, allowing for potential discovery of new species in poorly explored areas.

Habitat: Adults perch on overhanging leaves or snags in dappled shade at stream's edge in rainforest. When disturbed, adults often fly within branches to new perch sites making collecting difficult (Belle, 1982). Bright green translucent larvae of *A. furcatus* cling with elongate slender pretarsal claw to terrestrial veg-etation that trails in shallow streams over sand (Lou-ton, Garrison, and Flint, 1996).

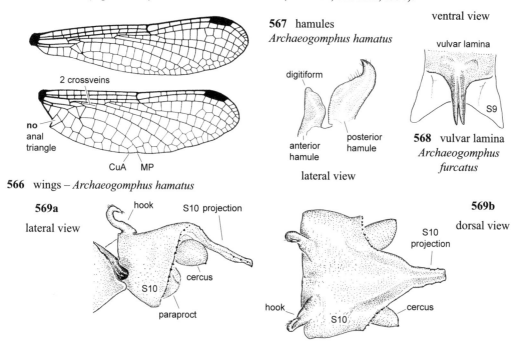

567 hamules
Archaeogomphus hamatus

digitiform

anterior
hamule

posterior
hamule

lateral view

ventral view

vulvar lamina

S9

568 vulvar lamina
Archaeogomphus furcatus

566 wings – *Archaeogomphus hamatus*

2 crossveins

no
anal
triangle

CuA   MP

569a

lateral view

hook   S10 projection

cercus

S10

paraproct

569b   dorsal view

S10
projection

hook   S10   cercus

male S10 – *Archaeogomphus hamatus*

Arigomphus Needham, 1897b: 181.

[♂ pp. 67, couplets 11-12; ♀ pp. 77-78, couplet 12]

Type species: *Gomphus pallidus* Rambur, 1842 [by original designation]

    syn *Orcus* Needham, 1897b: 166.

        Type species: *Gomphus pallidus* Rambur, 1842 [by original designation]

[NOTE: *Arigomphus* is a replacement name for *Or-cus* Needham, a junior primary homonym of *Orcus* Mulsant, 1850 in Coleoptera.]

7 species:

cornutus (Tough, 1900) [*Gomphus*]* – **L**
[Muttkowski and Whedon, 1915]
    syn *whedoni* (Muttkowski, 1913) [*Gomphus*]
furcifer (Hagen *in* Selys, 1878) [*Gomphus*]* – **L**
[Needham, Westfall, and May, 2000]
lentulus (Needham, 1902) [*Gomphus*]* – **L** [Bird,
1934 as *G. militaris*; Needham, Westfall, and May,
2000]
    syn *subapicalis* (Williamson, 1914) [*Gomphus*]
maxwelli (Ferguson, 1950) [*Gomphus*]* – **L**
[Needham, Westfall, and May, 2000]
pallidus (Rambur, 1842) [*Gomphus*]* – **L**
[Needham, Westfall, and May, 2000]
    syn *pilipes* (Hagen *in* Selys, 1858) [*Gomphus*]
submedianus (Williamson, 1914) [*Gomphus*]* – **L**
[Needham, Westfall, and May, 2000]
villosipes (Selys, 1854) [*Gomphus*]* – **L** [Needham,
1901; Needham, Westfall, and May, 2000]

References: Needham, Westfall, and May, 2000.
Distribution: SE Canada, E United States.

Medium gomphids (45–62 mm); postocellar ridge
complete; occiput with a transverse ridge along its
entire width; pterothorax (Figs. 573-574a-b) gray
to green or yellow with dark brown stripes, which
may be reduced or entirely absent; abdomen gray to
brown with greenish- or yellowish-gray markings.
Wings (Fig. 570) lacking basal subcostal crossveins;
1 crossvein in HW space between arculus branches
and point of branching of RP; subtriangles and supratriangles free; 1 cubito-anal crossvein; CuA and

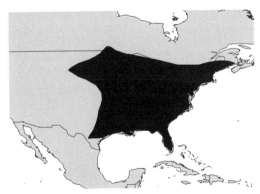

Map 31. Distribution of *Arigomphus* spp.

MP in HW divergent. Male cercus (Fig. 572) tapering
distally, usually with a branch not directed ventrally;
epiproct bifid. Vesica spermalis distal segment with 2
flagella of variable length. Female lacking auricles;
vulvar lamina bifid, extending to about a 1/3 of S9
length (Fig. 571). **Unique characters**: None known.
Status of classification: Good. Species easily recognizable. Carle (1986) included this genus in the
Gomphini, together with *Dromogomphus*, *Gomphus*,
Stylurus, and the Old World *Gastrogomphus* Needham, 1944.
Potential for new species: Unlikely.
Habitat: Breed in standing waters, like ponds, lakes,
ditches, or small slow streams, where the adults
perch on ground. Habits of all seven species given
by Dunkle (2000); Paulson (1999) provided detailed
information on the biology of *A. pallidus*.

571 vulvar lamina
        Arigomphus pallidus

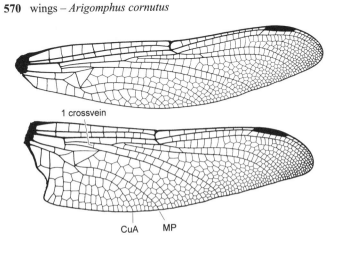

570 wings – *Arigomphus cornutus*

1 crossvein

CuA   MP

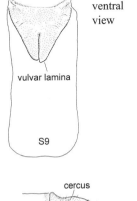

ventral
view

vulvar lamina

S9

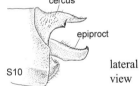

cercus

epiproct

S10

lateral
view

572 male S10 – *Arigomphus furcifer*

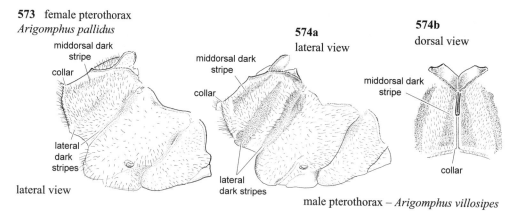

573 female pterothorax
Arigomphus pallidus

middorsal dark stripe

collar

lateral dark stripes

lateral view

574a
lateral view

middorsal dark stripe

collar

lateral dark stripes

574b
dorsal view

middorsal dark stripe

collar

male pterothorax – *Arigomphus villosipes*

Brasiliogomphus Belle, 1995: 23.
[♀ pp. 82, couplet 26]

Type species: *Brasiliogomphus uniseries* Belle, 1995
[by original designation]
1 species:

uniseries Belle, 1995

References: Belle, 1995; 1996.
Distribution: Known only from the holotype female
from São Paulo State, Brazil.

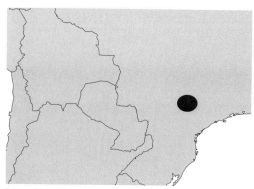

Map 32. Distribution of *Brasiliogomphus* sp.

Small gomphids (37 mm); occiput with a transverse
ridge interrupted medially (Fig. 576); pterothorax yellow to the sides of the medio-dorsal carina and with
2 broad dark brown mesanepisternal stripes, entirely
yellow laterally; abdomen predominantly yellow,
with a lateral brown stripe on S1-6. Wings (Fig. 575)
with basal subcostal crossveins; 2 or more crossveins
in HW space between arculus branches and point
of branching of RP; subtriangles and supratriangles
free; 1 cubito-anal crossvein; CuA and MP in HW divergent. Belle (1996) characterized this genus as having a single row of cells in the discoidal field of FW;
the wings from the holotype however show a double
row at the base of the discoidal field. Female vulvar
lamina bifid, extending to about 1/2 of S9 length (Fig.
577). **Unique characters**: None known.
Status of classification: Unknown. Belle (1996) established the tribe Agriogomphini under the Epigomphinae to receive *Agriogomphus*, *Brasiliogomphus*
and *Ebegomphus*. The genus, based on the holotype
female, will be difficult to assess until its male is discovered.
Potential for new species: Unknown.
Habitat: No data available. Larva unknown.

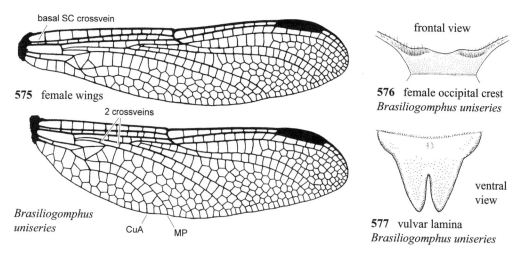

basal SC crossvein

575 female wings

2 crossveins

*Brasiliogomphus
uniseries*

CuA   MP

frontal view

576 female occipital crest
Brasiliogomphus uniseries

ventral
view

577 vulvar lamina
Brasiliogomphus uniseries

Cacoides Cowley, 1934a: 201.

[♂ pp. 74, couplet 34; ♀ pp. 87, couplet 43]

Type species: *Ictinus latro* Erichson, 1848 [by monotypy]

 syn *Cacus* Selys, 1854: 97 (78 reprint)

  Type species: *Ictinus latro* Erichson, 1848 [by monotypy]

[NOTE: *Cacoides* is a replacement name for *Cacus* Selys, 1854, a junior primary homonym of *Cacus* Gistl (1848) in Coleoptera.]

1 species:

latro (Erichson, 1848) [*Ictinus*]* – **L** [Belle, 1970]

References: Belle, 1986a.

Distribution: Atlantic slope of tropical South America, from Venezuela to Brazil.

Large gomphids (74–76 mm); postocellar ridge complete, with lateral edges projected into long horns (shared with *Mitragomphus*); occiput with a transverse ridge along its entire width, bearing a fringe of hairs; pterothorax gray to green or yellow with dark brown stripes, which may be reduced or entirely absent; abdomen gray to brown with greenish- or yellowish-gray markings. Subalar carina of HW projected into strong spine (Figs. 579a-b) shared with *Melanocacus* and *Zonophora*). Wings (Fig. 578) with basal subcostal crossveins; 2 or more crossveins in HW space between arculus branches and point of branching of RP; subtriangle of FW with 1 crossvein, of HW free; supratriangles crossed; 2 or more cubitoanal crossveins; CuA and MP in HW parallel; FW with a supplementary trigonal sector. Male cercus long (Figs. 584a-b), with upturned tip; epiproct small and entire, visible in lateral view (Fig. 584b); S8 with very wide lateral flanges, both in males and females.

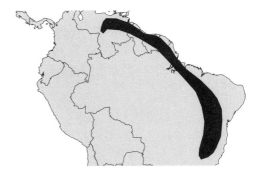

Map 33. Distribution of *Cacoides* sp.

Posterior hamule entire, lacking spines or hooks; vesica spermalis distal segment lacking flagella, and basal segment with an entire anterior rim (Figs. 583a-b). Female lacking auricles; vulvar lamina bifid, extending to about 1/2 of S9 length (Fig. 582) . **Unique characters**: RP2 with marked concavity proximal to pterostigma (Fig. 578); male auricle with a single tooth (Fig. 581).

Status of classification: Monotypic genus easily diagnosed. Carle (1986) included this genus in the tribe Lindeniini; Belle (1996) included it in his tribe Ictinogomphini, together with *Ictinogomphus* and *Melanocacus*.

Potential for new species: Unknown.

Habitat: Adults perch on tips of snags facing water at borders of stagnant ponds. Males are shy and unapproachable (Belle, 1986). Larvae inhabit vegetated ponds and ditches of sandy banks and lakes in the forest, where males defend territories from a perch in the marshy shores (Moore and Machado, 1992).

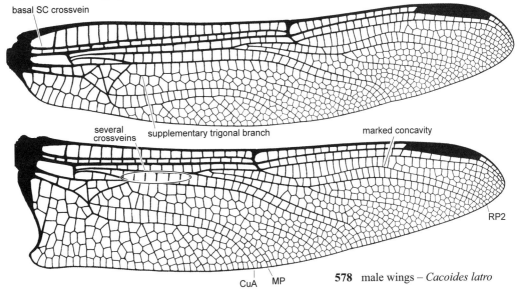

basal SC crossvein

several crossveins supplementary trigonal branch marked concavity

RP2

CuA MP **578** male wings – *Cacoides latro*

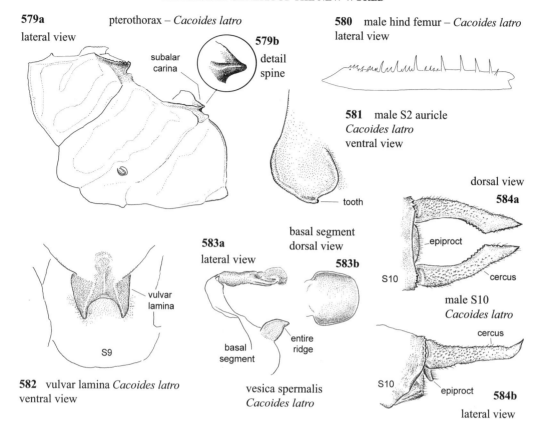

579a   pterothorax – *Cacoides latro*
lateral view

579b
subalar
carina   detail
spine

580   male hind femur – *Cacoides latro*
lateral view

581   male S2 auricle
Cacoides latro
ventral view

tooth

dorsal view
584a

epiproct

S10   cercus

male S10
Cacoides latro

583a
lateral view

basal segment
dorsal view
583b

vulvar
lamina

S9

basal
segment   entire
ridge

cercus

S10   epiproct   **584b**

lateral view

582   vulvar lamina *Cacoides latro*
ventral view

vesica spermalis
Cacoides latro

Cyanogomphus Selys, 1873: 753 (26 reprint).
[♂ pp. 71, couplet 26; ♀ pp. 81, couplet 24]
Type species: *Cyanogomphus waltheri* Selys, 1873
[by monotypy]
2 species:

comparabilis Belle, 1994
waltheri Selys, 1873*

References: Belle, 1966b; 1980; 1994a.
Distribution: S Brazil to NE Argentina.

Medium gomphids (39–45 mm); postocellar ridge
complete; occiput with a transverse ridge along its
entire width, bearing a fringe of hairs (Fig. 585);
pterothorax green with dark brown stripes; abdomen
brown becoming yellowish brown at apical segments.
Wings (Fig. 587) with basal subcostal crossveins;
2 or more crossveins in HW space between arculus
branches and point of branching of RP; subtriangles
and supratriangles free; 1 cubito-anal crossvein; CuA
and MP in HW divergent. Male cercus with branch directed ventrally (Fig. 589) between epiproct branches;
epiproct bifid, with tip of each branch spatulate. Anterior hamule subquadrate (Fig. 588), fused to anterior
lamina and with concave outer side; vesica spermalis
distal segment with 2 short flagella. Female lacking

Map 34. Distribution of *Cyanogomphus* spp.

auricles; vulvar lamina bifid, extending to about a 1/3
of S9 length (Fig. 590). **Unique characters**: None
known.

Status of classification: Five more northerly Amazonian species originally described under *Cyanogomphus* were recently removed and placed in *Ebegomphus* Needham by Belle (1996). Differentiable from
Tibiagomphus only by tibial armature (Figs. 586a-b);
possibly congeneric.

Potential for new species: Likely.
Habitat: No data available. Larva unknown.

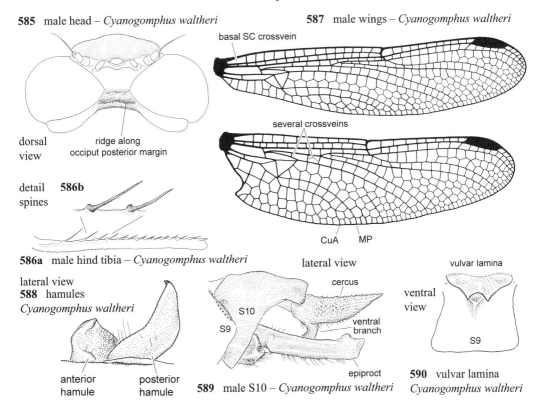

585 male head – *Cyanogomphus waltheri*

dorsal view

ridge along occiput posterior margin

detail spines **586b**

586a male hind tibia – *Cyanogomphus waltheri*

lateral view
588 hamules
Cyanogomphus waltheri

anterior hamule   posterior hamule

587 male wings – *Cyanogomphus waltheri*

basal SC crossvein

several crossveins

CuA   MP

lateral view

cercus

S10

S9

ventral branch

epiproct

589 male S10 – *Cyanogomphus waltheri*

vulvar lamina

ventral view

S9

590 vulvar lamina
Cyanogomphus waltheri

Desmogomphus Williamson, 1920: 1.
[♂ pp. 70, couplet 22; ♀ pp. 83–84, couplets 34–35]
Type species: *Desmogomphus tigrivensis* Williamson, 1920 [by original designation]
2 species:

paucinervis (Selys, 1873) [*Progomphus*?]* – **L** [Westfall, 1989]
tigrivensis Williamson, 1920* – **L** [Belle, 1970; 1977b]

References: Williamson, 1920; Belle, 1970; 1972a, b.
Distribution: Nicaragua south to Colombia, Venezuela, Guyana and Surinam.

Medium gomphids (41–46 mm); occiput with a transverse ridge along its entire width; pterothorax yellow to olive green with dark brown stripes; abdomen brown with whitish incomplete rings on anterior portions of S4-S7. Wings (Fig. 591) lacking basal subcostal crossveins; 2 or more crossveins in HW space between arculus branches and point of branching of RP; subtriangles and supratriangles free; 1 cubitoanal crossvein; CuA and MP in HW parallel or divergent. Male cercus with ventrally directed branch (Fig. 594a); epiproct bifid (Fig. 594c). Vesica spermalis distal segment lacking flagella, ending in a single spatulate projection. Female lacking auricles; vulvar

Map 35. Distribution of *Desmogomphus* spp.

lamina bifid, extending from about a 1/3 to almost entire length of S9. **Unique characters**: None known for adults; doubled dorsal hooks on abdominal segments 5-7 of larvae (Belle, 1970).
Status of classification: The male sex of only one species (*D. tigrivensis*) has been described. Belle (1996) considered it most closely related to *Diaphlebia*, and included both genera in his tribe Diaphlebiini.
Potential for new species: Good.
Habitat: Adults of *D. tigrivensis* have been taken while resting on leaves overhanging a creek, and larvae in moderately swift current in sand and gravel substrates in clear streams 5-10 m wide at about 300 m elevation

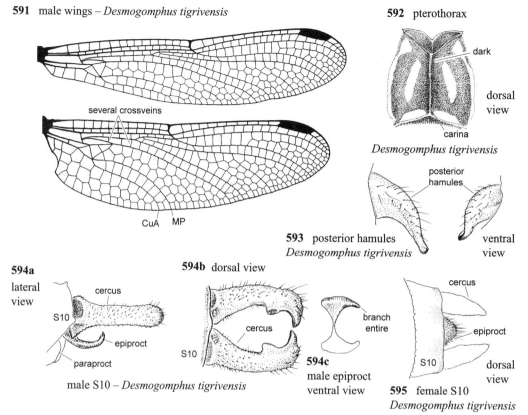

591 male wings – *Desmogomphus tigrivensis*

several crossveins

CuA   MP

592 pterothorax

dark

dorsal view

carina

Desmogomphus tigrivensis

posterior hamules

593 posterior hamules
Desmogomphus tigrivensis

ventral view

594a
lateral view

cercus

S10

epiproct

paraproct

male S10 – *Desmogomphus tigrivensis*

594b dorsal view

cercus

S10

594c
male epiproct
ventral view

branch entire

cercus

epiproct

S10

dorsal view

595 female S10
Desmogomphus tigrivensis

Diaphlebia Selys, 1854: 81 (62 reprint).
[♂ pp. 70, couplet 22; ♀ pp. 85, couplet 36]
Type species: *Diaphlebia angustipennis* Selys, 1854
[by monotypy]
2 species:

angustipennis Selys, 1854*
   syn *rokitanskyi* (St. Quentin, 1973) [*Zonophora*]
   syn *semilibera* Selys, 1869
nexans Calvert, 1903

References: Belle, 1977a.
Distribution: Amazonian forest from Venezuela to Brazil.

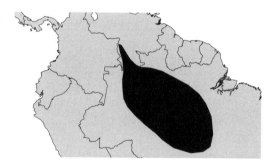

Map 36. Distribution of *Diaphlebia* spp.

Medium gomphids (50-53 mm); postocellar ridge projected laterally into short horns (Fig. 597); occiput with a transverse ridge along its entire width (Fig. 596), bearing a fringe of hairs along its summit; pterothorax and abdomen pale green with dark brown markings. Wings (Fig. 598) lacking basal subcostal crossveins; 2 or more crossveins in HW space between arculus branches and point of branching of RP; subtriangles and supratriangles free; 1 cubito-anal crossvein in FW, 2 in HW; CuA and MP in HW divergent. Male cercus long, straight, and unbranched (Figs. 601a-b); epiproct bifid (Fig. 601b). Posterior hamule ending in a single spine (Figs. 600a-b); ve-

sica spermalis distal segment lacking flagella. Female lacking auricles; vulvar lamina bifid, extending to almost entire length of S9 (Fig. 603). **Unique characters**: Female epiproct with an upturned hook (Figs. 602a-b).
Status of classification: Good. Species well defined. Belle (1996) considered it most closely related to *Desmogomphus*, and included both genera in his tribe Diaphlebiini.
Potential for new species: Likely.
Habitat: No information available; larva unknown. De Marmels (1989) indicated that numerous specimens were collected at night.

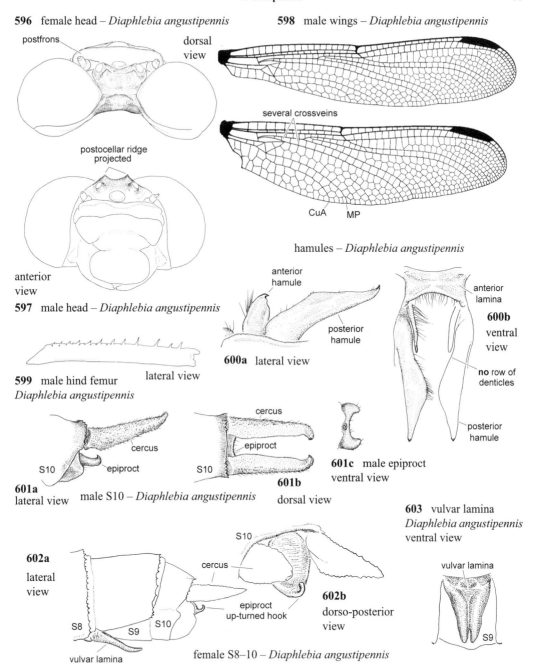

596 female head – *Diaphlebia angustipennis*

postfrons

dorsal view

postocellar ridge projected

anterior view

597 male head – *Diaphlebia angustipennis*

599 male hind femur
Diaphlebia angustipennis

lateral view

598 male wings – *Diaphlebia angustipennis*

several crossveins

CuA   MP

hamules – *Diaphlebia angustipennis*

anterior hamule

posterior hamule

600a lateral view

anterior lamina

600b ventral view

no row of denticles

posterior hamule

cercus

epiproct

601a lateral view   male S10 – *Diaphlebia angustipennis*

S10

cercus

cercus

epiproct

S10

601b dorsal view

601c male epiproct ventral view

603 vulvar lamina
Diaphlebia angustipennis
ventral view

602a lateral view

cercus

S10

epiproct up-turned hook

602b dorso-posterior view

S8   S9   S10

vulvar lamina

female S8–10 – *Diaphlebia angustipennis*

vulvar lamina

S9

Dromogomphus Selys, 1854: 58 (39 reprint).
[♂ pp. 67, couplet 10; ♀ pp. 77, couplet 11]
Type species: *Dromogomphus spinosus* Selys, 1854 [Kirby, 1890 by subsequent designation]
3 species:

armatus Selys, 1854* – **L** [Westfall and Tennessen, 1979; Louton, 1982b]
spinosus Selys, 1854* – **L** [Needham, 1901;

Westfall and Tennessen, 1979; Louton, 1982b]
spoliatus (Hagen *in* Selys, 1858) [*Gomphus*]* – **L** [Westfall and Tennessen, 1979; Louton, 1982b]

References: Westfall and Tennessen, 1979; Needham, Westfall, and May, 2000.
Distribution: SE Canada south to SE US and NE Mexico.

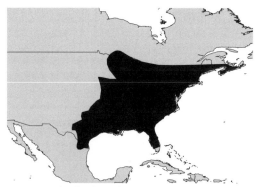

Map 37. Distribution of *Dromogomphus* spp.

Medium to large gomphids (53–74 mm); postocellar ridge complete; occiput with a transverse ridge along its entire width, with a fringe of hairs along its summit; hind femora distally with a row of very strong and long spines, about as long as femora width (Fig. 605, shared with *Mitragomphus* and *Perigomphus*), femora longer than head width; pterothorax green to yellow with dark brown stripes, which may be reduced laterally; abdomen greenish- or yellow with dark brown to reddish brown markings, distal segments orange in some species. Wings (Fig. 604) lack-ing basal subcostal crossveins; 1 crossvein in HW space between arculus branches and point of branching of RP; subtriangles and supratriangles free; 1 cubito-anal crossvein; CuA and MP in HW parallel or divergent. Male cercus unbranched, tapering distally (Fig. 607b); epiproct bifid (Fig. 607a). Anterior hamule cleft, posterior hamule quadrangular with a series of 2-3 small spines on anterior edge and a single spine on posterior edge (Fig. 606); vesica spermalis distal segment lacking flagella. Female with rudimentsary auricles; vulvar lamina bifid, extending to about a 1/3 of S9 length (Fig. 608). **Unique characters**: None known.

Status of classification: Good. Species easily diagnosed. Carle (1986) included this genus in the Gomphini, together with *Arigomphus*, *Gomphus*, *Stylurus*, and the Old World *Gastrogomphus*.

Potential for new species: Unlikely.

Habitat: Larvae inhabit variety of lotic environments, either turbid or clear, with sandy, muddy or rocky bottom, from slow running coastal plain streams to spring fed streams, rivers and glacial lakes (Louton, 1982b). Adults perch on the ground or on leaves of bushes or trees at forest edges and on banks of larval habitat; males patrol from a perch flying over open water and along shore (Dunkle, 2000).

604 male wings – *Dromogomphus spinosus*

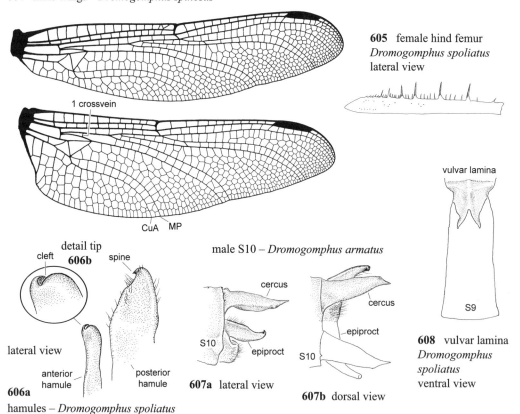

1 crossvein

CuA   MP

605 female hind femur
Dromogomphus spoliatus
lateral view

vulvar lamina

detail tip
cleft **606b** spine

male S10 – *Dromogomphus armatus*

cercus

cercus

epiproct

S9

lateral view

S10

epiproct

608 vulvar lamina
Dromogomphus spoliatus
ventral view

anterior
hamule

posterior
hamule

S10

606a

607a lateral view

607b dorsal view

hamules – *Dromogomphus spoliatus*

Ebegomphus Needham, 1944: 186.

[♂ pp. 71, couplets 25-26; ♀ pp. 82-83, couplets 28-29]

Type species: *Ebegomphus strumens* Needham, 1944 [by original designation]

   syn *Strumagomphus* Needham, 1944: 180 (*nomen nudum*)

5 species:

conchinus (Williamson, 1916) [*Cyanogomphus*]* – **L**? [Belle, 1966b]

demerarae (Selys, 1894) [*Cyanogomphus*?]* – **L** [Belle, 1966b; 1970]

   syn *strumens* Needham, 1944

minutus (Belle, 1970) [*Cyanogomphus*]* – **L**? [Belle, 1970]

pumilus (Belle, 1986) [*Cyanogomphus*] – **L** [Belle, 1986b]

schroederi (Belle, 1970) [*Cyanogomphus*]

References: Belle 1966b; 1970; 1986b; 1996.
Distribution: Colombia to NE Brazil.

Medium gomphids (35–45 mm); postocellar ridge complete; occiput broadly rounded (Fig. 609); pterothorax and abdomen green with dark brown markings. Wings (Fig. 611) with basal subcostal crossveins; 2 or more crossveins in HW space between arculus branches and point of branching of RP; subtriangles and supratriangles free; 1 cubito-anal crossvein; CuA and MP in HW parallel or divergent. Male cercus with branch directed ventrally (Fig. 612); epiproct bifid, each branch in turn branched. Anterior hamule quadrangular, ending in a spine, and fused to anterior lamina; posterior hamule with distal 1/3 bent antero-

Map 38. Distribution of *Ebegomphus* spp.

ventrally at a right angle (shared with *Octogomphus* and *Tibiagomphus*); vesica spermalis distal segment with a pair of short flagella. Female lacking auricles; vulvar lamina bifid, extending to about 1/2 the length of S9. **Unique characters**: None known.

Status of classification: Belle (1996) established the tribe Agriogomphini under the Epigomphinae to receive *Agriogomphus*, *Brasiliogomphus* and *Ebegomphus*.

Potential for new species: Unknown.

Habitat: Forest creeks; Belle (1970) found adult females of *E. minutus* hovering over sunny spots, low over the surface of quietly flowing water, at a heavily forested creek; and reported male flight as not swift. He also found an exuviae lying flat on a leaf of an aquatic plant. Larvae are not burrowers (Belle, 1966b), inhabiting leafy trash in pools and eddies of creeks. Larvae of *E. demerarae* were found emerging on large floating leaves at midday (Belle, 1970).

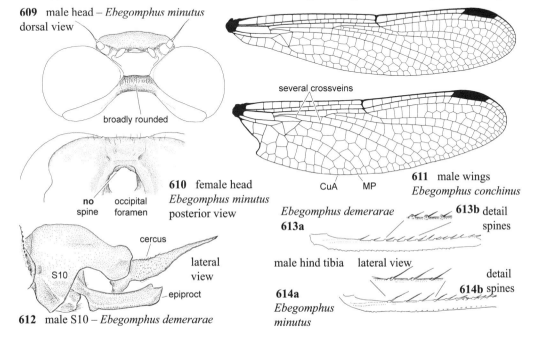

609 male head – *Ebegomphus minutus* dorsal view

broadly rounded

no spine   occipital foramen

610 female head *Ebegomphus minutus* posterior view

cercus

lateral view

S10

epiproct

612 male S10 – *Ebegomphus demerarae*

several crossveins

CuA   MP

611 male wings *Ebegomphus conchinus*

Ebegomphus demerarae
613a   **613b** detail spines

male hind tibia   lateral view

detail
614b spines
614a
Ebegomphus minutus

Epigomphus Hagen *in* Selys, 1854: 59 (40 reprint).
[♂ pp. 72, couplet 28; ♀ pp. 82-83, couplets 27, 32]
Type species: *Epigomphus paludosus* Hagen *in* Selys,
1854 [by monotypy]
     syn *Eugomphus* Kennedy, 1946: 666.
          Type species: *Epigomphus quadracies*
          Calvert, 1903 [by original designation]
28 species:

armatus Ris, 1918*
camelus Calvert, 1905*
clavatus Belle, 1980
compactus Belle, 1994
corniculatus Belle, 1989
crepidus Kennedy, 1936*
donnellyi González & Cook, 1988*
echeverrii Brooks, 1989* – **L** [Ramírez, 1996]
flinti Donnelly, 1989*
gibberosus Belle, 1988
houghtoni Brooks, 1989
hylaeus Ris, 1918* – **L** [Fleck, 2002b]
     syn *gracilis* Belle, 1970
jannyae Belle, 1993
llama Calvert, 1903*
maya Donnelly, 1989*
obtusus Selys, 1869*
occipitalis Belle, 1970*
paludosus Hagen *in* Selys, 1854* – **L**? [Costa,
1968]
paulsoni Belle, 1981
pechumani Belle, 1970
quadracies Calvert, 1903*
subobtusus Selys, 1878* – **L** [Ramírez, 1996]
subquadrices Kennedy, 1946*
subsimilis Calvert, 1920* – **L** [Ramírez, 1996]
sulcatistyla Donnelly, 1989
tumefactus Calvert, 1903*
verticicornis Calvert, 1908
westfalli Donnelly, 1986*

References: Calvert, 1920; Belle, 1970; 1988c; 1996;
Donnelly, 1986.
Distribution: Mexico south to NE Argentina.

Medium gomphids (43–55 mm); postocellar ridge
complete; rear of the head in the female with 2 lateral
depressions in some species, where male caudal appendages fit during copulation (shared with *Octogomphus*); pterothorax green to yellow with dark brown

Map 39. Distribution of *Epigomphus* spp.

or black stripes; abdomen largely black with green-
ish- or yellow markings, S7 usually pale. Wings (Figs.
617-618) with or without basal subcostal crossveins;
2 or more crossveins in HW space between arculus
branches and point of branching of RP; subtriangles
and supratriangles free; male HW triangle not defined
(shared with *Agriogomphus* and *Archaeogomphus*);
2 cubito-anal crossveins; CuA and MP in HW divergent. Spines of outer row of male hind tibiae approximately quadrangular (Fig. 615), sometimes projected
at distal edge (shared with *Tibiagomphus*); cercus
unbranched, tapering distally; epiproct bifid, each
branch entire or bifid (Figs. 620-621). Posterior hamule with small denticles on anterior edge and lacking
apical tooth (Figs. 619a-b); vesica spermalis distal
segment lacking flagella, and basal segment entire or
bifid. Female with rudimentary auricles; vulvar lamina bifid, extending to about 1/2 of S9 length. **Unique
characters**: Female epiproct polished and about as
long as cerci (Figs. 616a-b).
Status of classification: Species in general well defined, but reliance must be made on primary literature, since there is no synoptic work treating all of
them. Belle (1996) included only this genus in the
tribe Epigomphini.
Potential for new species: Likely.
Habitat: Adults have been taken over open brooks
in the forest (Belle, 1988c; Louton, Garrison, and
Flint, 1996). Mating adaptations described by Calvert
(1920). Larvae found less than 3 cm below surface of
substrate (organic matter, gravel, mud or sand) at bottom of creeks; emergence observed mostly in early
morning, taking place on firm substrate in horizontal
position (Ramírez, 1996).

615   male hind tibia – *Epigomphus paludosus*
lateral view

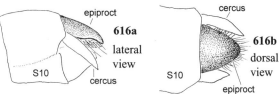

616a
lateral
view

616b
dorsal
view

female S10 – *Epigomphus armatus*

617 male wings – *Epigomphus flinti*

618 male wings – *Epigomphus paludosus*

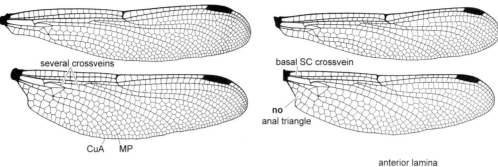

several crossveins

basal SC crossvein

no
anal triangle

CuA   MP

genital fossa – *Epigomphus paludosus*

anterior lamina

denticles

anterior
hamule

619a
lateral view

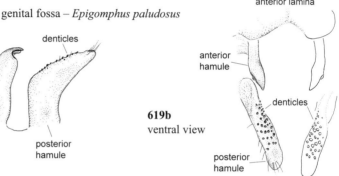

denticles

619b
ventral view

anterior
hamule

posterior
hamule

posterior
hamule

cercus

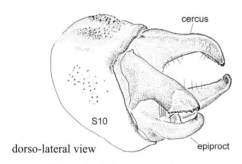

cercus

S10

dorso-lateral
view

S10

dorso-lateral view

epiproct

paraproct   epiproct

620 male S10 – *Epigomphus donnellyi*

621 male S10 – *Epigomphus paludosus*

Erpetogomphus Hagen *in* Selys, 1858: 329 (69 reprint).

[♂ pp. 68, couplet 13; ♀ pp. 76-79, couplet 5, 7, 13]

Type species: *Ophiogomphus crotalinus* Hagen *in* Selys, 1854 [Kirby, 1890 by subsequent designation]

syn *Herpetogomphus* Walsh, 1862: 388 (unjustified emendation of *Erpetogomphus*).

syn *Calogomphus* Carle, 1992: 148.

Type species: *Erpetogomphus eutainia* Calvert, 1905 [by original designation]

syn *Erpetocyclops* Carle, 1992: 148.

Type species: *Erpetogomphus ophibolus* Calvert, 1905 [by original designation]

22 species:

agkistrodon Garrison, 1994* – **L** [Novelo-Gutiérrez, 2002a]

boa Selys, 1859* – **L** [Novelo-Gutiérrez, 2002c]

bothrops Garrison, 1994* – **L** [Novelo-Gutiérrez, 2005]

compositus Hagen *in* Selys, 1858* – **L** [Needham, Westfall, and May, 2000]

syn *coluber* Williamson & Williamson, 1930

constrictor Ris, 1918* – **L** [Ramírez, 1996; Novelo-Gutiérrez, 2002a]

cophias Selys, 1858* – **L** [Novelo-Gutiérrez, 2002c]

crotalinus (Hagen *in* Selys, 1854) [*Ophiogomphus*]*

designatus Hagen *in* Selys, 1858* – **L** [Louton, 1982b]

elaphe Garrison, 1994*

elaps Selys, 1858* – **L** [Novelo-Gutiérrez, 2005]

erici Novelo-Gutiérrez 1999* – **L** [Novelo-Gutiérrez, 2002a]
eutainia Calvert, 1905* – **L** [Novelo-Gutiérrez, 2005]
  syn *diadophis* Calvert, 1905
heterodon Garrison, 1994*
lampropeltis lampropeltis Kennedy, 1918*
lampropeltis natrix Williamson & Williamson, 1930* – **L** [Needham, Westfall, and May, 2000; Novelo-Gutiérrez and González-Soriano, 1991]
leptophis Garrison, 1994
liopeltis Garrison, 1994* – **L** [Novelo-Gutiérrez, 2005]
ophibolus Calvert, 1905* – **L** [Novelo-Gutiérrez, 2002a]
sabaleticus Williamson, 1918* – **L** [Belle, 1992a; Ramírez, 1996; Novelo-Gutiérrez, 2002a]
schausi Calvert, 1919
sipedon Calvert, 1905*
tristani Calvert, 1912* – **L** [Ramírez, 1996; Novelo-Gutiérrez, 2002a]
viperinus Selys, 1868* – **L** [Novelo-Gutiérrez, 2005]
[NOTE: The status of *Ophiogomphus? menetriesii* Selys, 1854 was given by Garrison (1994), who argued that the name should be considered a *nomen dubium*.]

References: Garrison, 1994.
Distribution: US south to N Colombia and Venezuela.

Small to medium gomphids (38–56 mm); postocellar ridge complete or incomplete; occiput with a transverse ridge along its entire width, with a fringe of hairs along its summit; pterothorax gray to green to vivid blue green or yellow with dark brown stripes, which may be reduced or entirely absent; abdomen distal segments tan to dark reddish brown. Wings (Figs. 627-628) hyaline or colored at base; lacking basal subcostal crossveins; 1 crossvein in HW space between arculus branches and point of branching of RP; subtriangles and supratriangles free; 1 cubito-anal crossvein; CuA and MP in HW parallel or divergent. Male cercus (Figs. 629-630) entire or with a branch not directed ventrally; epiproct bifid; epiproct branches parallel to each other and upturned at about 180 degrees (Figs. 633a-b, shared with some species of *Ophiogomphus*). Posterior hamule (Fig. 631) variable, with or without a single end spine; vesica

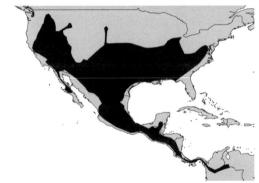

Map 40. Distribution of *Erpetogomphus* spp.

spermalis distal segment with 2 short flagella or lacking flagella, and with 2 ventro-lateral lobes (Figs. 632a-b, shared with most species of *Ophiogomphus*). Male auricles with small denticles in a row (Fig. 634, shared with *Ophiogomphus* and some *Agriogomphus* species); female auricles rudimentary. Vulvar lamina bifid (Figs. 622-623) or completely divided (Figs. 624-625), extending from about 1/4 to 1/2 of S9 length. **Unique characters**: None known.

Status of classification: Good. Complete revision by Garrison (1994) allows for the identification of all species but one, which was described later (Novelo-Gutiérrez and Garrison, 1999). Carle (1986) and Garrison (1994) considered this genus allied to *Ophiogomphus*, and Belle (1996) to *Onychogomphus*. All three genera have been included in the tribe Onychogomphini. *Calogomphus* and *Erpetocyclops* were proposed in a key as subgenera of *Erpetogomphus* by Carle (1992) based on sexual characters of the male, all of which we consider to represent no more than specific differences. *Calogomphus* and *Erpetocyclops* were not recognized by Garrison (1994) in the latest revision of the group and we follow him here in treating these names as synonyms.

Potential for new species: Likely. We know of one undescribed species from Mexico.

Habitat: Adults most commonly found near shores of streams and rivers, and sometimes away from water perching on ground or tips of weed stems, or foraging in agricultural stubble; females oviposit by swiftly tapping the water's surface or by hovering over moderately swift lotic environments (Garrison, 1994). Larvae found among gravel or leafy sediments, or in sand or mud banks of lotic environments (Louton, 1982b; Novelo-Gutiérrez and González-Soriano, 1991; Novelo-Gutiérrez, 2005).

vulvar lamina – ventral view

622 S9
Erpetogomphus sabaleticus

623 S9
Erpetogomphus tristani

S9 **624**
Erpetogomphus eutainia

S9 **625**
Erpetogomphus elaphe

Erpetogomphus agkistrodon
cercus **626**
female S10
epiproct
S10
dorsal view

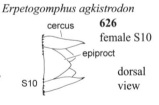

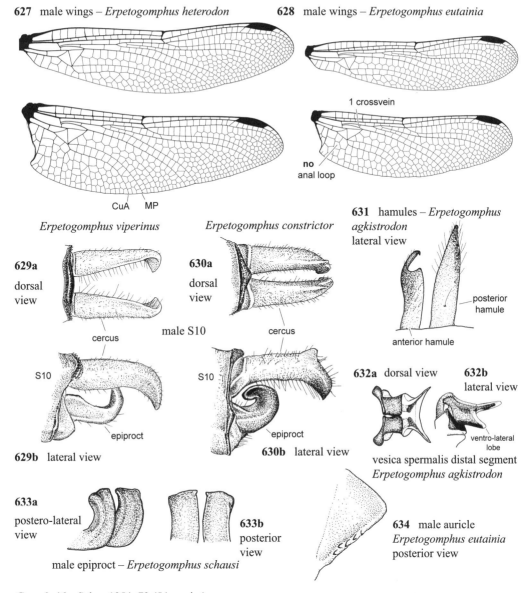

627 male wings – *Erpetogomphus heterodon*

628 male wings – *Erpetogomphus eutainia*

1 crossvein

no
anal loop

CuA   MP

Erpetogomphus viperinus

Erpetogomphus constrictor

631 hamules – *Erpetogomphus agkistrodon* lateral view

629a
dorsal
view

630a
dorsal
view

posterior
hamule

cercus

male S10

cercus

anterior hamule

S10

S10

632a dorsal view   **632b** lateral view

epiproct

epiproct

ventro-lateral
lobe

629b lateral view

630b lateral view

vesica spermalis distal segment
Erpetogomphus agkistrodon

633a
postero-lateral
view

633b
posterior
view

634 male auricle
Erpetogomphus eutainia
posterior view

male epiproct – *Erpetogomphus schausi*

Gomphoides Selys, 1854: 73 (54 reprint).

[♂ pp. 74, couplet 36; ♀ pp. 86, couplet 40]

Type species: *Diastatomma infumatum* Rambur, 1842
[Kirby, 1890 by subsequent designation]
   syn *Negomphoides* Muttowski, 1910: 81.
      Type species: *Diastatomma infumatum*
      Rambur, 1842 [by original designation]
   syn *Ammogomphus* Förster, 1914: 73.
      Type species: *Ammogomphus perditus*
      Förster, 1914 [by original designation]
3 species:

infumata (Rambur, 1842) [*Diastatomma*] – **L**
[Belle, 1992a]
perdita (Förster, 1914) [*Ammogomphus*]*
praevia St. Quentin, 1967*

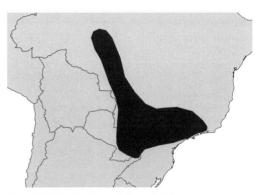

Map 41. Distribution of *Gomphoides* spp.

References: St. Quentin, 1967; Gloyd, 1973; 1974; Belle, 1984b.

Distribution: Brazil to Paraguay and NE Argentina.

Medium to large gomphids (62-68 mm); postocellar ridge complete or incomplete; occiput with a fringe of hairs along its summit; pterothorax green with dark brown stripes (Fig. 636); abdomen yellow to green, with brown markings. Wings (Fig. 635) with basal subcostal crossveins; 2 or more crossveins in HW space between arculus branches and point of branching of RP; subtriangle of FW with 2 or more crossveins, of HW free or with 1 or more crossveins; supratriangles free or crossed; male HW anal triangle reaching anal angle of wing; 1 cubito-anal crossvein; CuA and MP in HW divergent. Male cercus unbranched; epiproct bifid (Figs. 638a-b).

Posterior hamule with a single end spine; vesica spermalis distal segment with 2 long flagella (Figs. 637a-b). Female auricles absent; vulvar lamina bifid, extending to about 1/4 of S9 length; epiproct not polished, rudimentary and barely visible in dorsal view. **Unique characters**: Male epiproct with 1 outer dorso-basal tooth (Fig. 638b); S10 longer than S9 (Fig. 638b; unique for males, in females shared with *Peruviogomphus*).

Status of classification: Poorly known group. Species identification based on primary literature. Gloyd (1973; 1974) and Belle (1984b) redefined the genus as it is currently used. Belle (1996) considered it close to *Idiogomphoides* and *Phyllogomphoides*, including both genera in his tribe Gomphoidini.

Potential for new species: Likely.

Habitat: No information available.

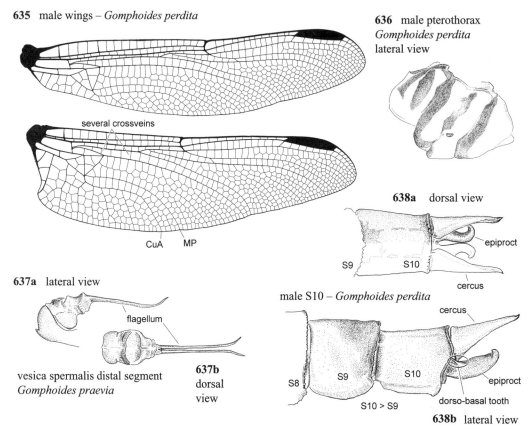

635 male wings – *Gomphoides perdita*

several crossveins

CuA   MP

636 male pterothorax *Gomphoides perdita* lateral view

638a dorsal view

S9   S10   epiproct   cercus

male S10 – *Gomphoides perdita*

637a lateral view

flagellum

vesica spermalis distal segment *Gomphoides praevia*

637b dorsal view

cercus   S8   S9   S10   epiproct   S10 > S9   dorso-basal tooth

638b lateral view

Gomphus Leach, 1815: 137.
[♂ pp. 67, couplets 9, 11-12; ♀ pp. 76-79, couplets 5, 7, 10, 12-13, 17]
Type species: *Libellula vulgatissima* Linnaeus, 1758 [by monotypy]
 syn *Thanatophora* Selys, 1850: 83.
  Type species: *Thanatophora egregia* Hansemann MS *in* Selys, 1850 [Cowley, 1934d, by subsequent designation]
 syn *Gomphurus* Needham, 1901: 446.

Type species: *Gomphus vastus* Walsh, 1862 [Cowley, 1934c by subsequent designation]
 syn *Hylogomphus* Needham, Westfall, and May, 2000: 332.
  Type species: *Gomphus adelphus* Selys, 1858 [Needham, Westfall, and May, 2000 by original designation]

[NOTE: *Hylogomphus* was originally proposed as a *nomen nudum* by Needham (1951: 23) and again by Needham and Westfall (1955: 224) without designation of type species. The correct introduction of *Hylogomphus* was by Needham, Westfall, and May (2000).]
    syn *Phanogomphus* Carle, 1986: 296.
        Type species: *Gomphus minutus* Rambur, 1842 [by original designation]
[NOTE: *Phanogomphus* was originally proposed as a subgenus of *Gomphus* by Carle (1986: 296), although he [incorrectly] stated that the name was "...to be credited to Carle and Cook (1984)" which, according to May and Carle (1996: 32), was an erroneous editorial substitution for "... Carle and Cook [in Carle, 1986]". We credit *Phanogomphus* to Carle, 1986.]
    syn *Stenogomphurus* Carle, 1996: 296.
        Type species: *Gomphus consanguis* Selys, 1879 [by original designation]
52 species; 38 New World species:

abbreviatus Hagen *in* Selys, 1878* – **L** [Needham, 1901; Louton, 1982b]
adelphus Selys, 1858* – **L**? [Hagen, 1885; Needham, 1901 as *G. brevis*; Louton, 1982b as *G. brevis*]
    syn *brevis* Hagen *in* Selys, 1878
apomyius Donnelly, 1966* – **L** [Needham, Westfall, and May, 2000]
australis (Needham, 1897) [*Arigomphus*]* – **L** [Westfall, 1950]
borealis Needham, 1901* – **L** [Needham, 1901]
cavillaris Needham, 1902* – **L** [Westfall, 1950]
    syn *brimleyi* Muttkowski, 1911
consanguis Selys, 1879* – **L** [Westfall and Trogdon, 1962; Louton, 1982b]
crassus Hagen *in* Selys, 1878* – **L** [Louton, 1982b; Tennessen and Louton, 1984]
    syn *walshii* Kellicott, 1899
descriptus Banks, 1896* – **L** [Walker, 1958]
    syn *argus* Needham, 1943
    syn *mortimer* Needham, 1943
dilatatus Rambur, 1842* – **L** [Westfall, 1974; Louton, 1982b]
diminutus Needham, 1950* – **L** [Needham, Westfall, and May, 2000]
exilis Selys, 1854* – **L** [Walker, 1958]
    syn *flavocaudatus* Walker, 1940
externus Hagen *in* Selys, 1858* – **L** [Needham and Hart, 1901; Walker, 1958; Louton, 1982b]
    syn *consobrinus* Walsh, 1863
fraternus fraternus (Say, 1840) [*Aeshna*]* – **L** [Garman, 1927; Walker, 1958; Louton, 1982b]
fraternus manitobanus Walker, 1958
geminatus Carle, 1979* – **L** [Carle, 1979]
gonzalezi Dunkle, 1992*
graslinellus Walsh, 1862* – **L** [Needham and Hart, 1901; Walker, 1958]
hodgesi Needham, 1950* – **L** [Westfall, 1965]

hybridus Williamson, 1902* – **L** [Broughton, 1928; Louton, 1982b]
kurilis Hagen *in* Selys, 1858* – **L** [Walker, 1958]
    syn *confraternus* Selys, 1873
    syn *donneri* Kennedy, 1917
    syn *sobrinus* Selys, 1873
lineatifrons Calvert, 1921* – **L** [Westfall, 1974; Louton, 1982b]
lividus Selys, 1854* – **L** [Garman, 1927; Byers, 1930; Walker, 1958; Louton, 1982b]
    syn *sordidus* Hagen *in* Selys, 1854
    syn *umbratus* Needham, 1897
lynnae Paulson, 1983* – **L** [Paulson, 1983; Needham, Westfall, and May, 2000]
militaris Hagen *in* Selys, 1858* – **L** [Landwer and Sites, 2003]
minutus Rambur, 1842* – **L** [Westfall, 1950]
modestus Needham, 1942* – **L** [Westfall, 1974; Louton, 1982b]
oklahomensis Pritchard, 1935* – **L** [Needham, Westfall, and May, 2000]
ozarkensis Westfall, 1975* – **L** [Louton, 1982b; Huggins and Harp, 1985]
parvidens Currie, 1917* – **L** [Carle, 1979; Louton, 1982b]
    syn *carolinus* Carle, 1979
quadricolor Walsh, 1863* – **L** [Walker, 1932; 1958; Louton, 1982b]
    syn *alleni* Howe, 1922
rogersi Gloyd, 1936* – **L** [Needham, 1943b as *G. consanguis*; Westfall and Trogdon, 1962; Louton, 1982b]
sandrius Tennessen, 1983*
septima delawarensis Donnelly & Carle, 2000* – **L** [Donnelly and Carle, 2000]
septima septima Westfall, 1956* – **L** [Donnelly and Carle, 2000]
    syn *secundus* Steinnmann, 1997
spicatus Hagen *in* Selys, 1854* – **L** [Needham, 1901; Stout, 1918]
vastus Walsh, 1862* – **L** [Walker, 1958; Westfall, 1974; Louton, 1982b]
ventricosus Walsh, 1863* – **L** [Louton, 1982b; 1983]

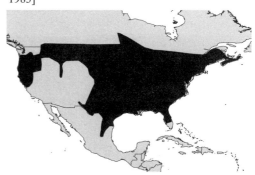

Map 42. Distribution of *Gomphus* spp.

viridifrons Hine, 1901* – **L** [Louton, 1982b]
westfalli Carle & May, 1987* – **L** [Needham, Westfall, and May, 2000]
[NOTE: *Gomphus williamsoni* Muttkowski, 1910, based on only one known specimen, is almost certainly a natural hybrid between *Gomphus graslinellus* and *G. lividus*.]

References: Needham, Westfall, and May, 2000.
Distribution: Holarctic; in the New World from S Canada south to US and N Mexico.

Small to large gomphids (34–73 mm); occiput with complete transverse ridge (Fig. 641), with a fringe of hairs along its summit; pterothorax largely greenish to yellow with brown stripes (Fig. 644); abdomen whitish to yellow or green with dark brown markings, apical segments black and yellow to orange brown. Wings (Figs. 639-640) lacking basal subcostal crossveins; 1 crossvein in HW space between arculus branches and point of branching of RP; subtriangles and supratriangles free; 1 cubito-anal crossvein; CuA and MP in HW parallel or divergent. Male cercus unbranched or with a branch not directed ventrally; epiproct bifid (Fig. 645). Posterior hamule with or without a cluster of denticles on anterior edge, ending in a single tooth (Fig. 646); vesica spermalis distal segment usually with 2 flagella, long or short, sometimes lacking (i.e. *G. brevis*), basal segment entire or bifid. Female with rudimentary auricles; vulvar lamina bifid, extending from a 1/4 (Fig. 647) to almost the entire length (Fig. 648) of S9. **Unique characters:**

None known.

Status of classification: Heterogeneous group in need of taxonomic revision and phylogenetic assessment. The North American species have traditionally been arranged in three subgenera, *Gomphus*, *Gomphurus*, and *Hylogomphus*, but controversy exists as to their status (Louton, 1982b; 1983; Carle, 1986; May and Carle, 1996). The relationship between the European type species *G. vulgatissimus* and the North American and Asian species is also uncertain; it is possible that Nearctic species of *G.* (*Hylogomphus*) are actually more closely related with the type species than are the Nearctic species of *G.* (*Gomphus*) (Louton, 1982b; Carle, 1986). Although the genus *Gomphus* as used here is very likely poly- or paraphyletic, we follow Needham, Westfall, and May (2000) in conserving its old usage until the relationships of all its species are resolved. Carle (1986) included this genus in the Gomphini, together with *Arigomphus*, *Dromogomphus*, *Stylurus*, and the Old World *Gastrogomphus*.

Potential for new species: Possible since many species are scarce and wary.

Habitat: Larvae burrow shallowly in sediment (silt, clay, sand, gravel) of rivers, streams, and lakes, and some species in ponds. Males patrol far out over rivers with bouncy flight and raised club (Dunkle, 2000). Mating takes place either on trees or on the ground, depending on the species. For a complete account on the habitat of all species see Dunkle (2000).

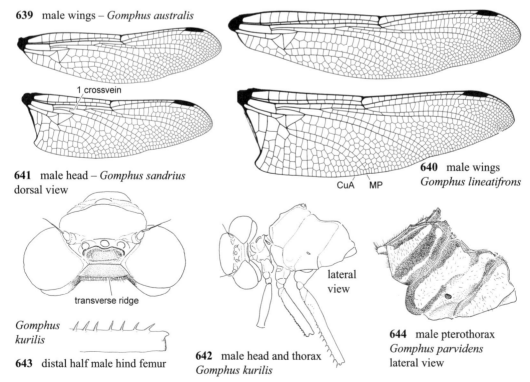

639 male wings – *Gomphus australis*

1 crossvein

641 male head – *Gomphus sandrius*
dorsal view

transverse ridge

640 male wings
Gomphus lineatifrons

CuA   MP

lateral view

Gomphus kurilis

643 distal half male hind femur

642 male head and thorax
Gomphus kurilis

644 male pterothorax
Gomphus parvidens
lateral view

645 male S10 – *Gomphus militaris*
dorso-lateral view

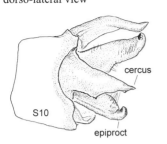

S10

cercus

epiproct

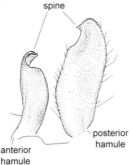

spine

anterior
hamule

posterior
hamule

646 hamules – *Gomphus kurilis*
lateral view

vulvar lamina   ventral view

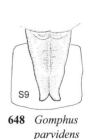

S9

647 *Gomphus
quadricolor*

S9

648 *Gomphus
parvidens*

Hagenius Selys, 1854: 82 (63 reprint).
[♂ pp. 69, couplet 18; ♀ pp. 83, couplet 30]
Type species: *Hagenius brevistylus* Selys, 1854 [by
monotypy]
1 species:

brevistylus Selys, 1854* – **L** [Needham, 1901;
Louton, 1982b]

References: Needham, Westfall, and May, 2000.
Distribution: SE Canada to SE United Sates.

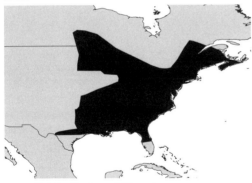

Map 43. Distribution of *Hagenius* sp.

Very large gomphids (73–90 mm); postocellar ridge
complete; occiput with complete transverse ridge in
male, broadly rounded in female, with a fringe of
hairs along its summit; pterothorax yellow with dark
brown stripes; abdomen yellow with black latero-
longitudinal stripes. Wings (Fig. 651) lacking basal
subcostal crossveins; 2 or more crossveins in HW
space between arculus branches and point of branching of RP; subtriangles and supratriangles free; 1 cubito-anal crossvein; CuA and MP in HW divergent;
FW with supplementary trigonal sector. Male cercus unbranched; epiproct bifid (Fig. 649). Anterior
lamina v-shaped (shared with some *Progomphus* and
Zonophora); posterior hamule with denticles on anterior edge, ending in a single tooth; vesica spermalis
distal segment lacking flagella, basal segment entire
(Fig. 650). Female with rudimentary auricles; vulvar
lamina bifid, extending to about 1/3 of S9. **Unique
characters**: None known.
Status of classification: Well defined species with
nearest relatives (*Sieboldius* Selys) in Asia.
Potential for new species: Unlikely.
Habitat: Larvae sprawl among leafy trash and wood-
chips along banks of forested streams and rivers.
Adults prey on other large insects, including dragonflies. Found flying at high speed along paths and forest edges, perching on rocks or vegetation at water.
Females oviposit while flying, splashing eggs from
the water onto the bank with the tip of the abdomen
(Dunkle, 2000).

649 male S10
Hagenius brevistylus
dorso-lateral view

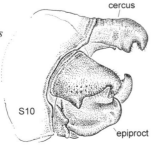

cercus

S10

epiproct

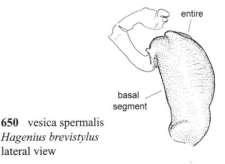

entire

basal
segment

650 vesica spermalis
Hagenius brevistylus
lateral view

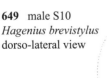

651   male wings – *Hagenius brevistylus*

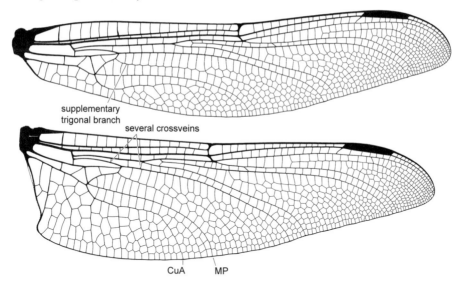

supplementary
trigonal branch
several crossveins

CuA   MP

Idiogomphoides Belle, 1984b: 106.
[♂ pp. 74, couplet 35; ♀ pp. 86, couplet 42]
Type species: *Gomphoides demoulini* St. Quentin, 1967 [by original designation]
3 species:

demoulini (St. Quentin, 1967) [*Gomphoides*]* – **L**?
[Needham, 1944 as *Gomphoides fuliginosus*]
emmeli Belle, 1995)*
ictinia (Selys, 1878) [*Gomphoides*] – **L**? [Belle, 1992a]

References: Belle, 1984b; Belle, 1995.
Distribution: Guyana and Surinam to S Brazil.

Map 44. Distribution of *Idiogomphoides* spp.

Large to very large gomphids (65–72 mm); postocellar ridge complete; occiput with complete transverse ridge, with a fringe of hairs along its summit; ptero-thorax and abdomen yellow with dark brown stripes and markings. Wings (Fig. 652) hyaline or colored at base; with basal subcostal crossveins; 2 or more crossveins in HW space between arculus branches and point of branching of RP; subtriangle of FW with 2 or more crossveins, of HW with 2 or more crossveins or free; supratriangles crossed; HW anal triangle reaching anal angle of wing; FW with 2 or more cubito-anal crossveins, HW with 1; CuA and MP in HW divergent. Male cerci with ventro-basal branch; epiproct small, visible in lateral view and bifid (Fig. 653). Posterior hamule with 2 subapical internal teeth (Fig. 654b) shared with some *Progomphus*. Vesica spermalis distal segment with 2 long flagella, basal segment bifid. Female lacking auricles; vulvar lamina bifid, almost completely divided, extending to almost distal margin of S9 (Figs. 655a-b); epiproct not polished, rudimentary and barely visible in dorsal view.
Unique characters: None known.
Status of classification: Belle (1984b; 1996) considered it allied to *Gomphoides* and *Phyllogomphoides*, including both genera in the tribe Gomphoidini.
Potential for new species: Likely since species are very rare and known from only a few specimens.
Habitat: No information available. Exuviae of possible *I. ictinia* collected on rocks of bank of the Frederick Willem Falls in Surinam; described by supposition by Belle (1992a) for this genus, adults of which have not been collected from either Guyana or Surinam.

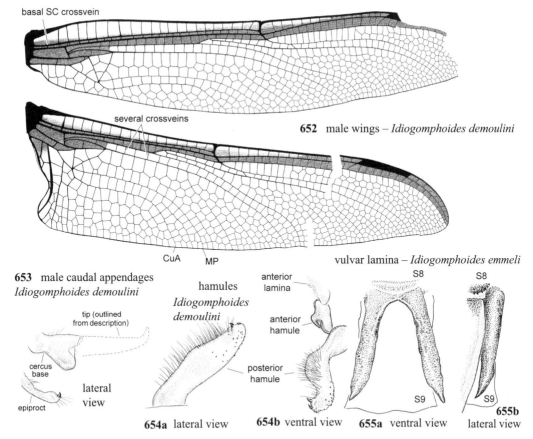

basal SC crossvein

several crossveins

652 male wings – *Idiogomphoides demoulini*

CuA MP

653 male caudal appendages
Idiogomphoides demoulini

hamules
Idiogomphoides demoulini

vulvar lamina – *Idiogomphoides emmeli*

S8 S8

tip (outlined from description)

anterior lamina

cercus base

anterior hamule

lateral view

epiproct

posterior hamule

lateral view

S9 S9

655b

654a lateral view **654b** ventral view **655a** ventral view lateral view

Lanthus Needham, 1897b: 166.

[♂ pp. 66, couplet 7; ♀ pp. 77, 79-80, couplets 8, 18]

Type species: *Gomphus parvulus* Selys, 1854 [by original designation]
3 species, 2 New World species:

parvulus (Selys, 1854) [*Gomphus*]* – **L** [Carle, 1980]
vernalis Carle, 1980* – **L** [Carle, 1980; Louton, 1982b]

References: Carle, 1980.
Distribution: Japan and North America; in the New World from E Canada to United States.

Small gomphids (29–40 mm); postocellar ridge complete; occiput broadly rounded; pterothorax yellow to green with dark brown stripes; abdomen mostly black, yellow or green basally. Wings (Fig. 657) lacking basal subcostal crossveins; 1 crossvein in HW space between arculus branches and point of branching of RP; subtriangles and supratriangles free; 1 cubito-anal crossvein; CuA and MP in HW divergent; FW with supplementary trigonal sector. Male cerci with ventro-basal branch; epiproct bifid (Figs. 658a-b). Posterior hamule with denticles on anterior edge.

Map 45. Distribution of *Lanthus* spp.

Vesica spermalis distal segment lacking flagella. Female with rudimentary auricles; vulvar lamina bifid, extending from 1/2 to 2/3 of S9 length (Fig. 659). **Unique characters**: None known.
Status of classification: Species well diagnosed. Carle (1980) considered this genus related to *Octogomphus* Selys, *Davidius* Selys, *Stylogomphus* Fraser, and *Sinogomphus* May, and later (1986) included it in the tribe Octogomphini together with *Octogomphus*, *Davidius*, and *Dubitogomphus* Fraser.

Potential for new species: Unlikely.

Habitat: Small spring fed streams and brooks, and small rivers of high quality water, in or near forests. Larvae burrow shallowly in sandy or silty places in beds of rocky streams (Needham, Westfall, and May, 2000). Males perch unwarily on rocks or trees over riffles (Dunkle, 2000).

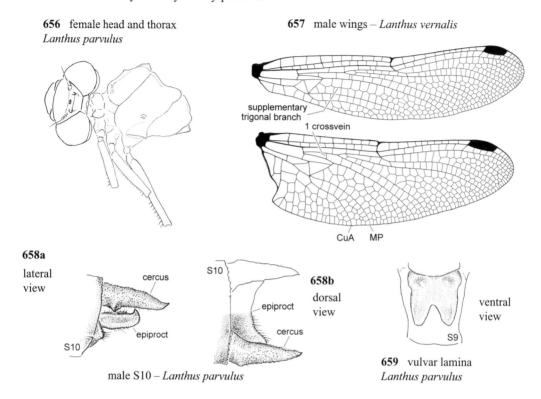

656 female head and thorax
Lanthus parvulus

657 male wings – *Lanthus vernalis*

supplementary trigonal branch
1 crossvein

CuA   MP

658a
lateral view

cercus

S10

658b
dorsal view

epiproct

cercus

epiproct

S10

ventral view

S9

male S10 – *Lanthus parvulus*

659 vulvar lamina
Lanthus parvulus

Melanocacus Belle, 1986a: 98.
[♂ pp. 74, couplet 34; ♀ pp. 87, couplet 43]
Type species: *Cacus mungo* Needham, 1940 [by original designation]
2 species:

interioris Belle, 1986a**
mungo (Needham, 1940) [*Cacus*]* – **L** [Needham, 1944; Belle, 1970]

References: Geijskes, 1964b; Belle, 1986a.
Distribution: Venezuela, Surinam, and Brazil.

Medium to large gomphids (60–66 mm); pterothorax and abdomen yellow with brown markings; subalar carina of HW projected into strong spine (shared with *Cacoides, Mitragomphus,* and *Zonophora*). Wings (Figs. 660-661) colored at base; with basal subcostal crossveins; 2 or more crossveins in HW space between arculus branches and point of branching of RP; subtriangle of FW with 2 or more crossveins, of HW free; supratriangles crossed; 2 or more cubito-anal crossveins; CuA and MP in HW parallel; FW with a

Map 46. Distribution of *Melanocacus* spp.

supplementary trigonal sector. Male cercus long, with upturned tip (Fig. 665a); epiproct small and entire, visible in lateral view (Figs. 665a-b). Posterior hamule lacking spines or hooks (Fig. 663); vesica spermalis distal segment lacking flagella, with 2 small apical spines (Figs. 664a-b), and basal segment with an entire anterior rim. Female vulvar lamina bifid,

extending beyond posterior margin of S9 (Fig. 666).
Unique characters: None known.

Status of classification: Good species resolution. Belle (1996) created the tribe Ictinogomphini to include both *Cacoides* and *Melanocacus*.

Potential for new species: Likely due to rarity and elusive habits of adults.

Habitat: Gallery forests of savannahs and evergreen forest. Adults have roosting sites high up in foliage of trees, and apparently only the females descend occasionally to oviposit, because the males are only known from reared larvae and newly hatched specimens; larvae resemble the decaying leaves among which they live in sandy bottomed streams (Belle, 1986a).

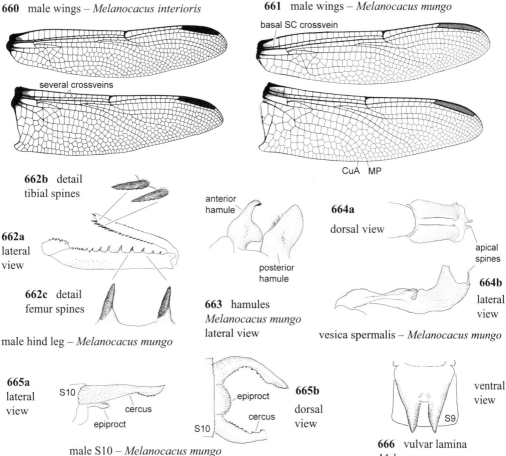

660 male wings – *Melanocacus interioris*

several crossveins

661 male wings – *Melanocacus mungo*

basal SC crossvein

CuA MP

662b detail tibial spines

662a lateral view

662c detail femur spines

male hind leg – *Melanocacus mungo*

anterior hamule

posterior hamule

663 hamules *Melanocacus mungo* lateral view

664a dorsal view

apical spines

664b lateral view

vesica spermalis – *Melanocacus mungo*

665a lateral view

S10

cercus

epiproct

epiproct

cercus

S10

665b dorsal view

male S10 – *Melanocacus mungo*

ventral view

S9

666 vulvar lamina *Melanocacus mungo*

Mitragomphus Needham, 1944: 215.

[♂ pp. 74, couplet 33]

Type species: *Mitragomphus ganzanus* Needham, 1944 [by original designation]
1 species:

ganzanus Needham, 1944*

References: Needham, 1944.
Distribution: Brazil.

Medium gomphids (about 61 mm); postocellar ridge complete with lateral edges projected into long horns (shared with *Cacoides*); occiput with a transverse

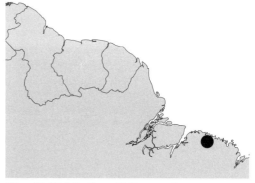

Map 47. Distribution of *Mitragomphus* sp.

ridge along its entire width, with a fringe of hairs along its summit; hind femora distally with a row of very strong and long spines, about as long as femora width (shared with *Dromogomphus* and *Perigomphus*), femora shorter than head width; pterothorax and abdomen yellow with black markings, subalar carina of HW (Fig. 668) projected into strong spine (shared with *Cacoides, Melanocacus,* and *Zonophora*). Wings (Fig. 667) colored at base; no basal subcostal crossveins; 2 or more crossveins in HW space between arculus branches and point of branching of RP; subtriangle of FW with 2 or more crossveins, of HW free; supratriangles crossed; 2 or more cubito-anal crossveins; CuA and MP in HW divergent. Male cercus (Figs. 671a, c-d) long, unforked; epiproct well developed and bifid (Figs. 671a-b). Posterior hamule lacking spines or hooks; vesica spermalis distal segment with 2 long flagella; basal segment with an entire anterior rim. **Unique characters**: S1 with a posteriorly directed medium spine (Fig. 670); male auricles with 2 teeth (Fig. 669).

Status of classification: Monotypic genus known only from holotype male. Belle (1996) included it together with *Zonophora* under the tribe Zonophorini.

Potential for new species: Unknown.

Habitat: Unknown.

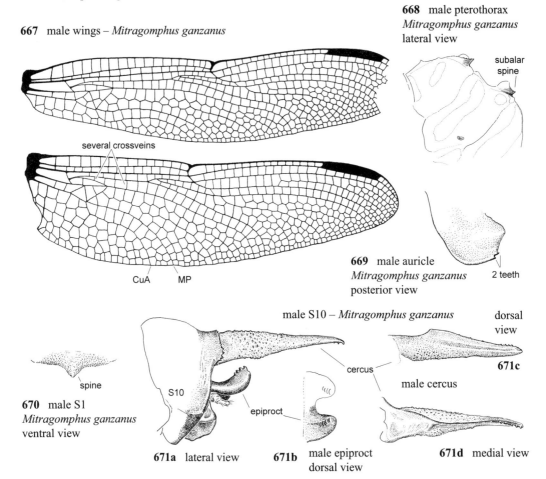

667 male wings – *Mitragomphus ganzanus*

668 male pterothorax *Mitragomphus ganzanus* lateral view

subalar spine

several crossveins

CuA   MP

669 male auricle *Mitragomphus ganzanus* posterior view   2 teeth

male S10 – *Mitragomphus ganzanus*   dorsal view

cercus   **671c**

spine   male cercus

670 male S1 *Mitragomphus ganzanus* ventral view   S10   epiproct

671a lateral view   **671b** male epiproct dorsal view   **671d** medial view

Neogomphus Selys, 1858: 679 (419 reprint)
[♂ pp. 66, couplet 5; ♀ pp. 79-80, couplets 15, 17-18]
Type species: *Hemigomphus molestus* Hagen *in* Selys, 1854 [Kirby, 1890 by original designation]
3 species:

bidens Selys, 1878* – L [Needham and Bullock, 1943]

edenticulatus Carle & Cook, 1984* – L [Belle, 1992a]
molestus (Hagen *in* Selys, 1854) [*Hemigomphus*]* – L [Needham and Bullock, 1943]

References: Carle and Cook, 1984.
Distribution: S Chile and SW Argentina.

Map 48. Distribution of *Neogomphus* spp.

Medium gomphids (39–47.5 mm); postocellar ridge complete; occiput broadly rounded; pterothorax yellow to green with dark brown stripes; abdomen mostly black, yellow or green basally. Wings (Fig. 672) lacking basal subcostal crossveins; 1 crossvein in HW space between arculus branches and point of branching of RP; subtriangles and supratriangles free; 1 cubito-anal crossvein; CuA and MP in HW divergent. Male cerci unbranched; epiproct bifid (Figs. 676a-b). Posterior hamule (Fig. 675) with denticles on anterior edge, and ending on a single tooth. Vesica spermalis distal segment with 2 long flagella, basal

segment bifid. Female with (Fig. 673) or without rudimentary auricles; vulvar lamina bifid (Fig. 677), extending from 2/3 of S9 length to beyond distal margin. **Unique characters**: None known.

Status of classification: Well defined species. Carle and Cook (1984) considered this genus allied to *Octogomphus* Selys, *Hemigomphus* Selys, *Davidius* Selys, *Lanthus* Needham, *Trigomphus* Bartenef, *Sinogomphus* May, *Stylogomphus* Fraser, and *Fukienogomphus* Chao, creating the tribe Octogomphini to encompass all of them. Belle (1996) considered it more closely related to *Hemigomphus* Selys and *Praeviogomphus* Belle, and included them in the tribe Hemigomphini. The status of *Hemigomphus elegans* Selys, 1854 from the interior of Brazil is unknown; according to Belle (1996: 302) the lost holotype and only known specimen may belong to a different (undescribed) genus within the Hemigomphini, although Carvalho (2000) suggested that *Praeviogomphus proprius* could be a synonym of *Hemigomphus elegans*. Potential for new species: Likely.

Habitat: Streams, rivers and lakes of clear and cold water, bottomed with gravel and cobble. Adults perch on stones along banks or on leaves of bushes overhanging the water (Carle and Cook, 1984). Larvae reported to emerge on stones in quiet and shallow sections of streams (Belle, 1992a).

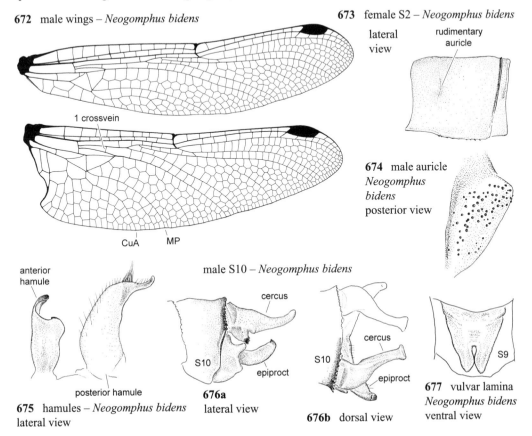

672 male wings – *Neogomphus bidens*

1 crossvein

CuA   MP

673 female S2 – *Neogomphus bidens*
lateral view   rudimentary auricle

674 male auricle *Neogomphus bidens* posterior view

anterior hamule

posterior hamule

675 hamules – *Neogomphus bidens* lateral view

male S10 – *Neogomphus bidens*

cercus

S10   epiproct

676a lateral view

cercus

S10   epiproct

676b dorsal view

S9

677 vulvar lamina *Neogomphus bidens* ventral view

Octogomphus Selys, 1873: 759.

[♂ pp. 67, couplet 12; ♀ pp. 75, couplet 2]

Type species: *Neogomphus? specularis* Selys, 1859 [by monotypy]
1 species:

specularis (Hagen *in* Selys, 1859) [*Neogomphus*?]*
– **L** [Kennedy, 1917]

References: Needham, Westfall, and May, 2000.
Distribution: SW Canada to W United States and NW Mexico.

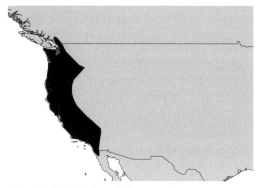

Map 49. Distribution of *Octogomphus* sp.

Medium gomphids (51–53 mm); occiput broadly rounded; rear of the head in the female with 2 lateral depressions, where male caudal appendages fit during copulation (Fig. 679a, shared with some *Epigomphus* species). Pterothorax yellow with dark brown stripes; abdomen mostly black, with yellow markings. Wings (Fig. 678) lacking basal subcostal crossveins; 1 crossvein in HW space between arculus branches and point of branching of RP; subtriangles and supratriangles free; 1 cubito-anal crossvein; CuA and MP in HW divergent. Male cercus with branch not directed ventrally (Fig. 680a); epiproct bifid with bifid branches (Fig. 680b). Posterior hamule with distal 1/3 bent antero-ventrally at a right angle (shared with *Ebegomphus* and *Tibiagomphus*), with denticles on anterior edge, and ending on a single tooth (Fig. 681). Vesica spermalis distal segment lacking flagella, basal segment bifid. Female with rudimentary auricles; vulvar lamina bifid, extending to about 1/2 of S9 length. **Unique characters**: Female postocellar ridge incomplete, with lateral and medial projections on each side (Fig. 679b).

Status of classification: Carle and Cook (1984) included it in the tribe Octogomphini, together with *Neogomphus* Selys, *Hemigomphus* Selys, *Davidius* Selys, *Lanthus* Needham, *Trigomphus* Bartenef, *Sinogomphus* May, *Stylogomphus* Fraser, and *Fukienogomphus* Chao.

Potential for new species: Unlikely.

Habitat: Rapid streams. Adults can be found on twigs of trees away from water, and males perch on sunny stones and leaves near riffles, and patrol through shady areas (Dunkle, 2000).

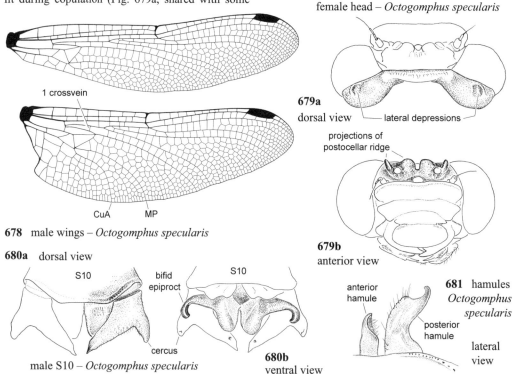

678 male wings – *Octogomphus specularis*

1 crossvein

CuA   MP

680a dorsal view

S10   bifid epiproct   S10

cercus

male S10 – *Octogomphus specularis*

680b ventral view

female head – *Octogomphus specularis*

679a
dorsal view   lateral depressions

projections of postocellar ridge

679b
anterior view

anterior hamule   **681** hamules
Octogomphus specularis

posterior hamule

lateral view

Ophiogomphus Selys, 1854: 39 (20 reprint).
[♂ pp. 67-68, couplets 8, 11, 13; ♀ pp. 77, 79, couplets 8, 18]
Type species: *Aeschna serpentina* Charpentier, 1825 [Selys, 1858 by subsequent designation]
  syn *Ophionuroides* Carle, 1986: 316.
    Type species: *Ophiogomphus anomalus* Harvey, 1898 [by original designation]
  syn *Ophionurus* Carle, 1986: 316.
    Type species: *Ophiogomphus alleghanensis* Carle, 1982b [by original designation]
27 species, 20 New World species:

acuminatus Carle, 1981* – **L** [Carle, 1981; Louton, 1982a as *bouchardi*]
  syn *bouchardi* Louton, 1982
anomalus Harvey, 1898* – **L** [Walker, 1933; 1958]
arizonicus Kennedy, 1917* – **L** [Dunkle, 1976]
aspersus Morse, 1895* – **L** [Needham, 1901; Walker, 1933; 1958]
australis Carle, 1992* – **L** [Carle, 1992]
bison Selys, 1873* – **L** [Needham, 1904]
  syn *sequoiarum* Butler, 1914
carolus Needham, 1897* – **L** [Needham, 1901; Walker, 1933; 1958]
colubrinus Selys, 1854* – **L** [Hagen, 1885]*
edmundo Needham, 1951
howei Bromley, 1924* – **L** [Kennedy and White, 1979; Louton, 1982b]
incurvatus Carle, 1982 * – **L** [Carle, 1982b]
  syn? *incurvatus alleghaniensis* Carle, 1982*
mainensis fastigiatus Donnelly, 1987*
mainensis mainensis Packard, 1863* – **L** [Garman, 1927; Walker, 1933; 1958; Louton, 1982b]
  syn *johannus* Needham, 1897
morrisoni Selys, 1879* – **L** [Kennedy, 1917]
  syn *morrisoni nevadensis* Kennedy, 1917
occidentis Hagen, 1882* – **L** [Hagen, 1885; Walker, 1933]
  syn *occidentis californicus* Kennedy, 1917
  syn *phaleratus* Needham, 1902
purepecha González & Villeda, 2000*
rupinsulensis (Walsh, 1862) [*Herpetogomphus*]* – **L** [Walker, 1933; 1958; Louton, 1982b]
  syn *carolinus* Hagen, 1885
  syn *pictus* (Needham, 1897) [*Herpetogomphus*]
severus montanus Selys, 1878*
severus severus Hagen, 1874* – **L** [Walker, 1933]
smithi Tennessen & Vogt, 2004* – **L** [Smith, 2005]?
susbehcha Vogt & Smith, 1993* – **L** [Vogt and Smith, 1993]
westfalli Cook & Daigle, 1985* – **L** [Cook and Daigle, 1985]

References: Carle, 1992; Needham, Westfall, and May, 2000.
Distribution: Holarctic; in the New World in Canada and United States to central Mexico.

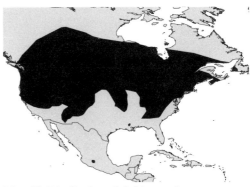

Map 50. Distribution of *Ophiogomphus* spp.

Small to medium gomphids (31–54 mm); pterothorax and abdomen yellow to green with dark brown or black stripes. Wings (Figs. 682-683) colored or hyaline at base; lacking basal subcostal crossveins; 1 crossvein in HW space between arculus branches and point of branching of RP; subtriangles and supratriangles free; 1 cubito-anal crossvein; CuA and MP in HW divergent. Male cercus unbranched; epiproct bifid (Figs. 687a-b). Vesica spermalis distal segment with 2 long or short flagella, and 2 ventro-lateral lobes in most species (Fig. 685; shared with *Erpetogomphus*), and basal segment bifid. Male auricles with small denticles in a row (shared with *Erpetogomphus* and some *Agriogomphus* species); auricles rudimentary in females. Vulvar lamina bifid, extending from about 1/2 of S9 length to beyond its distal margin. **Unique characters**: Posterior hamule with long and spatulate posterior tip (Figs. 686a-b).
Status of classification: Well diagnosed species. Considered to represent the sister group of *Erpetogomphus* (Garrison, 1994). Chao (1984) placed *Ophiogomphus* in his newly erected subfamily Onychogomphinae and Carle (1986) further assigned this genus to the tribe Onychogomphini.
Ophionuroides and *Ophionurus* were proposed in a key as subgenera of *Ophiogomphus* by Carle (1986) based on sexual characters of male and female, all of which we consider to represent no more than specific differences. *Ophionuroides* and *Ophionurus* have not been recognized by Needham, Westfall, and May (2000) and we follow them here in treating these names as synonyms.
Potential for new species: Possible.
Habitat: Gravel or rock, sometimes mud or sand, bottomed streams and rivers of clear water, some also on lakes, where larvae burrow shallowly. Adults perch on ground or vegetation (bushes, branches, tree canopies) in open woods away from water, or on rocks or sand bars at shore of water. Males patrol areas with riffles or pools, depending on the species (Dunkle, 2000).

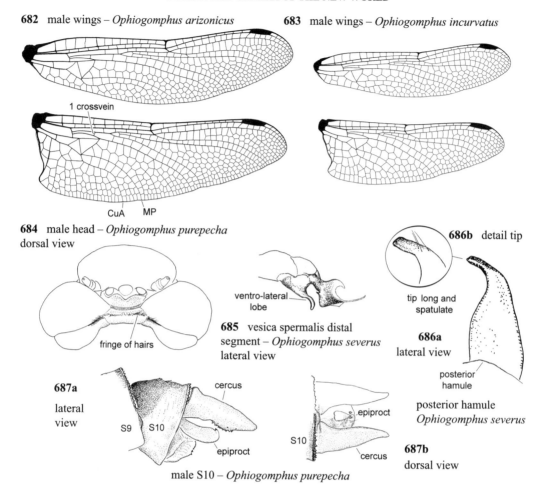

682 male wings – *Ophiogomphus arizonicus*   **683** male wings – *Ophiogomphus incurvatus*

1 crossvein

CuA   MP

684 male head – *Ophiogomphus purepecha*
dorsal view

686b detail tip

ventro-lateral
lobe

tip long and
spatulate

685 vesica spermalis distal
segment – *Ophiogomphus severus*
lateral view

686a
lateral view

fringe of hairs

posterior
hamule

687a

lateral
view

cercus

posterior hamule
Ophiogomphus severus

S9   S10

epiproct

epiproct

S10

cercus

687b
dorsal view

male S10 – *Ophiogomphus purepecha*

Perigomphus Belle, 1972b: 64.
[♂ pp. 69, couplet 20; ♀ pp. 84, couplet 35]
Type species: ***Perigomphus pallidistylus*** Belle, 1972
[by original designation]
1 species:

pallidistylus (Belle, 1972) [*Diaphlebia*
(*Perigomphus*)]* – **L** [Westfall, 1989]

References: Belle, 1972b.
Distribution: Costa Rica to Panama.

Map 50. Distribution of *Perigomphus* sp.

Small gomphids (34–42 mm); postocellar ridge complete; occiput broadly rounded (Fig. 689), covered with coarse hairs; pterothorax and abdomen green with dark brown markings; hind femora distally with a row of very strong and long spines, about as long as femora width (Fig. 690, shared with *Dromogomphus* and *Mitragomphus*), femora shorter than head width. Wings (Fig. 688) lacking basal subcostal crossveins; 2 or more crossveins in HW space between arculus branches and point of branching of RP; subtriangles and supratriangles free; 1 cubito-anal crossvein;

CuA and MP in HW parallel or divergent. Male cercus long, straight, and unbranched (Figs. 692a-b); epiproct bifid (Figs. 692a-c). Posterior hamule ending in a single spine, with small denticles on anterior edge; vesica spermalis distal segment with 2 short flagella, and basal segment bifid. Female auricles rudimentary; vulvar lamina bifid, extending to about 2/3

of S9 length. **Unique characters**: None known.
Status of classification: Belle (1972b) included his
new species *D. pallidistylus* originally in a new subgenus within *Diaphlebia*, *Perigomphus*, which was
later elevated to generic rank (Westfall, 1989), and is
now considered to be related to *Hemigomphus* Selys,
Neogomphus Selys *in* Hagen and *Praeviogomphus*

Belle within the tribe Hemigomphini (Belle, 1996).
Potential for new species: Unknown.
Habitat: A male of *P. pallidistylus* was observed land-
ing on a leaf about 1.5 m high in complete shade under canopy, at edge of a rocky and sandy to muddy
bottomed stream, of swift current and clear water,
mostly less than 30 cm deep (Paulson *pers. comm.*).

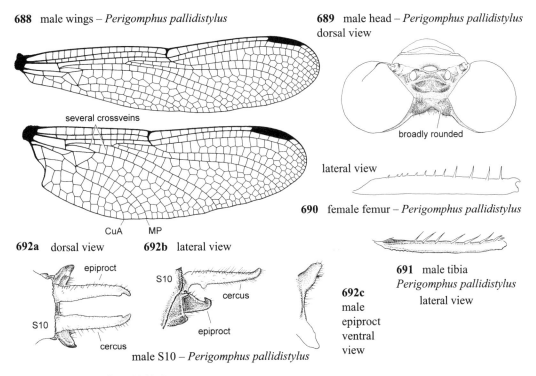

688 male wings – *Perigomphus pallidistylus*

several crossveins

CuA   MP

692a dorsal view   **692b** lateral view

epiproct

S10

S10   cercus

epiproct

cercus

male S10 – *Perigomphus pallidistylus*

689 male head – *Perigomphus pallidistylus*
dorsal view

broadly rounded

lateral view

690 female femur – *Perigomphus pallidistylus*

691 male tibia
Perigomphus pallidistylus
lateral view

692c
male
epiproct
ventral
view

Peruviogomphus Klots, 1944: 3.
[♂ pp. 69-70, couplets 19, 23; ♀ pp. 80, couplet 22]
Type species: *Peruviogomphus moyobambus* Klots,
1944 [by original designation]
3 species:

bellei Machado, 2005
moyobambus Klots, 1944*
pearsoni Belle, 1979*
spec. indet. 1992 – **L** [Belle, 1992a]

References: Klots, 1944; Belle, 1979; 1992a; Machado, 2005a.
Distribution: Ecuador and Peru to Brazil.

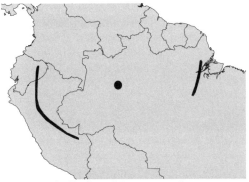

Map 51. Distribution of *Peruviogomphus* spp.

Medium gomphids (about 45–46 mm); postocellar
ridge complete; occiput with a transverse ridge along
its entire width; pterothorax and abdomen green with
dark brown markings. Wings (Fig. 693a) with basal
subcostal crossveins; 2 or more crossveins in HW
space between arculus branches and point of branching of RP; subtriangles free; supratriangles crossed;
male HW anal triangle reaching anal angle of wing;

1 cubito-anal crossvein; CuA and MP in HW divergent. Male cercus (Figs. 694a-b) with a branch not
directed ventrally; epiproct bifid. Posterior hamule
(Figs. 695a-b) ending in a single spine; vesica spermalis distal segment with 2 long flagella (Fig. 696a),
and basal segment bifid (Fig. 696b). Female vulvar
lamina bifid, extending to about 1/4 of S9 length; S10
longer than S9 in the female (shared with both sexes

of *Gomphoides*). **Unique characters**: Inner margin of male HW anal triangle with a series of closely set comb-like denticles (Fig. 693b).

Status of classification: Three or four species involved. Association of females to a certain species difficult, since all three known species were described based on a single male. Belle (1996) created the monotypic tribe Peruviogomphini for this genus, within the sub-family Gomphoidinae. Some of the characters mentioned as diagnostic for *P. bellei* by Machado (2005a),

such as inner margin of male HW anal triangle with a series of closely set comb-like denticles and S8-9 with lateral flanges (i.e. Figs. 693b, 694a), are found in all three described species.

Potential for new species: Likely. Specimens rare in collections.

Habitat: *Peruviogomphus pearsoni* is a rain-forest lake species of the tropical lowlands (Paulson *pers. comm.*). An adult of *P. moyobambus* was collected at an ox-bow lake in primary forest in SE Peru (Louton, Garrison, and Flint, 1996).

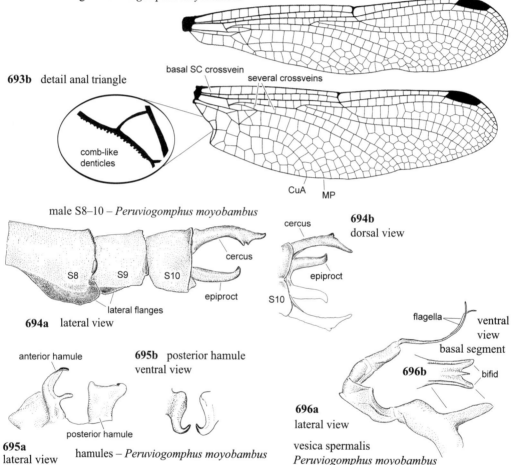

693a　male wings – *Peruviogomphus moyobambus*

693b　detail anal triangle

basal SC crossvein

several crossveins

comb-like denticles

CuA　MP

male S8–10 – *Peruviogomphus moyobambus*

cercus

694b dorsal view

cercus

S8　S9　S10

epiproct

S10

epiproct

lateral flanges

694a　lateral view

anterior hamule

695b　posterior hamule ventral view

flagella

ventral view

basal segment

696b

bifid

posterior hamule

696a

lateral view

695a lateral view

hamules – *Peruviogomphus moyobambus*

vesica spermalis

Peruviogomphus moyobambus

Phyllocycla Calvert, 1948a: 62.

[♂ pp. 72, couplet 30; ♀ pp. 85, couplet 37]

Type species: *Cyclophylla signata* Hagen *in* Selys, 1854 [replacement name for *Cyclophylla*].

　　syn *Cyclophylla* Selys, 1854: 76 (57 reprint).
　　　　Type species: *Cyclophylla signata* Hagen *in* Selys, 1854 [Kirby, 1890, by subsequent designation]

[NOTE: *Phyllocycla* is a replacement name for *Cyclophylla* Selys, a junior primary homonym of *Cyclophylla* Brandt, 1837 in Coelenterata.]

31 species:

anduzei (Needham, 1943) [*Cyclophylla*]*
argentina (Hagen *in* Selys, 1878) [*Cyclophylla*]*
　– **L** [Rodrigues Capítulo, 1983a]
　　syn *eugeniae* (Navás, 1927) [*Gomphoides*]
armata Belle, 1977
baria Belle, 1987
bartica Calvert, 1948
basidenta Dunkle, 1987*

brasilia Belle, 1988
breviphylla Belle, 1975*
diphylla (Selys, 1854) [*Cyclophylla*]*
elongata (Selys, 1858) [*Cyclophylla*] – L?
[Needham, 1940]*
foliata Belle, 1988
gladiata (Hagen *in* Selys, 1854) [*Cyclophylla*]
hamata Belle, 1990*
hespera (Calvert, 1909) [*Gomphoides*]*
malkini Belle, 1970*
medusa Belle, 1988*
modesta Belle, 1970* – L [Belle, 1970; Fleck, 2002b]
murrea Belle, 1988
neotropica Belle, 1970* – L [Needham, 1944 as *pachystyla*; Belle, 1970; Fleck, 2002b]
ophis (Selys, 1869) [*Cyclophylla*]* – L [Belle, 1970] syn *pachystyla* (Needham, 1944) [*Cyclophylla*]
pallida Belle, 1970
pegasus (Selys, 1869) [*Cyclophylla*]
propinqua Belle, 1972* – L? [Needham, 1940]
signata (Hagen *in* Selys, 1854) [*Cyclophylla*]
sordida (Selys, 1854) [*Cyclophylla*]
speculatrix Belle, 1975
titschacki (Schmidt, 1942) [*Gomphoides*]
uniforma Dunkle, 1987*
vesta Belle, 1972*
viridipleuris (Calvert, 1909) [*Gomphoides*]* – L [Belle, 1992a]
volsella (Calvert, 1905) [*Gomphoides*]* – L [Needham, 1940]

References: Belle, 1988b; 1990.
Distribution: Mexico south through N Argentina and Uruguay.

Medium gomphids (40–56 mm); postocellar ridge complete or incomplete; occiput with transverse ridge along posterior margin, with a fringe of hairs; pterothorax green or greenish yellow with brown stripes; abdomen mostly black or brown becoming paler distally (orange to pale brown), or mostly pale green. Wings (Figs. 699-700) with basal subcostal crossveins in HW; 2 or more crossveins in HW space between arculus branches and point of branching

Map 52. Distribution of *Phyllocycla* spp.

of RP; subtriangle of FW with 1 crossvein, of HW free; supratriangles crossed; male HW anal triangle reaching anal angle of wing; 1 cubito-anal crossvein; CuA and MP in HW parallel or divergent. Male cercus unbranched or with a dorso-distal branch; epiproct rudimentary, not visible in lateral view (Figs. 702a-b). Posterior hamule (Fig. 697) ending in a single tooth; vesica spermalis distal segment with 2 long flagella (Fig. 698a), basal segment bifid. Female auricles absent; vulvar lamina bifid, extending from 1/4 to about a 1/3 of S9; epiproct not polished, visible or not in dorsal view. Postero-dorsal rim of S10 distinguished from anterior part in both sexes by polished surface, less dense setation, different color and/or basal groove or ridge (shared with *Aphylla*).
Unique characters: Serrulation on outer side of vesica spermalis distal flagella (Fig. 698b).
Status of classification: Belle (1988b) revised the entire genus and included a key to males, but reference to primary literature is still necessary for illustrations; only one new species (*P. hamata* Belle, 1990) has been described since 1988. The genus is apparently allied to *Aphylla*, and Belle (1996) included both in the tribe Aphyllini.
Potential for new species: Very likely.
Habitat: Creeks of dense forest and rivers, where adults perch on leaves of overhanging trees (Belle, 1970), and larvae, with S10 developed as a tube as long as previous 4 or 5 abdominal segments, probably inhabit silt-covered backwaters (Needham, Westfall, and May, 2000).

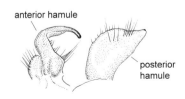

697 hamules – *Phyllocycla volsella* lateral view

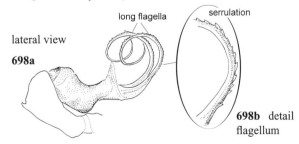

lateral view
698a

698b detail flagellum

vesica spermalis distal segment – *Phyllocycla volsella*

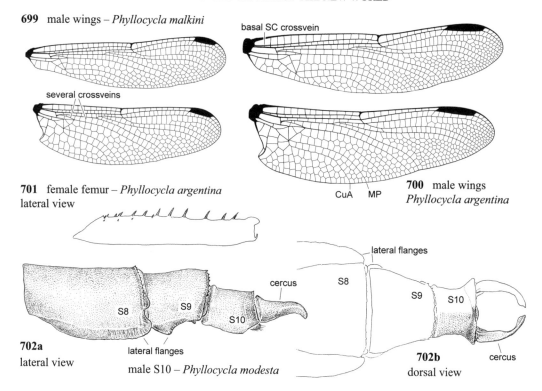

699   male wings – *Phyllocycla malkini*

basal SC crossvein

several crossveins

701   female femur – *Phyllocycla argentina*
lateral view

CuA   MP

700   male wings
Phyllocycla argentina

lateral flanges

cercus

S8

S9

S10

702a
lateral view

lateral flanges

male S10 – *Phyllocycla modesta*

S8

S9   S10

702b
dorsal view

cercus

Phyllogomphoides Belle, 1970: 112.

[♂ pp. 73, 75, couplets 31, 37; ♀ pp. 86, couplets 38, 41]

Type species: *Gomphoides fuliginosa* Hagen *in* Selys, 1854 [by original designation]

46 species:

aculeus Belle, 1982
albrighti (Needham, 1950) [*Gomphoides*]* – **L**?
[Needham, Westfall, and May, 2000]
andromeda (Selys, 1869) [*Cyclophylla*]* – **L** [Belle, 1970]
angularis Belle, 1982*
annectens (Selys, 1869) [*Gomphoides*?] – **L** [Costa, Santos, and Telles, 1999a]
apiculatus Cook & González, 1990*
appendiculatus (Kirby, 1899) [*Gomphoides*] – **L**?
[Needham, 1940 as *bifasciatus*]
atlanticus (Belle, 1970) [*Negomphoides*]
audax (Hagen *in* Selys, 1854) [*Gomphoides*]
bifasciatus (Hagen *in* Selys, 1878) [*Gomphoides*]*
– **L** [Ramírez, 1996]
brunneus Belle, 1981
burgosi Brooks, 1989
calverti (Kirby, 1897) [*Cyclophylla*]
camposi (Calvert, 1909) [*Gomphoides*]*
cassiopeia (Belle, 1975) [*Gomphoides*]
cepheus Belle, 1980*
cornutifrons (Needham, 1944) [*Aphylla*]
cristatus (Needham, 1944) [*Gomphoides*]* – **L**
[Belle, 1970]

Map 53. Distribution of *Phyllogomphoides* spp.

danieli González & Novelo, 1990*
duodentatus Donnelly, 1979* – **L** [Novelo Gutiérrez, 1993]
fuliginosus (Hagen *in* Selys, 1854) [*Gomphoides*]*
– **L** [Belle, 1970 as *audax*]
imperator Belle, 1976
indicatrix Belle, 1989
insignatus Donnelly, 1979
joaquini Rodrigues, 1992* – **L** [Muzón, Pessacq, and von Ellenrieder, 2006]
lieftincki (Belle, 1970) [*Negomphoides*]*
litoralis Belle, 1984 – **L** [Müller and Suhling, 2001]
luisi González & Novelo, 1990* – **L** [Novelo Gutiérrez ,1993]

major Belle, 1984* – **L** [Belle, 1970 as *fuliginosus*]
nayaritensis Belle, 1987*
pacificus (Selys, 1873) [*Gomphoides*]* – **L** [Novelo Gutiérrez, 1993]
pedunculus Belle, 1984*
praedatrix Belle, 1982
pseudangularis Belle, 1994
pseudoundulatus Belle, 1984
pugnifer Donnelly, 1979*
regularis (Selys, 1873) [*Gomphoides*]
selysi (Navás, 1924) [*Gomphoides*]
semicircularis (Selys, 1854) [*Gomphoides*]
singularis Belle, 1979
spiniventris Belle, 1994
stigmatus (Say, 1840) [*Aeschna*]* – **L** [Needham, 1904; Needham, Westfall, and May, 2000]
suasillus Donnelly, 1979
suasus (Selys, 1859) [*Gomphoides*]* – **L** [Novelo Gutiérrez, 1993]
    syn *perfidus* (Hagen, 1861) [*Gomphoides*]
suspectus Belle, 1994
undulatus (Needham, 1944) [*Gomphoides*]* – **L**? [Belle, 1970]

References: Belle, 1984c (South America); Donnelly, 1979 (Central America); Needham, Westfall, and May, 2000 (United States, N Mexico).
Distribution: South-central United States south through Bolivia, Paraguay, and N Argentina.

Medium to large gomphids (52–77 mm); postocellar ridge complete to absent; occiput with a transverse ridge, with a fringe of hairs along its summit; ptero-thorax green with dark brown to black stripes; abdomen mostly black, with pale sub-basal markings, in some species apical segments reddish brown. Wings (Figs. 703-704) with basal subcostal crossveins; 2 or more crossveins in HW space between arculus branches and point of branching of RP; subtriangle of FW with 1 or more crossveins, of HW free or with 2 or more crossveins; supratriangles crossed; male HW anal triangle reaching anal angle of wing; 1 cubito-anal crossvein; CuA and MP in HW parallel or divergent. Male cercus (Figs. 707a-b) unbranched, long and arcuate, or with a branch directed ventrally; epiproct bifid. Posterior hamule (Fig. 708) with a single end spine or lacking spine or tooth; vesica spermalis distal segment with 2 long flagella, and basal segment bifid. Female vulvar lamina bifid (Fig. 706), extending to about a 1/4 to a 1/3 of S9 length. **Unique characters**: None known.

Status of classification: Good species diagnoses in general but due to high number of new species described after main revisions (Belle 1984c, Donnelly 1979), reference to primary literature is still necessary for specific identification. Belle (1996) considered the genus related to *Gomphoides* and *Idiogomphoides*, including them in the tribe Gomphoidini. Larvae of *Phyllogomphoides* and *Gomphoides* are also very similar in structure and habits.
Potential for new species: Likely.
Habitat: Larvae inhabit small and large streams crossing forests (Belle 1984c), where they burrow in muddy beds. Males perch on stones or tip of branches of low vegetation on banks, and females visit the water only for oviposition.

703 male wings – *Phyllogomphoides pugnifer*   **704** male wings – *Phyllogomphoides andromeda*

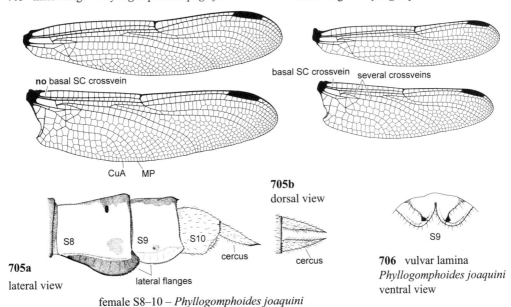

no basal SC crossvein

basal SC crossvein   several crossveins

CuA   MP

705b
dorsal view

S8   S9   S10
cercus
705a
lateral view
lateral flanges
cercus

female S8–10 – *Phyllogomphoides joaquini*

S9

706 vulvar lamina
Phyllogomphoides joaquini
ventral view

707a   lateral view

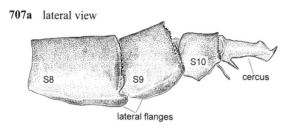

male S8–10 – *Phyllogomphoides pedunculus*

707b
dorsal view

708   hamules – *Phyllogomphoides joaquini*
ventral view

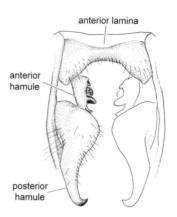

Praeviogomphus Belle, 1995: 21.
[♀ pp. 79, couplet 15]
Type species: *Praeviogomphus proprius* Belle, 1995
[by original designation]
1 species:

proprius Belle, 1995 – **L**? [Carvalho, 2000]

References: Belle, 1995; Carvalho, 2000.
Distribution: Known only from the holotype female
from Rio de Janeiro State, Brazil, and larval state by
supposition (determined as this species based on pharate adult morphology, Carvalho, 2000).

Medium gomphids (48 mm); postocellar ridge complete; occiput with a transverse ridge along its entire
width; pterothorax and abdomen yellow with dark
brown markings. Wings (Fig. 709) lacking basal subcostal crossveins; 1 crossvein in HW space between
arculus branches and point of branching of RP; FW
subtriangle free, HW subtriangle free or crossed; supratriangles free; 1 cubito-anal crossvein; CuA and
MP in HW divergent; FW with or without a supplementary trigonal sector. Female vulvar lamina bifid (Fig. 710), extending to about 3/4 of S9 length.
Unique characters: None known.
Status of classification: Unknown. The genus, based
on holotype female, will be difficult to assess until

Map 54. Distribution of *Praeviogomphus* sp.

its male is discovered. However, Carvalho (2000) illustrated pharate adult male characters from larvae
he assigned to this species. He suggests that this species may be found to be a synonym of *Hemigomphus*
(later assigned to *Neogomphus) elegans* Selys, whose
unique type from the interior of Brazil is lost.
Potential for new species: Unknown.
Habitat: No data available, however the type locality
is "Itatiaya" which probably refers to Parque Nacional do Itatiaia, an area in the mountain forest of NW
Rio de Janiero State. Carvalho (2000) collected larvae assigned to this species from primary rain forest
in Teresópolis, Rio de Janiero, Brazil.

709 female wings – *Praeviogomphus proprius*

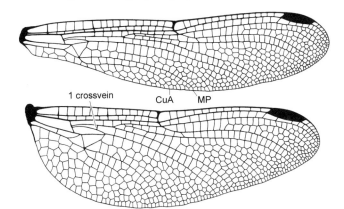

1 crossvein   CuA   MP

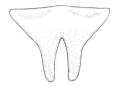

710 vulvar lamina
Praeviogomphus proprius
ventral view

Progomphus Selys, 1854: 69 (50 reprint).
[♂ pp. 70, 73, couplets 21, 31; ♀ pp. 81, 83, 86, couplets 25, 34, 38]
Type species: *Progomphus gracilis* Hagen *in* Selys, 1854 [Kirby, 1890, by subsequent designation]
   syn *Alloprogomphus* Carle, 1986: 322.
      Type species: *Progomphus complicatus* Selys, 1854 [by original designation]
   syn *Archaeoprogomphus* Carle, 1986: 322.
      Type species: *Progomphus geijskesi* Needham, 1944 [by original designation]
   syn *Eoprogomphus* Carle, 1986: 322.
      Type species: *Progomphus tibialis* Belle, 1973 [by original designation]
   syn *Neaprogomphus* Carle, 1986: 322.
      Type species: *Diastatomma obscurum* Rambur, 1842 [by original designation]
67 species:

abbreviatus Belle, 1973* – **L** [De Marmels, 1981a]
aberrans Belle, 1973*
adaptatus Belle, 1973
alachuensis Byers, 1939* – **L** [Needham, 1941]
amarillus Tennessen, 1992
amazonicus Belle, 1973
angeloi Belle, 1994
anomalus Belle, 1973 – **L** [Belle, 1991]
approximatus Belle, 1966 – **L** [Belle, 1966a]
auropictus Ris, 1911
australis Belle, 1973
basalis Belle, 1994
basistictus Ris, 1911*
bellei Knopf & Tennessen, 1980* – **L** [Tennessen, 1993]
belyshevi Belle, 1991* – **L** [Belle, 1991]
bidentatus Belle, 1994
boliviensis Belle, 1973*
borealis McLachlan *in* Selys, 1873* – **L** [Kennedy, 1917; Belle, 1991]

   syn ?*meridionalis (nomen nudum)* Hagen, 1885
brachycnemis Needham, 1944 – **L** [Belle, 1966a]
clendoni Calvert, 1905* – **L** [Belle, 1991]
   syn *williamsi* Needham, 1943
complicatus Selys, 1854* – **L**? [Santos, 1968a]
conjectus Belle, 1966 – **L** [Belle, 1992a]
costalis Hagen *in* Selys, 1854
delicatus Belle, 1973
dorsopallidus Byers, 1934 – **L** [Knopf, 1977]
elegans Belle, 1973
fassli Belle, 1973
flinti Belle, 1975
formalis Belle, 1973
geijskesi Needham, 1944 – **L**? [Belle, 1972a]
gracilis Hagen *in* Selys, 1854*
guyanensis Belle, 1966 – **L** [Belle, 1966a]
herrerae Needham & Etcheverry, 1956
incurvatus bivittatus De Marmels, 1991 – **L** [De Marmels, 1991a]
incurvatus incurvatus Belle, 1973
integer Hagen *in* Selys, 1878* – **L** [Needham, 1941]
intricatus Hagen *in* Selys, 1858* – **L** [Limongi, 1983]
joergenseni Ris, 1908*
kimminsi Belle, 1973
lepidus Ris, 1911* – **L**? [Needham, 1941]
longistigma Ris, 1918 – **L** [Needham, 1941]
maculatus Belle, 1984
mexicanus Belle, 1973* – **L**? [Belle, 1991]
microcephalus Belle, 1994
montanus Belle, 1973
nervis Belle, 1973
nigellus Belle, 1990
obscurus (Rambur, 1842) [*Diastatomma*]* – **L** [Needham and Hart, 1901; Louton, 1982b]
occidentalis Belle, 1983
perithemoides Belle, 1980

perpusillus Ris, 1918
phyllochromus Ris, 1918* – **L** [Limongi, 1983]
pijpersi Belle, 1966
polygonus Selys, 1879
pygmaeus Selys, 1873* – **L** [Limongi, 1983; Belle, 1991]
racenisi De Marmels, 1983 – **L** [De Marmels, 1990a]
recticarinatus Calvert, 1909*
recurvatus Ris, 1911
risi Williamson, 1920 – **L** [Belle, 1991]
serenus Hagen *in* Selys, 1878* – **L** [Needham, 1941; Needham, Westfall, and May, 2000]
superbus Belle, 1973
tantillus Belle, 1973
tennesseni Daigle, 1996
tibialis Belle, 1973 – **L** [Belle, 1973]
victor St. Quentin, 1973
virginiae Belle, 1973
zephyrus Needham, 1941 – **L**? [Belle, 1992a]
zonatus Hagen *in* Selys, 1854* – **L**? [Belle, 1991]

References: Belle, 1973; 1991.
Distribution: S Canada and United States south to Uruguay, S Argentina and N Chile.

Small to medium gomphids (25.5–55 mm); postocellar ridge complete; occiput with a transverse ridge, with a fringe of hairs along its summit; pterothorax yellow to gray to green with dark brown stripes; abdomen mostly black or brown, with pale sub-basal markings, or mostly pale. Wings (Figs. 711-714) with or without basal subcostal crossveins; 2 crossveins in HW space between arculus branches and point of branching of RP; subtriangles free or crossed; supratriangles free; 1 cubito-anal crossvein; CuA and MP in HW divergent or parallel. Female vulvar lamina bifid, extending to about three-fourths of S9 length. Male cercus unbranched (Figs. 717a-b); epiproct bifid (Fig. 717c), each branch bifid or entire. Anterior lamina usually v-shaped (Figs. 718-719) shared with *Hagenius* and *Zonophora*); posterior hamule ending in a single spine

Map 55. Distribution of *Progomphus* spp.

or 2 subapical spines; vesica spermalis distal segment with 2 long or short flagella, and basal segment bifid. Female lacking auricles; vulvar lamina bifid, extending to about 1/4 or less of S9 length. **Unique characters**: Posterior hamule anterior edge with denticles on a single row (Figs. 718-719) or with a chitinous ridge at base; epiproct in males either divided into two separate halves or deeply forked with each half independently movable (Dunkle 1984).

Status of classification: Good. Easily recognized genus but many new species have been described since Belle's revision (1973), making it necessary to refer to primary literature for specific identifications. *Alloprogomphus*, *Archaeoprogomphus*, *Eoprogomphus* and *Neaprogomphus* were proposed in a key as subgenera of *Progomphus* by Carle (1986) based on differences of wing venation, legs, and sexual characters of the male, all of which we consider to represent no more than specific differences. These subgenera have not been recognized by odonatologists and we treat these names as synonyms. *Progomphus* has been placed in its own tribe, Progomphini, within the subfamily Gomphoidinae (Belle, 1996).

Potential for new species: Very likely since 14 of the 67 known species have been described since 1973.

Habitat: Larvae burrow in sand-bottomed streams, rivers and lakes. Males perch on the sand, rocks or low vegetation along the shore.

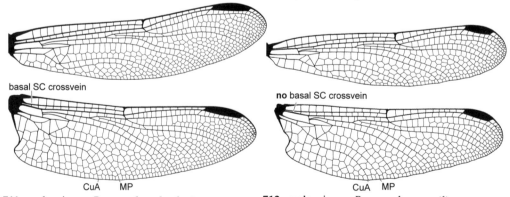

711   male wings – *Progomphus clendoni*

712   male wings – *Progomphus gracilis*

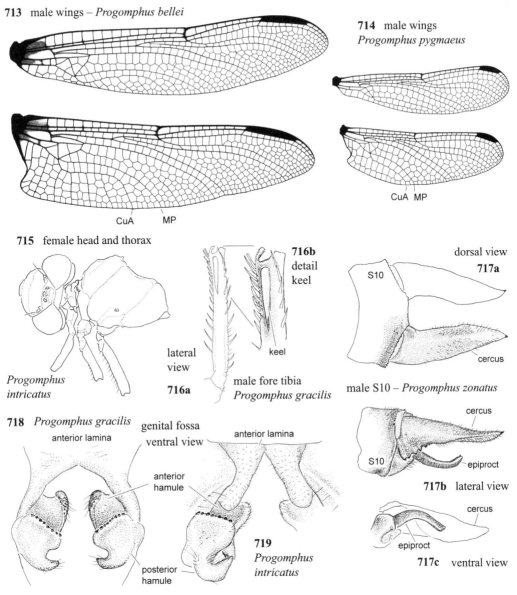

713 male wings – *Progomphus bellei*

714 male wings
Progomphus pygmaeus

CuA MP

CuA MP

715 female head and thorax

Progomphus intricatus

lateral view
716a

716b detail keel

keel

male fore tibia
Progomphus gracilis

dorsal view
717a

S10

cercus

male S10 – *Progomphus zonatus*

cercus

S10

epiproct

717b lateral view

cercus

epiproct

717c ventral view

718 *Progomphus gracilis*

anterior lamina

genital fossa
ventral view

anterior lamina

anterior hamule

posterior hamule

719 *Progomphus intricatus*

Stylogomphus Fraser, 1922: 69.
[♂ pp. 67, couplet 9; ♀ pp. 77, couplet 10]
Type species: *Stylogomphus inglisi* Fraser, 1922 [by original designation]
10 species, 2 New World species:

albistylus (Hagen *in* Selys, 1878) [*Gomphus*]* – **L** [Walker, 1958; Louton, 1982b]
  syn *naevius* (Hagen *in* Selys, 1878) [*Gomphus*]
sigmastylus Cook & Laudermilk, 2004* – **L** [Cook and Laudermilk, 2004]

References: Cook and Laudermilk, 2004.
Distribution: Oriental region and SE Canada to E United States in the New World.

Map 56. Distribution of *Stylogomphus* spp.

Small gomphids (31–36 mm); postocellar ridge incomplete; occiput broadly rounded; pterothorax yellow to green with dark brown stripes; abdomen mostly black, with narrow yellow bands at base of segments. Wings (Fig. 720) lacking basal subcostal crossveins; 1 crossvein in HW space between arculus branches and point of branching of RP; subtriangles and supratriangles free; 1 cubito-anal crossvein; CuA and MP in HW divergent. Male cercus unbranched (Figs. 721a-b); epiproct bifid. Posterior hamule with cluster of denticles on anterior edge, ending on a single tooth. Vesica spermalis distal segment lacking flagella, basal segment bifid. Female with rudimentary auricles; vulvar lamina bifid, extending to about a 1/3 of S9 length. **Unique characters**: None known.

Status of classification: A second species for the New World has been recently described (Cook and Laudermilk, 2004), which is very close to *S. albistylus*; evidence of hybridization suggests that it could represent just clinal variation. Closely related to *Octogomphus* and *Lanthus* (Needham, Westfall, and May, 2000).

Potential for new species: Unlikely.

Habitat: Clear streams, where males patrol from stones in riffles. Adults forage from tree leaves (Dunkle, 2000).

720   male wings – *Stylogomphus sigmastylus*

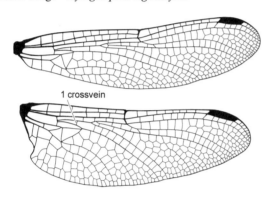

721a   dorsal view

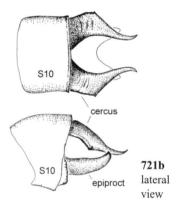

721b lateral view

male S10 – *Stylogomphus albistylus*

Stylurus Needham, 1897b: 166.

[♂ pp. 66, couplet 4; ♀ pp. 76, 78-79, couplets 5, 13]

Type species: *Gomphus plagiatus* Selys, 1854 [by original designation]

29 species, 12 New World species:

amnicola (Walsh, 1862) [*Gomphus*]* – **L** [Walker, 1928; Louton, 1982b]
   syn *abditus* (Butler, 1914) [*Gomphus*]
falcatus Gloyd, 1944*
intricatus (Hagen *in* Selys, 1858) [*Gomphus*]* – **L** [Walker, 1928; Louton, 1982b]
ivae Williamson, 1932* – **L** [Louton, 1982b; Needham, Westfall, and May, 2000]
laurae Williamson, 1932* – **L** [Louton, 1982b; Needham, Westfall, and May, 2000]
notatus (Rambur, 1842) [*Gomphus*]* – **L** [Walker, 1928]
   syn *fluvialis* (Walsh, 1862) [*Gomphus*]
   syn *jucundus* (Needham, 1943) [*Gomphus*]
olivaceus (Selys, 1873) [*Gomphus*]* – **L** [Walker, 1928]
   syn *olivaceus nevadensis* (Kennedy, 1917) [*Gomphus*]
plagiatus (Selys, 1854) [*Gomphus*]* – **L** [Walker, 1928; Louton, 1982b]

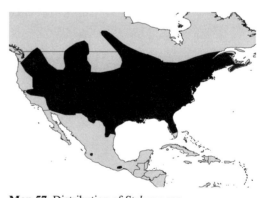

Map 57. Distribution of *Stylurus* spp.

   syn *elongatus* (Selys, 1854) [*Gomphus*]
potulentus (Needham, 1942) [*Gomphus*]* – **L** [Needham, Westfall, and May, 2000]
scudderi (Selys, 1873) [*Gomphus*]* – **L** [Needham, 1901; Walker, 1928; Louton, 1982b]
spiniceps (Walsh, 1862) [*Macrogomphus*?]* – **L** [Walker, 1928; 1958; Louton, 1982b]
   syn *segregans* Needham, 1897
townesi Gloyd, 1936*

References: Needham, Westfall, and May, 2000.
Distribution: S Canada to central Mexico.

Medium to large gomphids (41–68 mm); postocellar ridge complete; occiput with a transverse ridge along its entire width, with a fringe of hairs along its summit; pterothorax yellow to green with dark brown to black stripes; abdomen largely yellow or green with dark brown reduced to lateral areas, to mostly black with narrow yellow rings, to apical segments orange-brown. Wings (Fig. 723) lacking basal subcostal crossveins; 1 crossvein in HW space between arculus branches and point of branching of RP; subtriangles and supratriangles free; 1 cubito-anal crossvein; CuA and MP in HW divergent or parallel. Male cercus unbranched, tapering distally; epiproct bifid (Fig. 725). Anterior hamule digitiform (Fig. 724, shared with *Archaeogomphus*), lacking spines or hooks; posterior hamule narrow at base ending in a single spine;

vesica spermalis distal segment lacking flagella, and basal segment bifid. Female with rudimentary auricles; vulvar lamina bifid (Fig. 726), extending from a 1/5 to about a 1/3 of S9 length. **Unique characters**: None known.
Status of classification: Good. Carle (1986) elevated this group, originally considered as a subgenus of *Gomphus*, to generic rank. Relationships between Nearctic and certain European species remain uncertain (Louton, 1982b). Carle (1986) included this genus in the Gomphini, together with *Arigomphus*, *Dromogomphus*, *Gomphus*, and the Old World *Gastrogomphus*.
Potential for new species: Fair.
Habitat: Larvae burrow on sand, silt, mud, or gravel beds of rivers, streams, and occasionally lakes. Males patrol over water and perch on tree leaves or banks (Dunkle, 2000).

722 female head and thorax
Stylurus intricatus

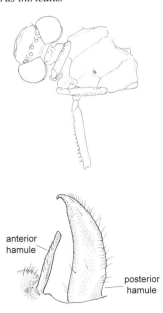

723 male wings – *Stylurus potulentus*

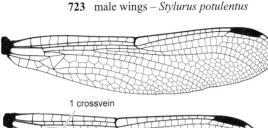

1 crossvein

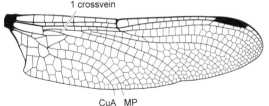

CuA MP

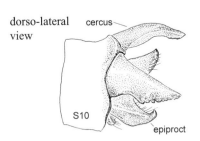

dorso-lateral view   cercus

S10
epiproct

724 hamules – *Stylurus falcatus*
lateral view

anterior hamule

posterior hamule

725 male S10 – *Stylurus falcatus*

S9

726 vulvar lamina
Stylurus intricatus
ventral view

Tibiagomphus Belle, 1992a: 2.
[♂ pp. 71, couplet 25; ♀ pp. 83, couplet 29]
Type species: *Cyanogomphus uncatus* Fraser, 1947 [by original designation]
2 species:

noval (Rodrigues Capítulo, 1985) [*Cyanogomphus*] – **L** [Rodrigues Capítulo, 1985]
uncatus (Fraser, 1947) [*Cyanogomphus*]*

References: Rodrigues Capítulo, 1985; Belle 1992a.
Distribution: SE Brazil, Paraguay, N Argentina and Uruguay.

Medium gomphids (37–41 mm); postocellar ridge complete; occiput with a transverse ridge along its entire width, bearing a fringe of hairs; pterothorax pale green to yellow or grey with dark brown stripes (Fig. 727); abdomen mostly brown with pale mark-

Map 58. Distribution of *Tibiagomphus* spp.

ings on basal segments. Wings (Fig. 728) with basal subcostal crossveins; 2 or more crossveins in HW space between arculus branches and point of branching of RP; subtriangles and supratriangles free; 1 cubito-anal crossvein; CuA and MP in HW divergent. Spines of outer row of male hind tibiae quadrangular and projected at distal edge (Figs. 729a-b, shared with some *Epigomphus*); male cercus (Figs. 7311-b) with branch directed ventrally; epiproct well developed,

bifid, with tip of each branch spatulate (shared with *Cyanogomphus*). Anterior hamule fused to anterior lamina, with outer side concave (Fig. 730); posterior hamule narrow with distal 1/3 bent antero-ventrally at a right angle, ending in a single spine; vesica spermalis distal segment with 2 short flagella, and basal segment bifid. Female lacking auricles; vulvar lamina bifid (Fig. 731), extending from about 1/2 to 2/3 of S9 length. **Unique characters**: None known.

Status of classification: Described by Belle in 1992a; further diagnosed by Belle in 1996. Belle considered it related to *Cyanogomphus*, including both genera under his tribe Cyanogomphini in the subfamily Epigomphinae (Belle, 1996). Differentiable from *Tibiagomphus* only by tibial armature (Figs. 729a-b); possibly congeneric.

Potential for new species: Unknown. The minor differences between the original description and illustrations of *Cyanogomphus noval* compared with the holotype of *Cyanogomphus uncatus* and material of the latter species we have examined are likely due to individual variation. We suspect that *Tibiagomphus noval* is a synonym of *T. uncatus*.

Habitat: Streams. No further information available.

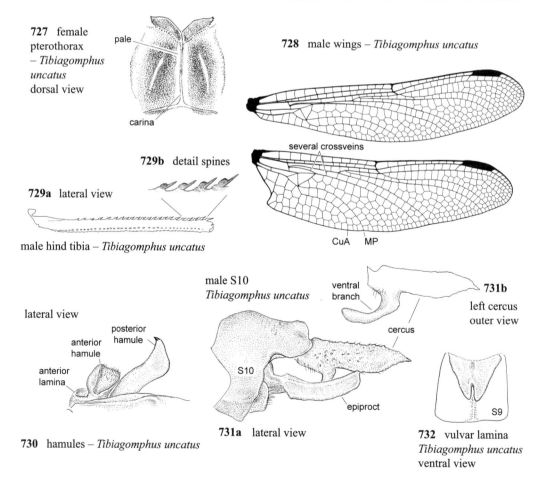

727 female pterothorax – *Tibiagomphus uncatus* dorsal view

pale / carina

728 male wings – *Tibiagomphus uncatus*

several crossveins

CuA   MP

729b detail spines

729a lateral view

male hind tibia – *Tibiagomphus uncatus*

lateral view

anterior hamule
posterior hamule
anterior lamina

730 hamules – *Tibiagomphus uncatus*

male S10
Tibiagomphus uncatus

ventral branch

cercus

S10

epiproct

731a lateral view

731b left cercus outer view

732 vulvar lamina *Tibiagomphus uncatus* ventral view

S9

Zonophora Selys, 1854: 80 (61 reprint).

[♂ pp. 68, couplet 17; ♀ pp. 85, couplet 36]

Type species: *Diastatomma campanulata* Burmeister, 1839 [by monotypy]

10 species:

batesi Selys, 1869* – **L** [Belle, 1966c]
  syn *bodkini* Campion, 1920
calippus calippus Selys, 1869* – **L**? [Belle, 1966c]
calippus klugi Schmidt, 1941*
calippus spectabilis Campion, 1920
campanulata annulata Belle, 1983
campanulata campanulata (Burmeister, 1839)
  [*Diastatomma*]
campanulata machadoi St. Quentin, 1973* – **L**
  [Belle, 1992a]
diversa Belle, 1983
nobilis Belle, 1983*
regalis Belle, 1976
solitaria obscura Belle, 1976
solitaria solitaria Rácenis, 1970
supratriangularis Schmidt, 1941*
surinamensis Needham, 1944
wucherpfennigi Schmidt, 1941*

Map 59. Distribution of *Zonophora* spp.

References: Schmidt, 1941a; Belle, 1963; 1966c; 1983b.

Distribution: Colombia and Venezuela south to Paraguay and NE Argentina.

Medium to large gomphids (50–70 mm); occiput with a transverse ridge along its entire width, with a fringe of hairs along its summit; postocellar ridge complete with lateral edges projected into short horns (Fig. 734); pterothorax and abdomen yellow with black markings, subalar carina of HW projected into strong spine (shared with *Cacoides, Melanocacus,* and *Mitragomphus*). Wings (Fig. 733) with or without basal subcostal crossveins; 2 or more crossveins in HW space between arculus branches and point of branching of RP; subtriangles free; supratriangles free or crossed; 2 or more cubito-anal crossveins; CuA and MP in HW divergent or parallel. Male cercus long and unforked (Figs. 735a-b); epiproct bifid (Fig. 735a). Anterior lamina v-shaped or entire; posterior hamule with or without a cluster of denticles on anterior edge, ending in a single tooth (Fig. 736a); vesica spermalis distal segment with 2 short to long flagella, and basal segment bifid. Female lacking auricles; vulvar lamina bifid, extending from about 1/2 of S9 length to beyond S9 posterior margin and strongly sclerotized (Fig. 737). **Unique characters**: None known.

Status of classification: Good species resolution. Belle (1996) considered this genus related to *Mitragomphus*, including both under the tribe Zonophorini in the subfamily Zonophorinae.

Potential for new species: Likely.

Habitat: Larvae found in the muddy or silt beds and banks of small creeks along gallery forests crossing the savannah (Belle, 1966c). Adults found in trees, but in sunny weather also in lower vegetation in partly shady areas of the creeks (Belle, 1963).

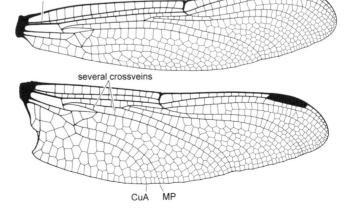

basal SC crossvein

733 male wings
Zonophora calippus klugi

several crossveins

CuA   MP

734 female head – *Zonophora calippus klugi*
dorsal view

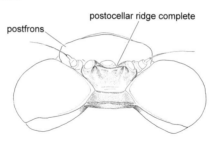

postfrons

postocellar ridge complete

735a dorsal view   male S10 – *Zonophora calippus klugi*

735b lateral view

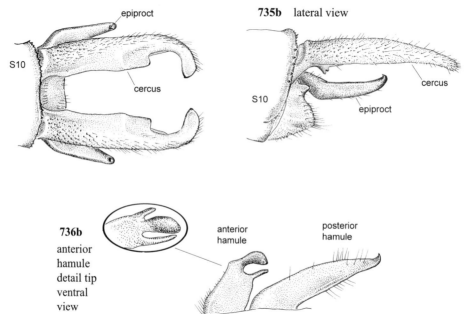

epiproct

S10

cercus

cercus

S10

epiproct

736b
anterior
hamule
detail tip
ventral
view

anterior
hamule

posterior
hamule

736a lateral view

hamules – *Zonophora campanulata*

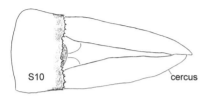

S10

cercus

737 female S10 – *Zonophora batesi*
dorsal view

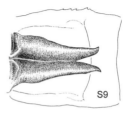

S9

738 vulvar lamina – *Zonophora batesi*
ventral view

7. Neopetaliidae

Subantarctic province of austral region: Monotypic, one species in one genus (endemic).

Diagnostic characters: Prementum cleft; eyes meeting on top of head at a single point; vertex forming a prominent tubercle between ocelli (Fig. 739); wings with five red spots along costal space (Fig. 740); FW triangle longitudinally elongated; pterostigma bracevein present; male with tibial keels in fore legs (Fig. 741); terga S5-8 in male and S2-8 in female with ventro-apical tufts of black hairs (Figs. 745a-b); male genitalia with u-shaped anterior hamule, posterior hamule with bilobed tip, genital ligula with wide lateral semihyaline extensions (Fig. 742), and vesica spermalis third segment with a pair of apical flap-like processes (Figs. 743a-b); female genitalia consisting on a short vulvar lamina from sternum S8, and two pairs of gonapophyses from sternum S9 (Figs. 745a-b); female sternum S10 expanded into a large circular, semihyaline plate (Figs. 745a-c).

Status of classification: Originally placed close to cordulegastrids by Selys (1854b). Later misplaced as an aeshnoid in the same family as austropetaliids (Selys, 1858; 1869a; 1878; Fraser, 1929; 1957; 1960; Tillyard and Fraser, 1940; Schmidt, 1941b; Davies and Tobin, 1985) until recently when Carle and Louton (1994) clearly showed that it belongs to the libelluloid lineage, based on larval (scoop-shaped labium, ventral proventricular lobe with an additional tooth) and adult (presence of male tibial keels) characters. Molecular data also support this relationship (Misof, Rickert, Buckley, Fleck, and Sauer, 2001).

Neopetalia Cowley, 1934a: 201.
Type species: *Petalia punctata* Hagen *in* Selys, 1854 [by monotypy]
punctata (Hagen *in* Selys, 1854)* – **L** [Carle and Louton, 1994]

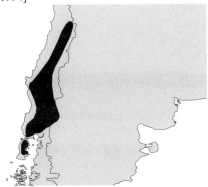

Map 60. Distribution of *Neopetalia* sp.
<u>References</u>: Schmidt, 1941b; Carle and Louton, 1994.
<u>Distribution</u>: S Chile and SW Argentina.

Medium anisopterans (57–58 mm). Frons quadrangular in dorsal view; eyes wider than long, meeting on top of head at a single point (Fig. 739). Frons yellowish, thorax brown with yellow spots and stripes and abundant whitish long silky hairs, abdomen reddish brown with paired dorsal yellow spots. Wings (Fig. 740) hyaline or tinged with brown, with five red spots located at the two primary antenodals, nodus and basal and distal ends of pterostigma; sectors of arculus separated; triangles with 2 cells; supratriangles free; discoidal field widening towards wing margin; 1-2 accessory crossveins in Cu-A space; 6-7 bridge crossveins; Mspl and Rspl lacking; HW anal loop small and rounded; male HW with 3-celled anal triangle. Male tibiae keeled (Fig. 741); male cerci flat in lateral view and curved medially in dorsal view (Figs. 744a-b); epiproct trifid (Figs. 744a-b). Female vulvar lamina from sternum S8 bifid and about as long as 1/3 of sternum S9 (Fig. 745a); two pairs of gonapophyses on sternum S9, the basal more external pair leaf-like, the inner-distal pair cylindrical (Figs. 745a-b); sternum S10 expanded into a large, circular, semihyaline plate, that completely hides the caudal appendages in ventral view (Figs. 754a-c); cerci cylindrical and pointed (Fig. 745c). **Unique characters**: Terga S5-8 in male and S2-8 in female with ventro-apical tufts of black hairs (Figs. 745a-b); male genitalia with u-shaped anterior hamule, posterior hamule with bilobed tip ('bone head'-shaped), genital ligula with wide lateral semihyaline extensions (Fig. 742), and vesica spermalis third segment with a pair of apical flap-like processes (Figs. 743a-b); female sternum S10 expanded into large circular semihyaline plate (Figs. 745a-c).
<u>Status of classification</u>: Good.
<u>Potential for new species</u>: Unlikely.
<u>Habitat</u>: Larvae breed in silt-bottomed pools of small streams and seepage areas, where they burrow leaving only eyes and terminalia over substrate; they are ambush hunters and their preferred prey are mayfly larvae (Carle and Louton, 1994). Adults fly close over spring seeps, and are most active when cloudy; both males and females patrol forest roads with a weak flight, occasionally soaring and rising above tree tops (Carle and Louton, 1994). Females hover momentarily close to water surface and dip abdomen in water to apparently release egg masses (Carle and Louton, 1994).

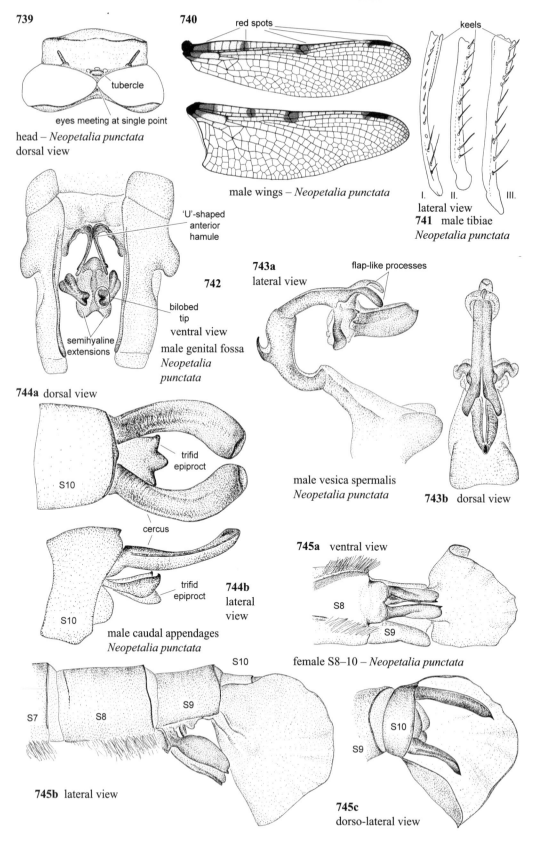

739

head – *Neopetalia punctata*
dorsal view

tubercle

eyes meeting at single point

740

red spots

male wings – *Neopetalia punctata*

keels

I. II. III.

lateral view
741 male tibiae
Neopetalia punctata

742

'U'-shaped anterior hamule

bilobed tip

ventral view
male genital fossa
Neopetalia punctata

semihyaline extensions

743a
lateral view

flap-like processes

male vesica spermalis
Neopetalia punctata

743b dorsal view

744a dorsal view

trifid epiproct

S10

cercus

trifid epiproct

S10

744b
lateral view

male caudal appendages
Neopetalia punctata

745a ventral view

S8

S9

female S8–10 – *Neopetalia punctata*

S10

S7 S8 S9

745b lateral view

S10

S9

745c
dorso-lateral view

8. Cordulegastridae

Holarctic, with one species extending south into Central America: About 45 spp. in 3 genera.

New World: 10 spp. in 1 genus.

Diagnostic characters: Prementum cleft; eyes almost contiguous dorsally or separated by less than distance between lateral ocelli; vertex forming a prominent tubercle between ocelli (Figs. 746-747); triangles elongated longitudinally and similar in shape in FW and HW; brace crossvein absent (Fig. 748); male mesotibial spines approximately quadrangular (as in Figs. 750a-b); male anterior and posterior hamules well developed (Fig. 752); female with sclerotized, straight and long gonapophyses of S8, extending to tip of abdomen or beyond (Fig. 751).

Status of classification: The higher classification within this family has undergone several revisions with resulting inflation of generic and suprageneric names. Carle (1983) recognized *Zoraena* and *Taeniogaster*, and Lohmann (1992) performed a phylogenetic analysis on the family resulting in eight of the nine New World species being allocated to six new genera (*Archegaster, Kalyptogaster, Lauragaster, Pangaeagaster, Taeniogaster, Zoraena*) and three tribes (Taeniogastrini, Cordulegastrini, and Thecagastrini) based primarily on specific differences. We follow Needham, Westfall, and May (2000) and Tennessen (2004) in considering all New World species in just one genus, *Cordulegaster*.

Cordulegaster Leach, 1815: 136.
Type species: *Libellula boltonii* Donovan, 1807 [Kirby, 1890, by subsequent designation]
  syn *Thecophora* Charpentier, 1840: 14.
    Type species: *Aeschna lunulata* Charpentier, 1825 [by monotypy]
  syn *Thecophora* Selys, 1854: 100 (81 reprint).
    Type species: *Thecophora diastatops* Selys, 1854 [by monotypy]
  syn *Thecagaster* Selys, 1854: 103 (84 reprint).
    Type species: *Cordulegaster brevistigma* Selys, 1854 [by monotypy]
  syn *Taeniogaster* Selys, 1854: 107 (88 reprint).
    Type species: *Aeschna obliqua* Say, 1840 [by monotypy]
  syn *Zoraena* Kirby, 1890: 79.
    Type species: *Thecophora diastatops* Selys, 1854 [replacement name for *Thecophora* Selys, a junior primary homonym of *Thecophora* Charpentier, 1840 in Odonata]
  syn *Kuldanagaster* Yousuf & Yunus, 1974: 143.
    Type species: *Kuldanagaster pakistanica* Yousuf & Yunus, 1974 [by original designation]
  syn *Archegaster* Lohmann, 1992: 6.
    Type species: *Cordulegaster sayi* Selys, 1854 [by original designation]
  syn *Kalyptogaster* Lohmann, 1992: 9.
    Type species: *Cordulegaster erroneus* Hagen *in* Selys, 1878 [by original designation]
  syn *Pangaeagaster* Lohmann, 1992: 10.
    Type species: *Cordulegaster maculatus* Selys, 1874 [by original designation]
  syn *Lauragaster* Lohmann, 1992: 10.
    Type species: *Cordulegaster diadema* Selys, 1868 [by original designation]
  syn *Sonjagaster* Lohmann, 1992: 11.

Type species: *Cordulegaster insignis* Schneider, 1845 [by original designation]
About 22 species; 10 New World species:

bilineata (Carle, 1983)*
diadema Selys, 1868* – **L**? [Needham, 1904]
diastatops (Selys, 1854)* – **L** [Needham, 1901; Walker, 1958]
  syn *lateralis* Scudder, 1866
dorsalis deserticola Cruden, 1969*
dorsalis dorsalis Hagen *in* Selys, 1858* – **L** [Needham, 1904]
erronea Hagen *in* Selys, 1878* – **L** [Louton, 1982b; Needham, Westfall, and May, 2000]
godmani McLachlan, 1878*
maculata Selys, 1854* – **L** [Needham, 1901; Byers, 1930; Walker, 1958; Louton, 1982b]
obliqua fasciata Rambur, 1842* – **L** [Byers, 1930; Needham and Westfall, 1955]
obliqua obliqua (Say, 1840)* – **L** [Needham, 1905a; Walker, 1958]
sayi Selys, 1854* – **L** [Needham, Westfall, and May, 2000]
talaria Tennessen, 2004*

References: Fraser, 1929; Lohmann, 1992; Needham, Westfall, and May, 2000; Tennessen, 2004.
Distribution: Holarctic region; in the New World from SE Canada to Costa Rica.

Medium to large cordulegastrids (55–88 mm); brown or black body with yellow stripes on pterothorax and yellow rings or spots on abdominal segments. Wings (Fig. 748) hyaline, lacking crossveins in median and subcostal spaces; triangles with 1-3 cells (usually 2);

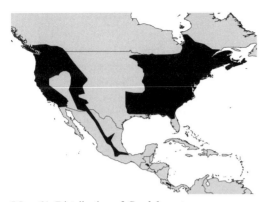

Map 61. Distribution of *Cordulegaster* spp.

subtriangles free; supratriangles crossed; 1-3 Cu-A crossveins; sectors of arculus separated; male with 3-4 celled anal triangle; anal loop defined and lacking a brace crossvein. Males with tibial keels in fore legs (Figs. 749a-b). Anterior hamule large, sub-quadrangular and foliate; posterior hamule narrow and arcuate (Fig. 752); vesica spermalis distal segment lacking flagella, with a medio-ventral lobe (Fig. 754a); basal segment with an anterior ridge (Fig. 754b); cercus

short, straight and pointed (Figs. 753a-b), sometimes with a sub-basal tooth (Fig. 753a); epiproct subquadrate (Fig. 753c). Female gonapophyses of S8 strongly sclerotized, forming a straight and long ovipositor, extending to tip of abdomen or beyond (Fig. 751). **Unique characters**: Male meso and hind tibiae with an outer row of quadrangular spines (Figs. 750a-b, shared with Old World genera, Lohmann 1992). Status of classification: Good species resolution. Evidence in favor of the validity for the recently described sibling species *Cordulegaster bilineata* was marshaled by Pilgrim, Roush, and Krane (2002). Potential for new species: Unlikely. However one new species similar to *C. bilineata* and *C. diastatops* was recently described (Tennessen, 2004). Habitat: Adults course slowly up and down meandering streams and rivulets within forests and seepages, where larvae live buried shallowly under the surface of silt or sand of their beds. Larvae bury themselves by kicking silt and sand over their backs with their hind legs. Females lay eggs by inserting their ovipositor into the soft bottom of these water bodies while hovering. Details of the biology for adults of all species (except *C. talaria*) were given by Dunkle (2000).

746 *Cordulegaster godmani*

vertex

head
dorsal
view

eyes almost
contiguous

747 *Cordulegaster sayi*

748 male wings – *Cordulegaster diadema*

sectors separated

keel

750b
detail

749b detail

quadrangular
spines

lateral view

749a
male mesotibia
Cordulegaster
dorsalis

750a male mesotibia
Cordulegaster
diadema

lateral view

S10

S8   S9

ovipositor

751 female S8–10
Cordulegaster diadema

752 male genital fossa – *Cordulegaster godmani*
lateral view

male caudal appendages – *Cordulegaster godmani*

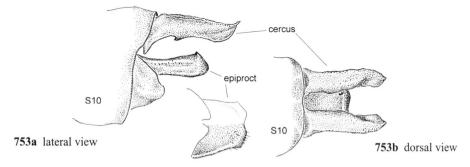

753a lateral view

753c male epiproct
ventral view

753b dorsal view

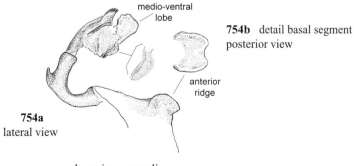

754b detail basal segment
posterior view

754a
lateral view

male vesica spermalis
Cordulegaster diadema

9. Libellulidae − Key to subfamilies

Cosmopolitan: About 1357 spp. in 191 genera.

New World: 490 spp. in 63 genera (51 endemic).

Diagnostic characters: Prementum of adult entire (Carle, 1995); eyes meeting on top of head for a considerable distance (*i.e.* Fig. 1380), except for *Diastatops* (Fig. 1379); primary antenodal veins indistinguishable from secondary antenodal veins (*i.e.* Fig. 759-762, Fraser, 1957), but visible in FW in *Antidythemis, Pantala, Tauriphila*, and *Tramea*, and in some specimens of *Zenithoptera*; antenodal veins aligned (Bechly, 1996; Needham, Westfall, and May, 2000), except for the last one in several libelluline genera (*i.e.* Fig. 761); triangles of FW and HW unequal in shape and in distance from arculus, with FW triangle transversely elongated and twice or more as far from arculus than longitudinally elongated HW triangle (*i.e.* Figs. 759-760, Needham, Westfall, and May, 2000); pterostigma bracevein absent (*i.e.* Figs. 761-762); terga of S2-3 with latero-longitudinal carinae; anterior hamules reduced, either short and erect or obsolete (Carle and Louton, 1994; Carle, 1995); posterior hamules well developed (*i.e.* Figs. 757-758).

Status of classification: We treat this family as comprising three subfamilies, Libellulinae, Macromiinae and Corduliinae. Most workers have followed Gloyd (1959) in treating Macromiinae as a separate family, and some workers consider Corduliinae as a separate family from Libellulinae following St. Quentin (1939). However, the characters traditionally used to define these groups are homoplastic. Other authors (Carle, 1995; Bechly, 1996) subdivide the corduliines even further, with several groups treated as distinct families (Austrocorduliidae, Corduliidae, Gomphomacromiidae, and Synthemistidae). We consider that the differences on which these subdivisions were based (*i.e.* basal space crossed and presence of an accessory antenodal crossvein in subcostal space, which are the distinguishing characters for Synthemistidae) are less significant than generic differences for other odonate families, and that a phylogenetic analysis is needed in order to establish if they are monophyletic. Some studies (*i.e.* Bechly, 1996; Jarzembowski and Nel, 1996; May, 1997) indicate that corduliines constitute a paraphyletic group, with macromiines and libellulines forming monophyletic groups; however, the relationships among libelluline and corduliine genera remain unsolved, and we consider that a phylogenetic analysis of the whole family is needed before updating its classification.

Key to subfamilies

1. Posterior margin of compound eye with evagination delimited from rest of eye (Fig. 755); distance between triangle and arculus about twice longer in FW than in HW (Fig. 759); stiff setae at anterior corners of male anterior lamina present (Fig. 757) .... **Macromiinae** (Page 141)

1'. Posterior margin of compound eye without evagination delimited from rest of eye (Fig. 756); distance between triangle and arculus longer than twice in FW than in HW (Fig. 760); stiff setae at anterior corners of male anterior lamina absent (Fig. 758) 2

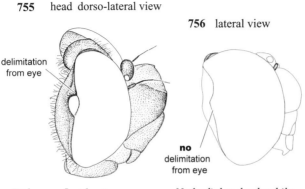

755 head dorso-lateral view

756 lateral view

delimitation from eye

Didymops floridensis

no delimitation from eye

Nothodiplax dendrophila

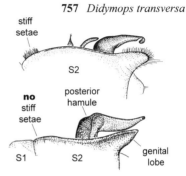

757 *Didymops transversa*

stiff setae

S2

no stiff setae

posterior hamule

S1   S2

genital lobe

758 *Navicordulia errans*
male genital fossa
lateral view

759 male wings – *Macromia annulata*   **760** male wings – *Rialla villosa*

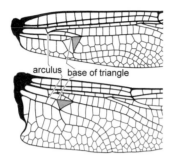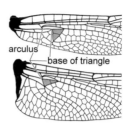

2(1). Second crossvein between RP1,2 and IRP2 developed as oblique vein (Fig. 761); tibial keels in male fore legs absent (Fig. 763); male S2 lacking auricles (Fig. 765) .... **Libellulinae** (Page 173)

2'. Second crossvein between RP1,2 and IRP2 not oblique (Fig. 762); tibial keels in male fore legs present (Fig. 764); male S2 with auricles (Fig. 766) .... **Corduliinae** (Page 145)

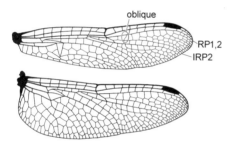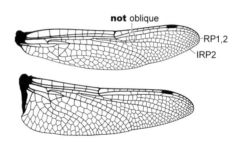

761 male wings – *Micrathyria aequalis*   **762** male wings – *Rialla villosa*

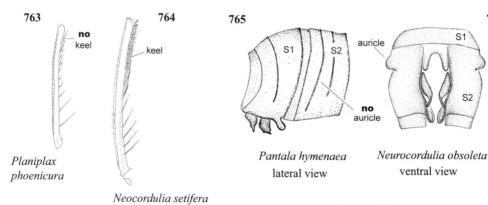

763 *Planiplax phoenicura*

764 *Neocordulia setifera* male fore tibia   lateral view

765 *Pantala hymenaea* lateral view

766 *Neurocordulia obsoleta* ventral view

male genital fossa

10. Macromiinae

Nearly worldwide; absent from neotropical region: About 136 spp. in 4 genera.

New World: 9 spp. in 2 genera (1 endemic).

Diagnostic characters: Ventral pretarsal tooth as long as or longer than dorsal tooth (shared with Old World *Macromidia* and some libelluline members of *Macrothemis*) (Gloyd, 1959; May, 1997); armature of larval proventriculus (Gloyd, 1959); frontal plate of larval head with an upcurved pyramidal horn (Tillyard, 1917; Needham, Westfall, and May, 2000). They differ from all New World corduliines by the distance between triangle and arculus about twice longer in FW than in HW (longer than twice in corduliines); hind margin of compound eye with evagination delimited from rest of eye (Figs. 767-768); presence of stiff setae at anterior corners of male anterior lamina (Figs. 769-770), and from all New World corduliines except *Lauromacromia* and *Neocordulia* by the rounded basal segment of vesica spermalis, lacking lateral projections (Fig. 774b).

Status of classification: According to May (1997) this subfamily is monophyletic, although the relationships among its genera are not yet resolved.

Key to males

1. Eyes contiguous for a short distance (shorter than vertex length) (Fig. 767); vertex more or less rounded and larger than occiput; rear of occiput with posterior bulge (Fig. 767); anterior lamina not longer (**A**) than wide (**B**) (Fig. 769) .... ***Didymops*** (Page 142)

1.' Eyes contiguous for a long distance (longer than vertex length) (Fig. 768); vertex with prominent tubercles and smaller than occiput; rear of occiput lacking posterior bulge (Fig. 768); anterior lamina longer (**A**) than wide (**B**) (Fig. 770) .... ***Macromia*** (Page 143)

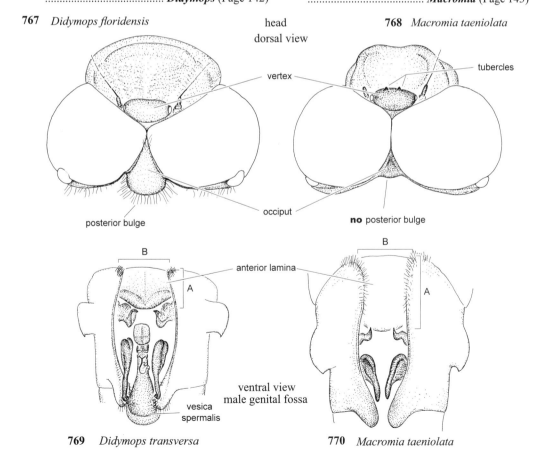

767 *Didymops floridensis*

head
dorsal view

768 *Macromia taeniolata*

tubercles

vertex

posterior bulge

occiput

no posterior bulge

anterior lamina

ventral view
male genital fossa

vesica
spermalis

769 *Didymops transversa*

770 *Macromia taeniolata*

Key to females

1. Eyes contiguous for a short distance (shorter than vertex length) (as in Figs. 767, 772a); vertex more or less rounded and larger than occiput; rear of occiput with posterior bulge (as in Figs. 767, 772a-b) .... **Didymops** (Page 142)

1.' Eyes contiguous for a long distance (longer than vertex length) (as in Fig. 768); vertex with prominent tubercles and smaller than occiput; rear of occiput lacking posterior bulge (as in Fig. 768) .... **Macromia** (Page 143)

Didymops Rambur, 1842: 142.

[♂ pp. 141, couplet 1; ♀ pp. 142, couplet 1]

Type species: *Didymops servillii* Rambur, 1842 [by monotypy]

2 species:

floridensis Davis, 1921* – **L** [Needham and Westfall, 1955; Needham, Westfall, and May, 2000]

transversa (Say, 1840) [*Libellula*]* – **L** [Cabot, 1890; Needham, 1901; Louton, 1982b]
   syn *cinnamomea* (Burmeister, 1839)
   Epophthalmia]
   syn *servillii* Rambur, 1842

References: Needham, Westfall, and May, 2000.
Distribution: SE Canada to SE United States.

Large macromiines (56–68 mm); reddish brown with lateral yellow stripes on pterothorax and yellow bands or dorsal spots on abdomen. Wings (Fig. 771) hyaline or with a small basal brown spot; 12-13 antenodal crossveins in FW; FW subtriangle and triangle free; 2 rows of cells at base of FW discoidal field. Male tibial keels on all legs. **Unique characters**: Eyes contiguous for a short distance (shorter than vertex length); vertex smoothly rounded and larger than occiput (Fig. 772a); posterior surface of occiput bulging (Figs. 772a-b); anterior lamina not longer than wide (Fig. 774b).

Status of classification: Good species resolution.
Potential for new species: Unlikely.
Habitat: Streams and slow flowing rivers with sandy bottoms in forests, also in lakes, where larvae sprawl on the bottom of backwater areas and cover themselves with a thin layer of silt (Louton, 1982b). Adults fly and perch low and hover while hunting (Dunkle, 2000).

Map 62.   Distribution of *Didymops* spp.

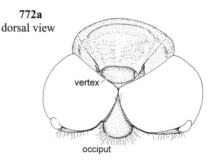

772a
dorsal view

vertex

occiput

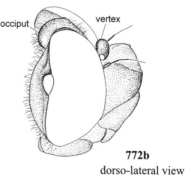

occiput   vertex

772b
dorso-lateral view

head – *Didymops floridensis*

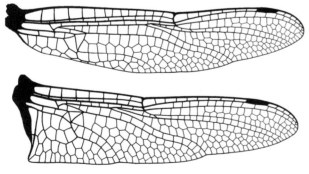

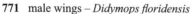

771   male wings – *Didymops floridensis*

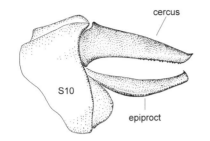

773

male S10
Didymops floridensis
lateral view

774a

lateral view

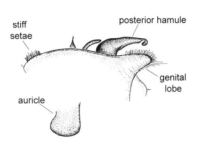

774b

ventral
view

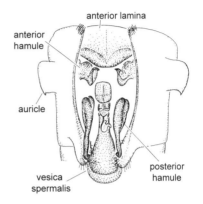

male genital fossa – *Didymops transversa*

Macromia Rambur, 1842: 137.

[♂ pp. 141, couplet 1; ♀ pp. 142, couplet 1]
Type species: *Macromia cingulata* Rambur, 1842
[Kirby, 1890 by subsequent designation]
84 species; 7 New World species:

alleghaniensis Williamson, 1909* – **L** [Needham
and Westfall, 1955; Louton, 1982b]
annulata Hagen, 1861* – **L** [Needham, 1950 as
caderita]
   syn *caderita* Needham, 1950
illinoiensis georgina (Selys, 1878) [*Epophthalmia*]*
– **L** [Louton, 1982b]
   syn *australensis* Williamson, 1909
illinoiensis illinoiensis Walsh, 1862* – **L** [Cabot,
1890; Walker and Corbet, 1975; Louton, 1982b]
magnifica McLachlan *in* Selys, 1874* – **L**
[Kennedy, 1915; Walker, 1937]
   syn *rickeri* Walker, 1937
margarita Westfall, 1947* – **L** [Needham, Westfall,
and May, 2000]
pacifica Hagen, 1861* – **L** [Needham, Westfall, and
May, 2000]
   syn *flavipennis* Walsh, 1862
taeniolata Rambur, 1842* – **L** [Cabot, 1890;

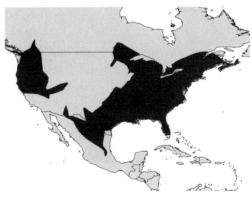

Map 63. Distribution of *Macromia* spp.

Louton, 1982b]

References: May, 1997; Needham, Westfall, and
May, 2000.
Distribution: Worldwide except Africa and neotropi-
cal region; in nearctic region from S Canada to N
Mexico.

Large to very large macromiines (65–91 mm); reddish brown with lateral yellow stripes and metallic green or greenish-blue reflections on pterothorax, and yellow bands or dorsal spots on abdomen. Wings (Fig. 776) hyaline or with a small basal brown spot; 13-20 antenodal crossveins in FW; FW subtriangle and triangle free or crossed; 2-3 rows of cells at base of FW discoidal field. Male tibial keels absent or short on middle legs, well developed on fore and hind legs. **Unique characters**: Vertex (Fig. 775) with prominent tubercles and smaller than occiput; posterior surface of occiput lacking bulge (Fig. 775); anterior lamina longer than wide (Fig. 778b).

Status of classification: Some names have been proposed for varieties and hybrids (*i.e. M. wabashensis*, most likely a hybrid between *M. pacifica* and *M. taeniolata*).

Potential for new species: Unlikely.

Habitat: Adults course up and down clean, sand-bottomed streams and rivers, some species also at lakes. Patrol flights encompass long stretches, usually at high speed and with a straight beat.

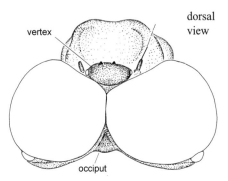

775   head – *Macromia taeniolata*

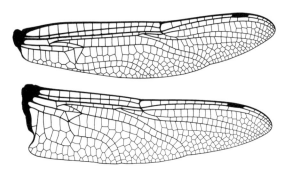

776   male wings – *Macromia annulata*

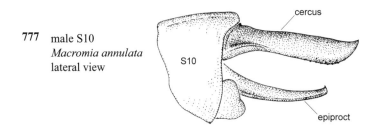

777   male S10
Macromia annulata
lateral view

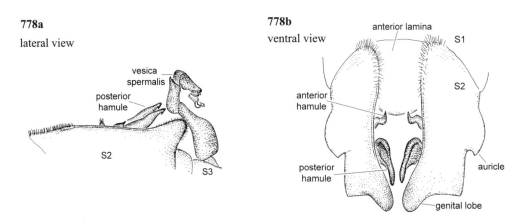

778a
lateral view

778b
ventral view

male genital fossa – *Macromia taeniolata*

11. Corduliinae

Cosmopolitan: About 276 spp. in 44 genera.

New World: 89 spp. in 15 genera (12 endemic).

Diagnostic characters: All New World corduliines share with macromiines the presence of male tibial keels (Figs. 862, 871, 891), presence of male auricles on S2 (Figs. 786-787), and HW anal angle angulated (Figs. 779-782, with the exception of *Williamsonia*);

and with libellulines presence of midrib (Aspl) in anal loop (Figs. 781-782, with the exception of *Gomphomacromia* and *Lauromacromia*, Figs. 779-780). For further differences with macromiines see under diagnostic characters for that subfamily (Page 141).

Status of classification: See discussion under Libellulinae (Page 173).

Key to males

1. HW anal loop polygonal; midrib of anal loop (Aspl) indistinct (Figs. 779-780) **2**

1'. HW anal loop elongate; midrib of anal loop (Aspl) distinct (Figs. 781-782) **3**

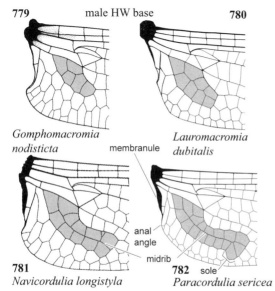

779 male HW base **780**

Gomphomacromia nodisticta

membranule

Lauromacromia dubitalis

anal angle

midrib

781 *Navicordulia longistyla*

782 sole *Paracordulia sericea*

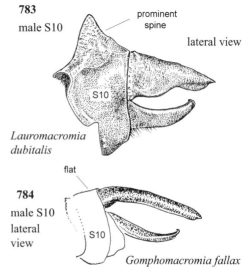

783 male S10

prominent spine

lateral view

S10

Lauromacromia dubitalis

flat

784 male S10 lateral view

S10

Gomphomacromia fallax

2(1). S10 with prominent dorsal spine or keel (Fig. 783); genital lobe present (as in Fig. 787); epiproct as seen in ventral view longer than wide (as in Fig. 785b); sternum S8 with protuberance (as in Fig. 785a) ... *Lauromacromia* (Page 160)

2'. S10 flat or with low medio-longitudinal carina (Fig. 784); genital lobe absent (Fig. 786); epiproct as seen in ventral view about as long as wide (Fig. 788); sternum S8 lacking protuberance .... *Gomphomacromia* (Page 158)

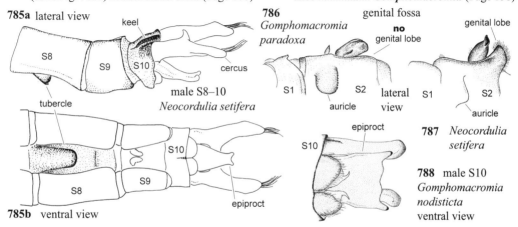

785a lateral view

keel

S8

S9

S10

cercus

tubercle

male S8–10
Neocordulia setifera

S8

S9

S10

epiproct

785b ventral view

786 genital fossa
Gomphomacromia paradoxa

no genital lobe

genital lobe

S1

S2 lateral view

auricle

S1

S2

auricle

epiproct

S10

787 *Neocordulia setifera*

788 male S10
Gomphomacromia nodisticta
ventral view

3(1). FW cubito-anal crossveins numbering 1 (Fig. 789) **4**

3'. FW cubito-anal crossveins numbering 2 or more (Fig. 790) **15**

789 male FW base
Navicordulia
longistyla

1 CuA crossvein

790 male FW base
Aeschnosoma
forcipula

2 CuA crossveins

4(3). HW anal margin with an excavation after distal end of anal triangle (Fig. 791) .... ***Navicordulia*** (Page 161)

4'. HW anal margin approximately linear (Figs. 792-793) **5**

triangle
margin
at angle

anal
margin
excavated

triangle
margin
at angle

anal
margin
linear

no sole

793 male
HW base
Williamsonia
litneri

anal loop
with sole

triangle
margin
in line

792 male HW base
Paracordulia sericea

791 male HW base – *Navicordulia longistyla*

5(4). Outer margin of HW anal triangle meeting anal margin of HW at an angle (as in Figs. 791-792) **6**

5'. Outer margin of HW anal triangle more or less in line with anal margin of HW (Fig. 793) .... ***Williamsonia*** (Page 170)

6(5). HW anal loop with sole (Fig. 792) **7**

6'. HW anal loop without sole (as in Fig. 791) .... ***Neocordulia*** (Page 163)

7(6). HW membranule not reaching end of anal triangle (Fig. 796); genital lobe approximately triangular (Fig. 798) ..... ***Paracordulia*** (Page 165)

7'. HW membranule surpassing end of anal triangle (Fig. 797) genital lobe rounded or quadrangular (Figs. 800-802) **8**

796 male HW – *Helocordulia selysii*

794 male S10

dorso-apical tooth

S10

epiproct

Cordulia shurtleffii
ventral view

795 male S10

S10

epiproct

Neocordulia setifera
ventral view

spots

membranule

797 male HW – *Rialla villosa*

8(7). Epiproct deeply bifurcate and with a dorso-apical tooth on each side (Fig. 794); lateral margin of genital fossa with an angulated lobe at anterior third (Fig. 800) .... ***Cordulia*** (Page 154)

8'. Epiproct not deeply bifurcate and lacking a dorso-apical tooth on each side (as in Fig. 795); lateral margin of genital fossa approximately straight (Figs. 801-802) **9**

PLATE 1

Tanypteryx hageni (Petaluridae) male from Siskiyou-Trinity County, California, United States, basking on granite bolder, photographed by R.W. Garrison

Anax walsinghami (Aeshnidae) male from Stanislaus County, California, United States, eating a nymphalid butterfly (*Vanessa cardui*), photographed by R.W. Garrison

Aeshna interrupta (Aeshnidae) male and female from British Columbia, Canada, in copula, photographed by R.W. Garrison

PLATE 2

Rhionaeschna variegata
(Aeshnidae) male from
Parque Nacional Queulat,
Chile, freshly emerged from
larval exuvia, photographed
by R.W. Garrison

Rhionaeschna variegata
(Aeshnidae) mature male
from Ñireguao, Chile,
photographed by R.W.
Garrison

Epigomphus paludosus
(Gomphidae) male
from Rio de Janeiro,
Brazil, photographed
by R.W. Garrison

PLATE 3

Progomphus gracilis (Gomphidae) male from Parque Nacional Itatiaia, Rio de Janeiro, Brazil, photographed by R.W. Garrison

Gomphus lynnae (Gomphidae) male from Wheeler County, Oregon, United States, photographed by R.W. Garrison

Phyllocycla breviphylla (Gomphidae) male from San Luis Potosi, Mexico, photographed by R.W. Garrison

PLATE 4

Erpetogomphus lampropeltis (Gomphidae) male from Ventura County, California, United States, photographed by R.W. Garrison

Cordulegaster diadema (Cordulegastridae) male from Greenlee County, Arizona, United States, photographed by R.W. Garrison

Didymops floridensis (Libellulidae: Macromiinae) juvenile male from Clay County, Florida, United States, photographed by R.W. Garrison

PLATE 5

Gomphomacromia paradoxa (Libellulidae: Corduliinae) male from Puerto Cisnes, Chile, photographed by R.W. Garrison

Diastatops obscura (Libellulidae: Libellulinae) male from Bolivar, Venezuela, photographed by R.W. Garrison

Dythemis fugax (Libellulidae: Libellulinae) male from Graham County, Arizona, United States, photographed by R.W. Garrison

PLATE 6

Elasmothemis cannacrioides (Libellulidae: Libellulinae) male from Madre de Dios, Peru, photographed by R.W. Garrison

Erythrodiplax juliana (Libellulidae: Libellulinae) male from Rio de Janeiro, Brazil, photographed by R.W. Garrison

Erythrodiplax castanea (Libellulidae: Libellulinae) male from Madre de Dios, Peru, photographed by J.A. Louton

PLATE 7

Perithemis intensa
(Libellulidae: Libellulinae) male from
Maricopa County,
Arizona, United States,
photographed by R.W.
Garrison

Libellula comanche
(Libellulidae: Libellulinae) male from Riverside
County, California, United States, photographed
by R.W. Garrison

Macrothemis imitans
(Libellulidae: Libellulinae) male from
Rio de Janeiro, Brazil,
photographed by R.W.
Garrison

PLATE 8

Scapanea frontalis
(Libellulidae: Libelluli-
nae) male from La Vega,
Dominican Republic,
photographed by R.W.
Garrison

Tramea onusta
(Libellulidae: Libel-
lulinae) male from
Riverside County, Cali-
fornia, United States,
photographed by R.W.
Garrison

Zenithoptera fasciata
(Libellulidae:
Libellulinae) male
from Pista de Kaw
Road, French Guiana,
photographed by R.W.
Garrison

9(8). HW with dark brown or reddish brown spots at antenodal crossveins (Fig. 806) **10**

9'. HW lacking dark spots at antenodal crossveins (Fig. 805) **11**

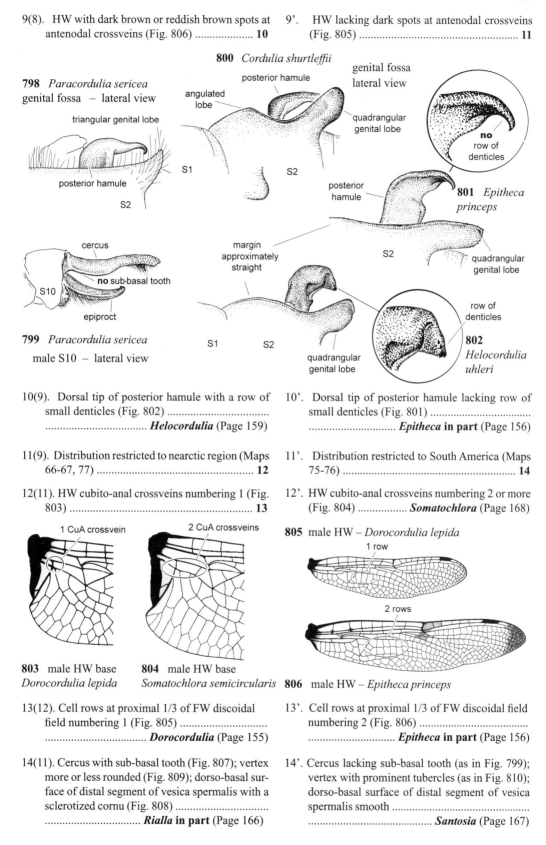

800 *Cordulia shurtleffii*

posterior hamule

genital fossa
lateral view

798 *Paracordulia sericea*
genital fossa – lateral view

angulated lobe

triangular genital lobe

quadrangular genital lobe

posterior hamule

no row of denticles

posterior hamule

S1

S2

S2

801 *Epitheca princeps*

cercus

margin approximately straight

S2

quadrangular genital lobe

no sub-basal tooth

S10

row of denticles

epiproct

799 *Paracordulia sericea*
male S10 – lateral view

S1

S2

quadrangular genital lobe

802 *Helocordulia uhleri*

10(9). Dorsal tip of posterior hamule with a row of small denticles (Fig. 802) ....
.... ***Helocordulia*** (Page 159)

10'. Dorsal tip of posterior hamule lacking row of small denticles (Fig. 801) ....
.... ***Epitheca* in part** (Page 156)

11(9). Distribution restricted to nearctic region (Maps 66-67, 77) **12**

11'. Distribution restricted to South America (Maps 75-76) **14**

12(11). HW cubito-anal crossveins numbering 1 (Fig. 803) **13**

12'. HW cubito-anal crossveins numbering 2 or more (Fig. 804) .... ***Somatochlora*** (Page 168)

1 CuA crossvein

2 CuA crossveins

805 male HW – *Dorocordulia lepida*

1 row

2 rows

803 male HW base
Dorocordulia lepida

804 male HW base
Somatochlora semicircularis

806 male HW – *Epitheca princeps*

13(12). Cell rows at proximal 1/3 of FW discoidal field numbering 1 (Fig. 805) ....
.... ***Dorocordulia*** (Page 155)

13'. Cell rows at proximal 1/3 of FW discoidal field numbering 2 (Fig. 806) ....
.... ***Epitheca* in part** (Page 156)

14(11). Cercus with sub-basal tooth (Fig. 807); vertex more or less rounded (Fig. 809); dorso-basal surface of distal segment of vesica spermalis with a sclerotized cornu (Fig. 808) ....
.... ***Rialla* in part** (Page 166)

14'. Cercus lacking sub-basal tooth (as in Fig. 799); vertex with prominent tubercles (as in Fig. 810); dorso-basal surface of distal segment of vesica spermalis smooth ....
.... ***Santosia*** (Page 167)

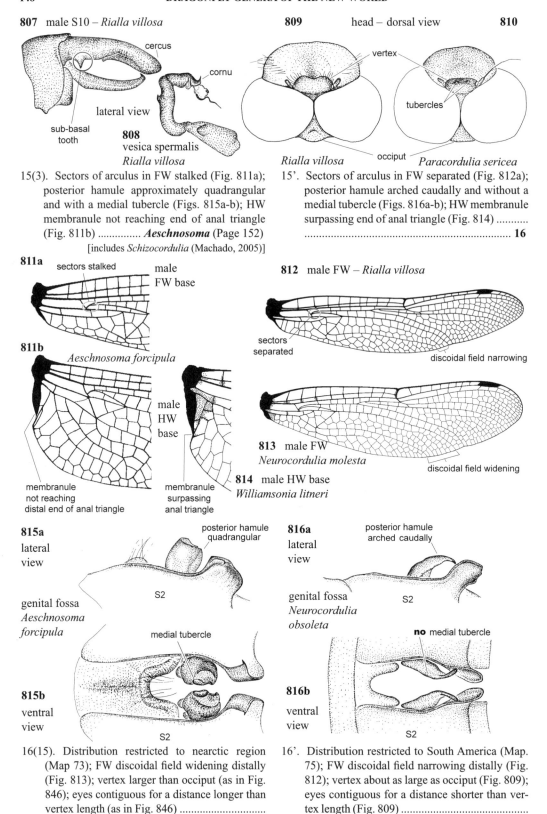

807   male S10 – *Rialla villosa*

cercus

cornu

lateral view

sub-basal tooth

808 vesica spermalis *Rialla villosa*

809   head – dorsal view   **810**

vertex

tubercles

Rialla villosa   occiput   *Paracordulia sericea*

15(3). Sectors of arculus in FW stalked (Fig. 811a); posterior hamule approximately quadrangular and with a medial tubercle (Figs. 815a-b); HW membranule not reaching end of anal triangle (Fig. 811b) .... ***Aeschnosoma*** (Page 152)
[includes *Schizocordulia* (Machado, 2005)]

15'. Sectors of arculus in FW separated (Fig. 812a); posterior hamule arched caudally and without a medial tubercle (Figs. 816a-b); HW membranule surpassing end of anal triangle (Fig. 814) ....
.... **16**

811a   sectors stalked   male FW base

811b   *Aeschnosoma forcipula*

male HW base

membranule not reaching distal end of anal triangle

membranule surpassing anal triangle

812   male FW – *Rialla villosa*

sectors separated

discoidal field narrowing

813   male FW *Neurocordulia molesta*

discoidal field widening

814   male HW base *Williamsonia litneri*

815a lateral view

posterior hamule quadrangular

S2

genital fossa *Aeschnosoma forcipula*

medial tubercle

815b ventral view

S2

816a lateral view

posterior hamule arched caudally

S2

genital fossa *Neurocordulia obsoleta*

no medial tubercle

816b ventral view

S2

16(15). Distribution restricted to nearctic region (Map 73); FW discoidal field widening distally (Fig. 813); vertex larger than occiput (as in Fig. 846); eyes contiguous for a distance longer than vertex length (as in Fig. 846) ....
.... ***Neurocordulia*** (Page 164)

16'. Distribution restricted to South America (Map. 75); FW discoidal field narrowing distally (Fig. 812); vertex about as large as occiput (Fig. 809); eyes contiguous for a distance shorter than vertex length (Fig. 809) ....
.... ***Rialla* in part** (Page 166)

Key to females

1.   HW anal loop polygonal; midrib of anal loop (Aspl) indistinct (as in Figs. 817-818) **2**

1'.   HW anal loop elongate; midrib of anal loop (Aspl) distinct (as in Figs. 819-820) **3**

817 male HW base
Gomphomacromia nodisticta

818 male HW base
Lauromacromia dubitalis

821

female sterna S8–9

Cordulia shurtleffii
vulvar lamina ventral view

822a
ventral view

Gomphomacromia paradoxa

cercus

vulvar lamina lobes

gonapophysis of S9

822b
lateral view

vulvar lamina lobes

membranule

anal angle

midrib

sole

819 male HW base
Navicordulia longistyla

820 male HW base
Paracordulia sericea

2(1). Vulvar lamina projected distally beyond tip of cerci, more than 1/3 of sternum S9, not bifid, with distal 1/2 divided into 2 long parallel-sided lobes (Figs. 822a-b) ....
.... ***Gomphomacromia*** (Page 158)

2'.   Vulvar lamina not reaching tip of cerci, and less than 1/3 of sternum S9, bifid, with distal 1/2 not divided into 2 long parallel-sided lobes (as in Fig. 821) ....
.... ***Lauromacromia*** (Page 160)

3(1). FW cubito-anal crossveins numbering 1 (as in Fig. 823) **4**

3'.   FW cubito-anal crossveins numbering 2 or more (as in Fig. 824) **18**

823 male FW base
Navicordulia longistyla

1 CuA crossvein

subtriangle free

824 male FW base
Aeschnosoma forcipula

2 CuA crossveins

crossed subtriangle

4(3). HW anal loop with sole (as in Fig. 820) ....
.... **5**

4'.   HW anal loop without sole (as in Fig. 819) ....
.... **17**

5(4). HW with dark brown or reddish brown spots at antenodal crossveins (as in Fig. 825) **6**

5'.   HW lacking dark spots at antenodal crossveins (as in Fig. 826) **7**

spots

membranule

825 male HW – *Helocordulia selysii*

826 male HW – *Rialla villosa*

6(5).   Vulvar lamina 1/3 to as long as 1/2 of sternum S9 (as in Figs. 827-828) ....
.... *Helocordulia* (Page 159)

6'.   Vulvar lamina as long as sternum S9 or longer (Fig. 829) .... ***Epitheca* in part** (Page 156)

827 *Cordulia shurtleffii*   **828** *Dorocordulia lepida*   **829** *Epitheca spinigera*

female sterna S8–9 ventral view

vulvar lamina

vulvar lamina

S8   S9   S8   S9   S8   S9

7(5).   FW subtriangle free (as in Fig. 823) ....
.... **8**

7'.   FW subtriangle crossed (as in Fig. 824) ....
.... **10**

8(7).   Vulvar lamina projected distally beyond tip of cercus, and as long as sternum S9 or longer; distal margin of vulvar lamina not bifid, and sternum S9 projected distally beyond tip of cerci Figs. 830a-b) ....
.... *Navicordulia* **in part** (Page 161)

8'.   Vulvar lamina not reaching tip of cerci, and 1/3 to as long as 1/2 of sternum S9; distal margin of vulvar lamina bifid (as in Figs. 827-828), and sternum of S9 not reaching posterior margin of S10 .... **9**

830a lateral view

female S8–10 *Navicordulia errans*

830b dorsal view

cercus   S10   S9   S8   vulvar lamina   S10   S8   S9   cercus   vulvar lamina

9(8).   FW discoidal field widening distally (as in Fig. 832) .... *Williamsonia* **in part** (Page 170)

9'.   FW discoidal field narrowing distally (as in Fig. 831) .... *Dorocordulia* **in part** (Page 155)

831 male FW *Dorocordulia lepida*

discoidal field narrowing

832 male FW *Williamsonia litneri*

discoidal field widening

1 CuA crossvein   2 CuA crossveins

male HW base

833 *Dorocordulia lepida*   **834** *Somatochlora semicircularis*

10(7).   Distribution restricted to nearctic region (Maps 65-67, 77) **11**

10'.   Distribution restricted to South America (Maps 71, 74-76) **13**

11(10). HW cubito-anal crossveins numbering 1 (as in Fig. 833) **12**

11'.   HW cubito-anal crossveins numbering 2 or more (as in Fig. 834) ....
.... ***Somatochlora*** (Page 168)

12(11). Vulvar lamina 1/3 to as long as 1/2 of sternum S9 (Figs. 827-828) ....
.... *Dorocordulia* **in part/** *Cordulia*

12'.   Vulvar lamina as long as sternum S9 or longer (Fig. 829) .... ***Epitheca* in part** (Page 156)

Dorocordulia (Page 155): Vulvar lamina narrowly cleft, with lobes separated by a space much narrower than each lobe's width (Fig. 828)

Cordulia (Page 154): Vulvar lamina broadly cleft, with lobes separated by a space about as wide as each lobe's width (Fig. 827)

13(10). Distal end of HW anal loop with 2 cells (as in Fig. 835) .... **Navicordulia** in part (Page 161)

13'. Distal end of HW anal loop with 3 or more cells (as in Fig. 836) **14**

male HW base

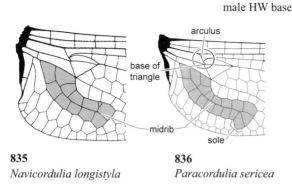

835
Navicordulia longistyla

836
Paracordulia sericea

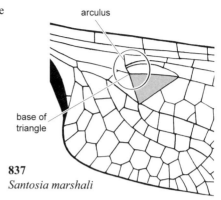

837
Santosia marshali

14(13). Base of HW triangle opposite to arculus (as in Fig. 837) **15**

14'. Base of HW triangle slightly proximal to arculus (as in Fig. 836) **16**

15(14). Vertex more or less rounded (as in Fig. 838) .... **Rialla** in part (Page 166)

15'. Vertex with prominent tubercles (as in Fig. 839) .... **Santosia** (Page 167)

head – dorsal view

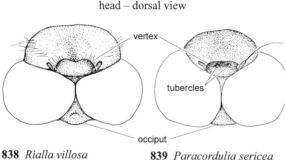

838 *Rialla villosa*   **839** *Paracordulia sericea*

female S8–10 – *Paracordulia sericea*

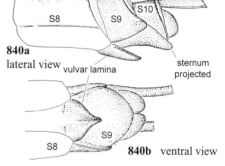

840a lateral view

840b ventral view

16(14). Vulvar lamina not triangular with concave sides, and with distal margin bifid (as in Figs. 827-828); sternum of S9 not reaching posterior margin of S10; vertex more or less rounded (as in Fig. 838) .... **Rialla** in part (Page 166)

16'. Vulvar lamina triangular with concave sides, and distal margin not bifid (Fig. 840a); sternum of S9 projected distally to about posterior margin of S10 (Fig. 840b); vertex with prominent tubercles (Fig. 839) .... **Paracordulia** (Page 165)

17(4). Vulvar lamina less than 1/3 of sternum S9 (as in Fig. 827); HW cubito-anal crossveins numbering 2 or more (as in Fig. 834); lateral carinae on at least some abdominal segments absent; cell rows at proximal 1/3 of FW discoidal field numbering 2 (as in Fig. 841) .... **Neocordulia** (Page 163)

17'. Vulvar lamina more than 1/3 of sternum S9 (as in Fig. 828); HW cubito-anal crossveins numbering 1 (as in Fig. 833); lateral carinae on at least some abdominal segments present; cell rows at proximal 1/3 of FW discoidal field numbering 1 (as in Fig. 842) .... **Williamsonia** in part (Page 170)

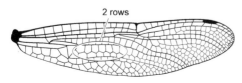

841 male FW – *Neocordulia setifera*

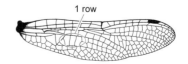

842 male FW – *Williamsonia litneri*

18(3). Sectors of arculus in FW stalked (as in Fig. 843) .... *Aeschnosoma* (Page 152)
[includes *Schizocordulia* (Machado, 2005)]

18'. Sectors of arculus in FW separated (as in Fig. 844) **19**

844 male FW – *Rialla villosa*

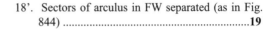

843 male FW base – *Aeschnosoma forcipula*

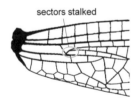

sectors stalked

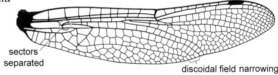

sectors separated

discoidal field narrowing

discoidal field widening

845 male FW – *Neurocordulia molesta*

19(18). Vulvar lamina less than 1/3 of sternum S9 (as in Fig. 827); distribution restricted to nearctic region (Map 73); FW discoidal field widening distally (as in Fig. 845); vertex larger than occiput (as in Fig. 846) ....
.... *Neurocordulia* (Page 164)

19'. Vulvar lamina more than 1/3 of sternum S9 (as in Fig. 828); distribution restricted to South America (Map 75); FW discoidal field narrowing distally (as in Fig. 844); vertex about as large as occiput (as in Fig. 838) ....
.... *Rialla* **in part** (Page 166)

Aeschnosoma Selys, 1870: 7 (4 reprint).
[♂ pp. 148, couplet 15; ♀ pp. 152, couplet 18]
Type species: *Aeschnosoma elegans* Selys, 1870 [by original designation]
[NOTE: The citation commonly used for this species name is Selys, 1871, but Selys mentioned *A. elegans* as type species of *Aeschnosoma* in 1870, and although he provided the specific description in 1871a, since it was the only species included in 1870, the year of the generic description applies to the species also.]
[NOTE: Machado, 2005c erected a new genus, *Schizocordulia* for *Aeschnosoma rustica* Hagen in Selys, 1871, based on the bifid male epiproct, long internal branch of the hamule, pilose plate on male sternum S7, and large female vulvar lamina. We consider these differences to be specific rather than generic.]
5 species:

auripennis Geijskes, 1970 – **L** [Geijskes, 1970]
elegans Selys, 1870
forcipula Hagen *in* Selys, 1871* – **L** [Geijskes, 1970]
  syn *peruviana* Cowley, 1934
marizae Santos, 1981 – **L** [Costa and Santos, 2000]
rustica Hagen *in* Selys, 1871

References: Geijskes, 1970; Santos, 1981; Machado, 2005c.
Distribution: Guiana and Venezuela south to Brazil and Peru.

Medium corduliines (38–60 mm); emerald green

Map 64. Distribution of *Aeschnosoma* spp.

eyes in live males; pterothorax brown or brown with yellow stripes, with metallic green and copper reflections; abdomen brown to black, with variable development of yellow spots. Wings (Fig. 847) hyaline or with a basal golden spot; sectors of arculus in FW stalked (shared with *Lauromacromia*); 2-4 Cu-A crossveins in FW; FW discoidal field widening distally; anal loop elongated with well-defined sole and midrib. Male tibial keels on all legs; sub-basal tooth on male cercus present or absent (Fig. 851). Vulvar lamina bifid, shorter than 1/3 of sternum S9 and not projected ventrally (Fig. 852). **Unique characters**: Distal segment of vesica spermalis narrow and distally projected (Figs. 850a-b); posterior hamule with a medial tubercle (Fig. 848b).
Status of classification: Species resolution seems to

be good.

Potential for new species: Likely. Most species are rare in collections, inhabiting streams within barely penetrable jungle; *A. forcipula* is however a common species of slow running creeks in savannah lowlands.

Habitat: Adults fly swiftly in an irregular pattern along forest trails and over creeks with water rich in tannins, hovering from time to time; females oviposit by rapidly striking the abdomen several times in the middle of the flowing creek to and fro in irregular flight (Geijskes, 1970).

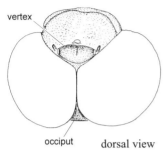

846 head – *Aeschnosoma forcipula*

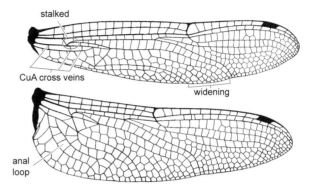

847 male wings – *Aeschnosoma forcipula*

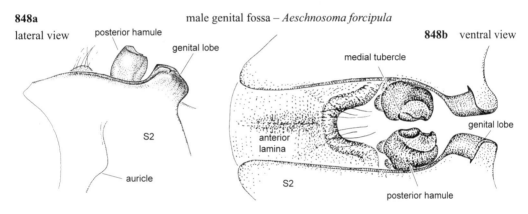

848a lateral view

male genital fossa – *Aeschnosoma forcipula*

848b ventral view

849 male S5–10 – *Aeschnosoma forcipula*

dorsal view

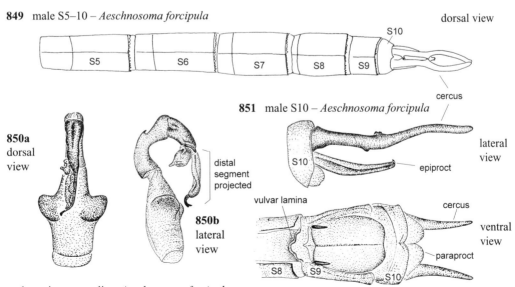

851 male S10 – *Aeschnosoma forcipula*

850a dorsal view

850b lateral view

male vesica spermalis – *Aeschnosoma forcipula*

852 female S8–10 – *Aeschnosoma forcipula*

Cordulia Leach, 1815: 136.

[♂ pp. 146, couplet 8; ♀ pp. 150, couplet 12]

Type species: *Libellula aenea* Linnaeus, 1758 [Jödicke and van Tol, 2003 by subsequent designation; validated by Commission Opinion 2110 (ICZN, 2005)]

> syn *Chlorosoma* Charpentier, 1840 [*nec Chlorosoma* Wagler, 1830 in Reptilia]
>> Type species: *Libellula aenea* Linnaeus, 1758 [by monotypy]

3 species; 1 New World species:

shurtleffii Scudder, 1866* – **L** [Needham, 1901]
> syn *bifurcata (nomen nudum)* Hagen, 1861

References: Needham, Westfall, and May, 2000.
Distribution: Holarctic; in nearctic region from Alaska and Canada to N United States.

Medium corduliines (43–50 mm); brown to black with metallic green-violet reflections on head and pterothorax; body hairy, especially on pterothorax. Wings (Fig. 853) hyaline or with golden basal spot; anal loop with well-developed midrib and sole. Male tibial keels on all legs; posterior hamule bent caudally from its base and with recumbent tip directed externally (Figs. 855a-b); distal segment of vesica spermalis with two apical flagella and a dorso-basal cornu. Vulvar lamina bifid and about as long as 1/2 of sternum S9 (Fig. 854, shared with *Rialla*). **Unique**

Map 65. Distribution of *Cordulia* sp.

characters: Male epiproct deeply bifid and with a dorso-apical tooth on each side (Figs. 856a-b); margin of genital fosa with triangular lobe at anterior 1/3 (Fig. 855a).
Status of classification: Well defined species.
Potential for new species: Unlikely in the New World (see Jödicke, Langhoff, and Misof, 2004).
Habitat: Bogs, marshes, fens, ponds, and lakes with boggy edges, where the larvae sprawl. Adults common; dart and hover while flying, found foraging in shade areas along margins of forests, or basking flat on leaves. Males have brief moving territories (Dunkle, 2000). Reproductive behavior studied by Hilton (1983a).

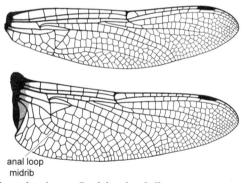

anal loop
midrib

853   male wings – *Cordulia shurtleffii*

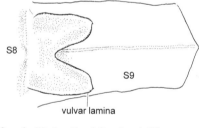

S8

S9

vulvar lamina

854   female S8–9 – *Cordulia shurtleffii* ventral view

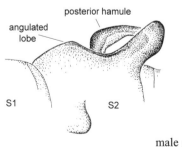

posterior hamule

angulated lobe

S1   S2

855a
lateral view

male genital fossa
Cordulia shurtleffii

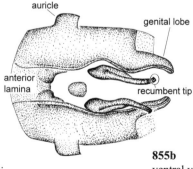

auricle

genital lobe

anterior lamina

recumbent tip

855b
ventral view

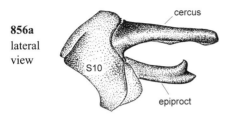

856a
lateral
view

cercus

S10

epiproct

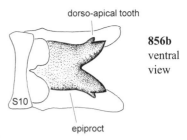

dorso-apical tooth

856b
ventral
view

S10

epiproct

male S10 – *Cordulia shurtleffii*

Dorocordulia Needham, 1901: 504.

[♂ pp. 147, couplet 13; ♀ pp. 150, couplets 9, 12]

Type species: *Cordulia libera* Selys, 1871a [by original designation]

2 species:

lepida (Hagen *in* Selys, 1871) [*Cordulia*]* – **L** [Garman, 1927]
libera (Selys, 1871) [*Cordulia*]* – **L** [Needham, 1901]

References: Needham, Westfall, and May, 2000.
Distribution: SE Canada and NE United States.

Small to medium corduliines (28–43 mm); brown to black with metallic green-copper reflections; body hairy, especially on pterothorax. Wings (Fig. 857) hyaline or with small golden basal spots; anal loop elongate with well-developed midrib and sole. Male tibial keels short to absent on middle legs, well developed on fore and hind legs; abdomen considerably widened (*D. libera*) or not (*D. lepida*); sub-basal tooth on male cercus (Fig. 860). Posterior margin of anterior lamina (Fig. 859b) with deep u-shaped cleft (shared with *Helocordulia*, *Neurocordulia* and *Somatochlora*); posterior hamule caudally bent with pointed tip (Figs. 859a-b); distal segment of vesica

Map 66. Distribution of *Dorocordulia* spp.

spermalis with two apical flagella. Vulvar lamina bifid and about as long as 1/3-1/2 of sternum S9 (Fig. 858). **Unique characters**: None known.
Status of classification: Good species resolution.
Potential for new species: Unlikely.
Habitat: Bogs, marshes, ponds, lakes and small streams of acid water and boggy or marshy margins. Feed along margins of forests or grassy forest clearings. Perch hanging from twigs and bask flat on leaves; males hover while patrolling (Dunkle, 2000).

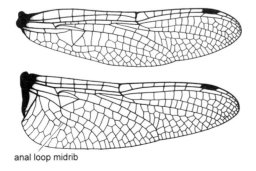

anal loop midrib

857 male wings – *Dorocordulia lepida*

vulvar lamina

S8   S9

858 female S8–9
Dorocordulia lepida
ventral view

859a lateral view

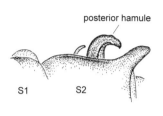

posterior hamule

S1 S2

859b ventral view

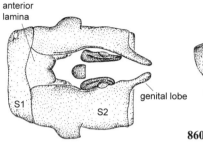

anterior
lamina

S1

S2

genital lobe

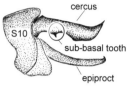

cercus

S10

sub-basal tooth

epiproct

860 male S10
Dorocordulia libera
lateral view

male genital fossa – *Dorocordulia lepida*

Epitheca Charpentier, 1840: 11.
[♂ pp. 147, couplets 10, 13; ♀ pp. 150, couplets 6, 12]
Type species: *Libellula bimaculata* Charpentier, 1825
[by monotypy]
12 species; 10 New World species:
  syn. *Tetragoneuria* Hagen, 1861
    Type species: *Libellula semiaquea* Burmeister,
    1839 [Kirby, 1890 by subsequent designation]
  syn. *Epicordulia* Selys, 1871
    Type species: *Epitheca princeps* Hagen, 1861
    [Kirby, 1890 by subsequent designation]
canis McLachlan, 1886* – **L** [Needham, 1901 as *T.
spinosa*]
costalis (Selys, 1871) [*Cordulia*]* – **L** [Tennessen,
1977]
  syn *williamsoni* Muttkowski, 1911
cynosura (Say, 1840) [*Libellula*]* – **L** [Cabot, 1890]
  syn *complanata* (Rambur, 1842) [*Cordulia*]
  syn *cynosura* race? *basiguttata* (Selys, 1871)
  [*Cordulia*]
  syn *cynosura simulans* Muttkowski, 1911
  syn *lateralis* (Burmeister, 1839) [*Epophthalmia*]
  syn *morio* Muttkowski, 1911
petechialis Muttkowski, 1911* – **L** [Needham and
Westfall, 1955; Needham, Westfall, and May, 2000]
princeps (Hagen, 1861) [*Epitheca*]* – **L** [Cabot,
1890; Needham, 1901; Broughton, 1928]
  syn *regina* (Hagen *in* Selys, 1871) [*Cordulia*]
semiaquea (Burmeister, 1839) [*Libellula*]* – **L**
[Needham and Westfall, 1955; Needham, Westfall,
and May, 2000]
  syn *diffinis (nomen nudum)* Hagen, 1861
  [*Tetragoneuria*]
  syn *semiaquea calverti* Muttkowski, 1915
sepia Gloyd, 1933* – **L** [Westfall, 1951]
spinigera Selys, 1871 [*Cordulia*]* – **L** [Needham,
1901; Walker, 1913]
  syn *indistincta* Morse, 1895
spinosa (Hagen *in* Selys, 1878) [*Cordulia*]* – **L**
[Tennessen, 1994]
stella Williamson *in* Muttkowski, 1911*
[NOTE: Geijskes (1970: 16) examined what he stated to be
the holotype of *Libellula tomentosa* Fabricius 1775, prob-

ably after the listing by Zimsen (1964) and determined that
it was a libellulid. After examining photos of this specimen
we determined that it is a species of *Macrothemis*. Pending
examination of the genitalia, its identity will remain unresolved. However, the true holotype of *Libellula tomentosa*
is stated by Selys (1871a) and Martin (1907) to be in the Sir
Joseph Banks collection in the Linnean Society of London.
Pending study of this specimen, we have not listed this name
here.]

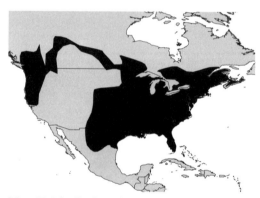

Map 67. Distribution of *Epitheca* spp.

References: Walker, 1966; Needham, Westfall, and
May, 2000 (as *Epicordulia* and *Tetragoneuria*).
Distribution: Holarctic; nearctic species in SE Cana-
da, United States and N Mexico.

Medium (32–48 mm, subgenus *Tetragoneuria*) to
large (58-78 mm, subgenus *Epicordulia*) corduliines;
brown with yellowish markings and metallic
greenish-copper reflections; body hairy, especially
pterothorax. Wings (Figs. 864-865) hyaline or with
dark spots variable in size and position (basal, basal
1/2, along costal and subcostal space, at nodus,
at tip); anal loop with well-developed midrib and
sole. Male tibial keels (Fig. 862) on all legs; male
abdomen constricted at S3; auricles small (Fig.
863b); male cercus with distal ventral keel but no
sub-basal tooth (Figs. 866-868). Posterior hamule
entire, bent caudally with tip pointed (Fig. 863a);

distal segment of vesica spermalis with two apical flagella. Vulvar lamina bifid along distal 5/6 of its length. **Unique characters**: Vulvar lamina as long as sternum S9 or slightly longer (Fig. 861).

Status of classification: Some authors (*i.e.* Needham, Westfall, and May, 2000) consider *Epitheca* to include three different genera: *Epicordulia* Selys, 1871 and *Tetragoneuria* Hagen, 1861 in North America and *Epitheca* in Eurasia. The differences are however slight and we prefer to follow Walker (1966) in treating them under a single genus. The status of some species is problematic (Donnelly, 1992b; 2003; May, 1995; Needham, Westfall, and May, 2000), and the identification of females should be made with caution when not associated with males.

Potential for new species: Possible.

Habitat: Ponds, lakes, marshes, swamps, temporary ponds and ditches, slow streams and rivers, with clear or muddy water. Perch obliquely on twigs or stems or hanging from twigs. Often forage over tree tops, and can form feeding swarms. Males hover periodically while patrolling. Female holds an egg-mass with bifid vulvar lamina, which is released with a double dip on a submerged plant, unfolding in the water into a gelatinous rope (Dunkle, 2000). Mating starts in flight and is completed on a perch. Several aspects of its biology have been studied, *i.e.* oviposition (Williamson, 1905), temporal isolation (Paulson, 1973), hatching pattern (Tennessen and Murray, 1978; Tennessen, 1979), seasonal regulation (Lutz, 1963a; b; 1970; Lutz and Jenner, 1964), and temperature regulation (May, 1987).

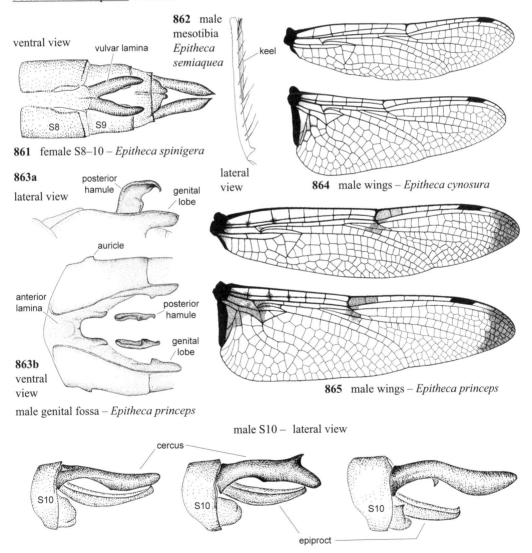

862 male mesotibia *Epitheca semiaquea*

keel

ventral view

vulvar lamina

S8   S9

861 female S8–10 – *Epitheca spinigera*

863a lateral view

posterior hamule

genital lobe

auricle

anterior lamina

posterior hamule

genital lobe

863b ventral view

male genital fossa – *Epitheca princeps*

lateral view

864 male wings – *Epitheca cynosura*

865 male wings – *Epitheca princeps*

male S10 – lateral view

cercus

S10

S10

S10

epiproct

866 *Epitheca sepia*   **867** *Epitheca spinosa*   **868** *Epitheca spinigera*

Gomphomacromia Brauer, 1864: 163.

[♂ pp. 145, couplet 2; ♀ pp. 149, couplet 2]

Type species: *Gomphomacromia paradoxa* Brauer, 1864 [by monotypy]

4 species:

chilensis Martin, 1921*
   syn *mexicana* Needham, 1933
fallax McLachlan, 1881*
nodisticta Ris, 1928*
paradoxa Brauer, 1864* − **L** [Theischinger and Watson, 1984]
   syn *etcheverryi* Fraser, 1957
   syn *chilensis (nomen nudum)* (Hagen, 1861) [*Cordulia*]
   syn var. *effusa* Navás, 1918
   syn var. *tincta* Navás, 1918

[NOTE: The name *Gomphomacromia fuliginosa* Martin, 1921 was introduced in error by Bridges (1994), Steinmann (1997), and Tsuda (2000), since Martin's (1921) record was of *Gomphoides fuliginosa*.]

References: von Ellenrieder and Garrison, 2005a.
Distribution: Ecuador to S Chile and Argentina along the Andes.

Map 68. Distribution of *Gomphomacromia* spp.

Small corduliines (35–42 mm); reddish brown to black, usually with yellow spots or stripes on pterothorax, and yellow spots on abdomen. Wings hyaline or with small basal brown spots; supratriangle usually free; subtriangle of HW usually present (absent in some *G. paradoxa*); anal loop polygonal, lacking sole and midrib (Figs. 869-870, shared with *Lauromacromia*). Male tibial keels absent or very short in middle legs (Fig. 871), well developed on fore and hind legs; male S7-9 widened (Fig. 873); male auricles with numerous small denticles on inner surface (Fig. 872a). Vulvar lamina projected caudally beyond tip of cerci; female S10 prolonged caudally to about tip of cerci (Figs. 877a-b, shared with *Navicordulia*). **Unique characters**: Postero-ventral corners of tergum 1 with denticulate ventral projections (Figs. 872a-b; absent in *G. fallax*); posterior margin of anterior lamina (Fig. 872b) with laminar lateral projections (possibly fused anterior hamules); genital lobe absent; large posterior hamule with a medio-ventral finger-like projection (Figs. 872a-b); distal segment of vesica spermalis with one dorso-apical cornu, bifid at tip (Figs. 876a-b); male epiproct quadrangular, about as wide as long in ventral view (Fig. 875). Vulvar lamina distal 1/2 divided into two flap-like rectangular or cylindrical lobes; female sternum S9 with two caudally directed cylindrical or ribbon-shaped projections, the lateral gonapophyses, approximately as long as cerci and surpassing the posterior margin of S10 (Figs. 877a-b).

Status of classification: Good. Well-diagnosed species.

Potential for new species: Likely especially in northern portion of distribution range.

Habitat: Mountain streams, seepages and bogs. In *G. nodisticta* and *G. paradoxa*, males defend small territories from other males, patrolling stream margins or forests paths with a low flight, occasionally perching on stones or low vegetation; mating pairs land on vegetation along stream margins. *G. paradoxa* is an exception among cordulines by being found in large numbers. At least some of the species in this genus seem to have semi-terrestrial larvae; larvae were found under stones at a distance of about three meters from a moist, moss cover rocky area in Osorno province, Chile, and on a moist, moss covered slope on a dirt trail in Pakitza, Peru (von Ellenrieder and Garrison, 2005a).

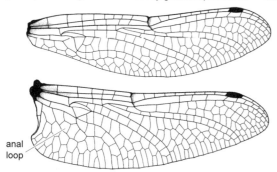

anal loop

869 male wings − *Gomphomacromia nodisticta*

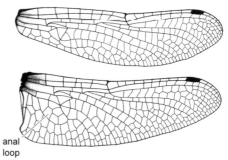

anal loop

870 male wings − *Gomphomacromia paradoxa*

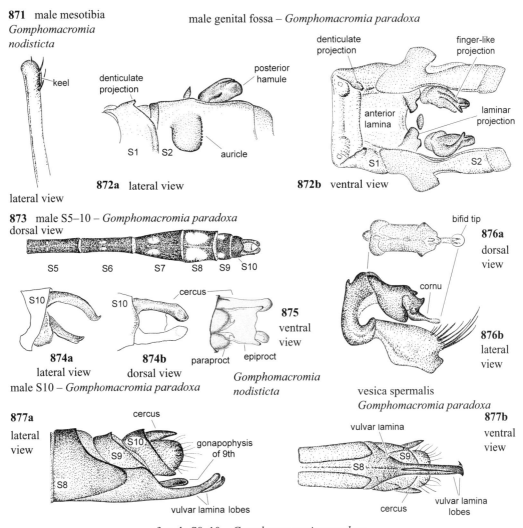

871 male mesotibia
*Gomphomacromia
nodisticta*

keel

lateral view

male genital fossa – *Gomphomacromia paradoxa*

denticulate
projection

posterior
hamule

S1   S2   auricle

872a lateral view

denticulate
projection

finger-like
projection

anterior
lamina

laminar
projection

S1   S2

872b ventral view

873 male S5–10 – *Gomphomacromia paradoxa*
dorsal view

S5   S6   S7   S8   S9   S10

bifid tip

876a
dorsal
view

cornu

876b
lateral
view

S10

S10

cercus

875
ventral
view

paraproct   epiproct

874a   **874b**
lateral view   dorsal view

male S10 – *Gomphomacromia paradoxa*

*Gomphomacromia
nodisticta*

vesica spermalis
Gomphomacromia paradoxa

877a
lateral
view

cercus

S10

S9

S8

gonapophysis
of 9th

vulvar lamina lobes

vulvar lamina

S9

S8

cercus

877b
ventral
view

vulvar lamina
lobes

female S8–10 – *Gomphomacromia paradoxa*

Helocordulia Needham, 1901: 484, 495.
[♂ pp. 147, couplet 10; ♀ pp. 150, couplet 6]

Type species: *Cordulia uhleri* Selys, 1871 [Mutt-
kowski, 1910 by subsequent designation]
2 species:

selysii (Hagen *in* Selys, 1878) [*Cordulia*?]* – **L**
[Kennedy, 1924; Wright, 1946c; Louton, 1982b]
uhleri (Selys, 1871) [*Cordulia*]* – **L** [Needham,
1901; Wright, 1946c; Louton, 1982b]

References: Needham, Westfall, and May, 2000.
Distribution: E Canada and United States.

Medium coruliines (38–46 mm); dark brown, with
orange markings on abdomen; body hairy, especially
pterothorax. Wings (Fig. 878) hyaline with small
dark spots at base and at antenodal crossveins; anal
loop with well-developed midrib and sole; Mspl in-

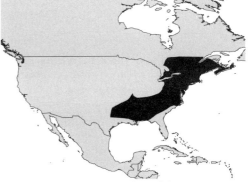

Map 69. Distribution of *Helocordulia* spp.

distinct or weakly developed. Male tibial keels on all
legs. Posterior hamule caudally bent (Fig. 881); distal
segment of vesica spermalis with two apical flagella.

Vulvar lamina bifid and about as long as 1/3-1/2 of sternum S9. **Unique characters**: Dorsal tip of posterior hamule with a row of small denticles (Fig. 881). Status of classification: Good species resolution. This genus is similar to *Epitheca*, and diagnostic wing characters given in previous keys (Needham and Westfall, 1955; Needham, Westfall, and May, 2000) are incorrect; in addition to the unique character mentioned here, males can be distinguished by the external carina on their posterior hamule (Fig. 881), which is lacking in *Epitheca* (Fig. 863a).
Potential for new species: Unlikely.
Habitat: Clean sand-bottomed streams, rivers, and lakes in forested areas. Males patrol along stream banks hovering briefly, perching occasionally on stems or on the ground.

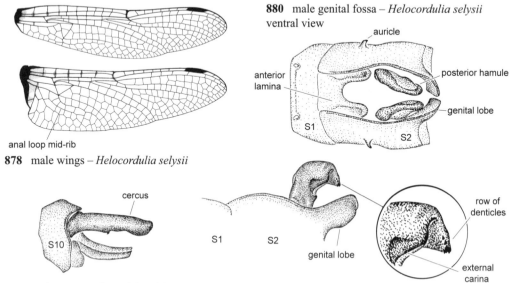

878 male wings – *Helocordulia selysii*

anal loop mid-rib

880 male genital fossa – *Helocordulia selysii* ventral view

auricle
anterior lamina
posterior hamule
genital lobe
S1
S2

cercus
S10
879 male S10 – *Helocordulia uhleri* lateral view

S1
S2
genital lobe
row of denticles
external carina
881 male genital fossa – *Helocordulia uhleri* lateral view

Lauromacromia Geijskes, 1970: 9.
[♂ pp. 145, couplet 2; ♀ pp. 149, couplet 2]
Type species: *Gomphomacromia dubitalis* Fraser, 1939 [by original designation]
5 species:

bedei Machado, 2005
dubitalis (Fraser, 1939) [*Gomphomacromia*]** – **L** [Fleck, 2002a]
luismoojeni (Santos, 1967) [*Neocordulia*]
flaviae Machado, 2002
picinguaba Carvalho et al., 2004

References: Geijskes, 1970; May, 1992b; Machado, 2002, 2005d; Carvalho, Salgado, and Werneck-de-Carvalho, 2004.
Distribution: Venezuela and French Guiana south to Brazil.

Medium corduliines (about 49-54 mm); pterothorax brown with yellow lateral stripes, metallic green iridescence and copper or blue reflections on intervening areas; abdomen dark brown with yellow markings. Wings (Fig. 882) hyaline, tinged with golden brown basally; sectors of arculus in FW stalked

Map 70. Distribution of *Lauromacromia* spp.

(shared with *Aeschnosoma*) or not (in *L. picinguaba*); FW discoidal field widening distally; anal loop polygonal, lacking sole and midrib (shared with *Gomphomacromia*). Male tibial keels on all legs; male auricles bare or with numerous small denticles on inner surface (shared with *Gomphomacromia*, as in Fig. 872a); sternum S8 with a protuberance near its base (shared with several *Neocordulia*, as in Figs. 894a-b); male dorsum of S10 with a prominent spine or keel (Fig. 884a, shared with some *Neocordulia*). Triangular genital lobe ending in a tubercle (Fig. 883, shared with *Paracordulia*); posterior hamule entire; basal segment of vesica spermalis rounded (shared only with *Neocordulia* among New World cordulines, as in Fig. 895); distal segment of vesica spermalis with a pair of very long apical flagella. Vulvar lamina bifid, shorter than 1/3 of sternum S9 and not projected ventrally. **Unique characters**: None known.

Status of classification: Good species resolution.
Potential for new species: Likely. Rare species of secretive habits in forests.
Habitat: The holotype male of *L. luismoojeni* was found perching on forest understory at the margin of a small river at 10 am (Santos, 1967b), and that of *L. flavia* hovering over the narrow, rock-bottomed side stream of a river with remnants of a gallery forest at 11 am (Machado, 2002). The last larval instar was described by supposition by Fleck (2002a) based on a larva from French Guiana.

882 male wings – *Lauromacromia dubitalis*

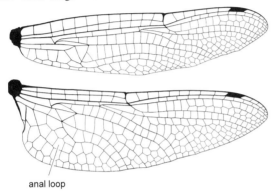

anal loop

883 male genital fossa – *Lauromacromia dubitalis*
lateral view

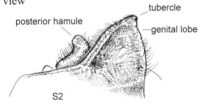

posterior hamule

tubercle

genital lobe

S2

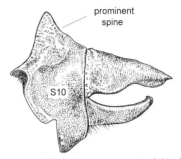

prominent spine

S10

884a male S10 – *Lauromacromia dubitalis*
lateral view

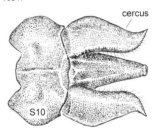

cercus

S10

884b male S10 – *Lauromacromia dubitalis*
dorsal view

Navicordulia Machado and Costa, 1995: 188.
[♂ pp. 146, couplet 4; ♀ pp. 150-151, couplets 8, 13]
Type species: *Dorocordulia errans* Calvert, 1909 [by original designation]
10 species:

amazonica Machado & Costa, 1995
atlantica Machado & Costa, 1995
errans (Calvert, 1909) [*Dorocordulia*]*
kiautai Machado & Costa, 1995
leptostyla Machado & Costa, 1995*
longistyla Machado & Costa, 1995*
mielkei Machado & Costa, 1995
miersi Machado & Costa, 1995
nitens De Marmels, 1991 [*Dorocordulia*] – **L** [De Marmels, 1991b]
vagans De Marmels, 1989 [*Dorocordulia*]

Map 71. Distribution of *Navicordulia* spp.

References: Machado and Costa, 1995.
Distribution: Tepuis in Venezuela south to SE Brazil and Paraguay.

Medium corduliines (36-47.5 mm); brown with metallic green, blue or copper reflections. Wings (Figs. 887a-b) hyaline or tinged with brown or yellow in females; triangles usually free (crossed in FW of *N. nitens* and *N. atlantica*); supratriangles free; subtriangles absent in HW (except for *N. miersi*); anal loop elongated with well-developed sole and midrib. Male tibial keels short or absent on middle legs, well developed on fore and hind legs. Posterior hamule caudally directed from its base, long, straight, with recumbent, pointed tip directed externally (Figs. 886a-b); distal segment of vesica spermalis with two apical flagella. Vulvar lamina entire, projected distally beyond tip of cerci; female S10 prolonged caudally to about tip of cerci (Figs. 890a-b). **Unique characters**: Male HW anal margin excavated posterior to distal end of anal triangle (Fig. 887b); sternum S7 of males with a pilose complex, consisting of a medio-longitudinal

ridge and one or two transverse carinae or protuberances associated with a medial and two lateral patches of hairs (Fig. 888); female sternum S9 entire, tongue, scoop or dish-shaped and projected caudally beyond tip of cerci (Figs. 890a-b).

Status of classification: Well defined species.

Potential for new species: Likely. *Navicordulia errans* and *N. leptostyla* from the Brazilian Cerrado (savannah) are abundant in the field; species from forested areas are elusive, and new species are likely to be found in the largely unexplored Amazonian forest.

Habitat: Roosting on tree canopies or bushes away from water; breeding in swamps and rivers (Machado and Costa, 1995), or in rock-bottomed forest pools with tannin rich-water (De Marmels, 1991b). The long vulvar lamina seems to indicate that oviposition occurs by digging eggs into mud or sediments in shallow water bodies (Machado and Costa, 1995).

885 head – *Navicordulia errans*
dorsal view

vertex

occiput

886a
ventral
view

auricle

anterior
lamina

posterior hamule

recumbent
tip directed
externally

886b
lateral
view

genital
lobe

male genital fossa – *Navicordulia errans*

887a male wings – *Navicordulia longistyla*

anal loop

triangle
margin
at angle

anal
margin
excavated

887b male HW base
Navicordulia longistyla

888
male S8

ventral
view

sternum
S8

pilose
complex

Navicordulia errans

889 male S10
Navicordulia errans

cercus

S10

epiproct

lateral
view

890a
lateral
view

cercus

S10

S9

S8

vulvar lamina

female S8–10
Navicordulia errans

S10

S8   S9

cercus

vulvar lamina

890b ventral view

Neocordulia Selys, 1882: 169 (6 reprint).
[♂ pp. 146, couplet 6; ♀ pp. 151, couplet 17]

Type species: *Gomphomacromia androgynis* Selys, 1871 [Cowley, 1934c by subsequent designation]
  syn. *Neocordulia* (*Mesocordulia*) May, 1992b: 24
    Type species: *Gomphomacromia batesi* Selys, 1971 [May, 1992b by original designation]
10 species:

androgynis (Selys, 1871) [*Gomphomacromia*] – **L** [Costa and Santos, 2000]
batesi batesi (Selys, 1871) [*Gomphomacromia*]*
batesi longipollex Calvert, 1909* – **L** [Novelo-Gutiérrez and Ramírez, 1995]
biancoi Rácenis, 1970* – **L** [De Marmels, 1990a]
campana May & Knopf, 1988*
carlochagasi Santos, 1967
griphus May, 1992*
mambucabensis Costa & Santos, 2000
matutuensis Machado, 2005
setifera (Hagen *in* Selys, 1871)
[*Gomphomacromia*]* – **L** [Costa and Santos, 2000]
  syn *valga (nomen nudum)* (Hagen, 1861) [*Cordulia*]

Map 72. Distribution of *Neocordulia* spp.

volxemi (Selys, 1874) [*Gomphomacromia*]

References: May, 1992b; Machado, 2005b.
Distribution: S Mexico south to Paraguay and SE Brazil.

Medium corduliines (43-58.5 mm); brown with metallic green and violet reflections on pterothorax. Wings (Fig. 892) hyaline or tinged with golden or brown; FW discoidal field parallel sided or narrowing distally; anal loop elongate and with midrib, lacking sole (shared only with some *Williamsonia* specimens). Male tibial keels on all legs (Fig. 891); sternum S8 with biconical process at base (Figs. 894a-b) in the species of the subgenus *Neocordulia* (shared with *Lauromacromia*, absent in the species of the subgenus *Mesocordulia*: *N. batesi*, *N. campana* and *N. griphus*); male dorsum of S10 with a prominent spine or keel (Fig. 894a. shared with *Lauromacromia* and *Paracordulia*) or approximately flat. Basal segment of vesica spermalis rounded (shared only with *Lauromacromia* among New World cordulines); distal segment of vesica spermalis with one apical flagellum (Fig. 895). Vulvar lamina bifid and less than 1/3 of sternum S9. **Unique characters**: Posterior margin of anterior lamina with medial arched rim bearing a tuft of setae on each lateral end, which gives the margin a tri-concave contour (Fig. 893a); posterior hamule bifid (Fig. 893a).
Status of classification: Good species resolution.
Potential for new species: Likely as most species are rare.
Habitat: Specimens collected patrolling along dirt roads in secondary forest and in forest trails. Breeding of known larvae occurs in sand-bottomed mountain streams of clear, slow-moving water (Costa and Santos, 2000), small creeks and ditches within forest (De Marmels, 1990a).

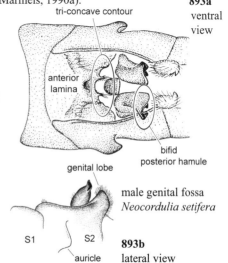

893a ventral view

tri-concave contour

anterior lamina

bifid posterior hamule

genital lobe

male genital fossa
Neocordulia setifera

S1   S2

auricle

893b lateral view

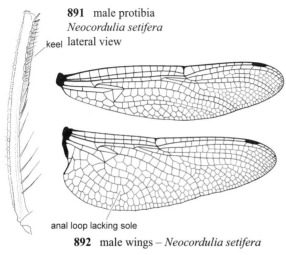

891 male protibia
Neocordulia setifera
keel   lateral view

anal loop lacking sole

892 male wings – *Neocordulia setifera*

894a latero-dorsal view

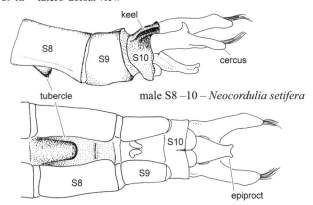

male S8 –10 – *Neocordulia setifera*

894b ventral view

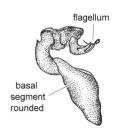

895 vesica spermalis
Neocordulia setifera
lateral view

Neurocordulia Selys, 1871a: 278 (44 reprint).
[♂ pp. 148, couplet 16; ♀ pp. 152, couplet 19]
Type species: *Libellula obsoleta* Say, 1840 [by monotypy]
    syn. *Platycordulia* Williamson, 1908: 431
        Type species: *Platycordulia xanthosoma* Williamson, 1908 [by original designation]
[NOTE: *Rostrocordulia* Hodges *in* Needham and Westfall, 1955: 355, was mentioned as a subgenus but that name is a *nomen nudum*.]
7 species:

alabamensis Hodges *in* Needham & Westfall,
1955* – **L** [Hodges *in* Needham and Westfall, 1955;
Louton, 1982b]
michaeli Brunelle, 2000* – **L** [Brunelle, 2000]
molesta (Walsh, 1863) [*Cordulia*]* – **L** [Needham
and Westfall, 1955; Louton, 1982b]
    syn *clara* Muttkowski, 1910
obsoleta (Say, 1840) [*Libellula*]*– **L** [Needham,
1901; Louton, 1982b; Westfall and Tennessen, 1996]
    syn *polysticta* (Burmeister, 1839) [*Libellula*]
virginiensis Davis, 1927* – **L** [Byers, 1937; Louton,
1982b]
xanthosoma (Williamson, 1908) [*Platycordulia*]*
– **L** [Williams and Dunkle, 1976; Louton, 1982b]
yamaskanensis (Provancher, 1875) [*Aeshna*]* – **L**
[Walker, 1913; Louton, 1982b]
    syn *jamascurensis* (Hagen *in* Selys, 1878)
    [*Epitheca*?]

References: Needham, Westfall, and May, 2000.
Distribution: SE Canada to SE United States.

Medium corduliines (40–55 mm); pale brown with
yellowish and orange markings lacking metallic reflections. Wings (Fig. 896) hyaline or tinged with reddish brown, with variable extension of dark macula-

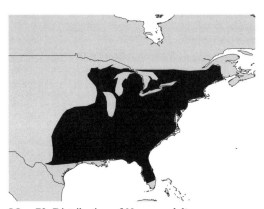

Map 73. Distribution of *Neurocordulia* spp.

tion (at base, antenodal crossveins, nodus, postnodal
crossveins); FW discoidal field widening distally;
anal loop with well-developed midrib and sole; 3-6
bridge crossveins; 2-3 cubito-anal crossveins in FW.
Male tibial keels on all legs, limited to apical tibial
tip in some species; sub-basal tooth on male cercus
present or absent (Fig. 899). Posterior hamule caudally bent, with pointed tip (Figs. 897a-b); distal
segment of vesica spermalis with two apical flagella
(Fig. 898). Vulvar lamina bifid and shorter than 1/3
of sternum S9. **Unique characters**: Distal segment
of vesica spermalis with a dorso-basal flap directed
anteriorly (Fig. 898).
Status of classification: Good species resolution.
Potential for new species: Likely due to elusive habits.
Habitat: Perch hanging from twigs in shaded areas of
forests during the day; active over water (clean forest
streams, rivers and lakes) during short periods of time
at dusk and dawn. Larvae inhabit large turbulent rivers, wave-beaten areas of large reservoirs, and spring-
fed streams (Louton, 1982b).

896 male wings – *Neurocordulia molesta*

male genital fossa – *Neurocordulia obsoleta*

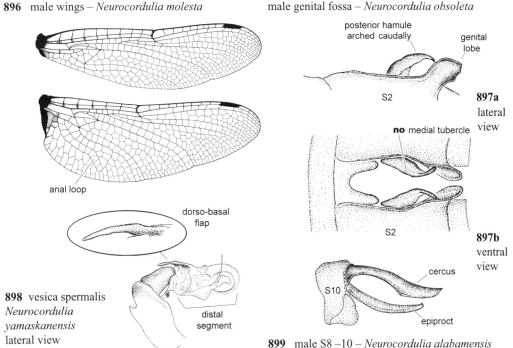

anal loop

dorso-basal flap

898 vesica spermalis
Neurocordulia yamaskanensis
lateral view

distal segment

posterior hamule arched caudally

genital lobe

S2

897a lateral view

no medial tubercle

S2

897b ventral view

cercus

S10

epiproct

899 male S8 –10 – *Neurocordulia alabamensis*
lateral view

Paracordulia Martin, 1907: 11, 33.
[♂ pp. 146, couplet 7; ♀ pp. 151, couplet 16]
Type species: *Cordulia sericea* [Cowley, 1934c by subsequent designation]
1 species:

sericea (Selys, 1871) [*Cordulia*]**

References: Geijskes, 1970.
Distribution: Surinam and Venezuela south to

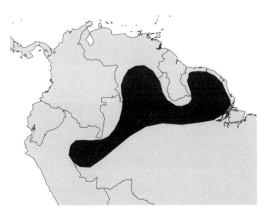

Map 74. Distribution of *Paracordulia* sp.

Amazonian region in Peru and Brazil.

Medium corduliines (45–50 mm); brown to black, with metallic greenish-blue reflections on frons and vertex, metallic green on pterothorax and copper on abdomen, lateral pale spots on last three segments of male abdomen. Wings (Fig. 901) hyaline or with yellow streaks in costal and cubital spaces; FW discoidal field narrowing to margin; anal loop elongate with midrib and sole well-developed. Vertex with prominent tubercles (Fig. 900, shared with *Santosia*); male tibial keels on all legs; male genital lobe ending in a tubercle (Fig. 903, shared with *Lauromacromia*); posterior hamule entire; vulvar lamina about as long as 1/2 of sternum S9 length. **Unique characters**: Vulvar lamina triangular; female sternum S9 projected distally to about posterior margin of S10 (Fig. 902a).
Status of classification: Very rare in collections, known so far from only six specimens. More than one species could be included under *P. sericea*, the known Venezuelan specimen probably representing a different species (*Paracordulia* sp. 2 of De Marmels, 1983).
Potential for new species: Likely due to secretive habits in forests.
Habitat: Flies swiftly and low over streams and ground in densely vegetated areas; female oviposits by dipping the abdomen on the water surface in the middle of creeks in shadowed areas (Geijskes, 1970; Rácenis, 1970).

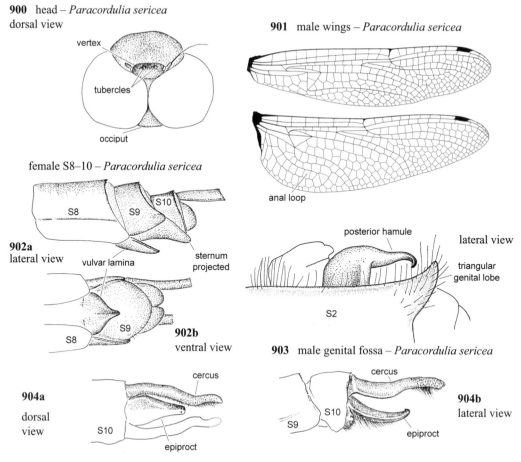

900   head – *Paracordulia sericea*
dorsal view

901   male wings – *Paracordulia sericea*

female S8–10 – *Paracordulia sericea*

902a
lateral view

902b
ventral view

903   male genital fossa – *Paracordulia sericea*

904a
dorsal
view

904b
lateral view

male S10 – *Paracordulia sericea*

Rialla Navás, 1915: 148.
[♂ pp. 147-148, couplets 14, 16; ♀ pp. 151-152, couplets 15-16, 19]
Type species: *Rialla membranata* Navás, 1915 [by
original designation]
   syn *Anticordulia* Needham & Bullock, 1943: 367
      Type species: *Cordulia villosa* Rambur, 1842
      [by original designation]
1 species:

villosa (Rambur, 1842) [*Cordulia*]* – **L** [Needham
and Bullock, 1943]
   syn *membranata* Navás, 1915

References: Needham and Bullock, 1943.
Distribution: S Chile and Argentina.

Map 75. Distribution of *Rialla* sp.

Medium corduliines (44–50 mm); eyes emerald green
in life; pale brown with metallic greenish-blue reflections in pterothorax and black margins and lateral yellow spots on abdominal segments; body hairy, especially pterothorax, which is covered with long golden

hairs. Wings (Fig. 906) hyaline or tinged with golden
brown; sectors of arculus in FW stalked (shared with
Aeschnosoma and *Lauromacromia*); triangle usually
crossed (occasionally free); supratriangle usually free
(occasionally crossed); 1-2 cubito-anal crossveins in

FW; FW discoidal field narrowing to wing margin; anal loop with well-developed midrib and sole. Occipital triangle bulging and as large as vertex (Fig. 905); male tibial keels on all legs; abdomen wider than high and triquetal; sub-basal tooth on male cercus present. Genital lobe lower than posterior hamule and bent caudally at a straight angle (Fig. 907a, shared only with some species of *Epitheca*); posterior hamule curved caudally from its base and ending in a recumbent tip directed dorsally (Fig. 907b); distal segment of vesica spermalis with two apical flagella

and a dorso-basal cornu (Figs. 908a-b). Vulvar lamina about as long as 1/2 of sternum S9 and bifid. **Unique characters**: None known.

Status of classification: Well defined species.

Potential for new species: Unlikely.

Habitat: Ponds and lakes with open water in antarctic forest and steppe areas, where males patrol following the banks with a swift and straight flight close over water. Copulation begins in flight and the mating pair lands in tree-tops (Jurzitza, 1975); adults perch occasionally on grasses or bushes, usually hanging vertically (Jurzitza, 1989).

905 head – *Rialla villosa*

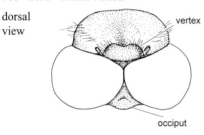

dorsal view

906 male wings – *Rialla villosa*

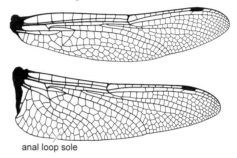

anal loop sole

907a lateral view

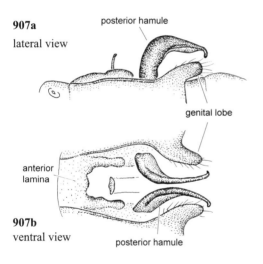

posterior hamule

genital lobe

anterior lamina

907b ventral view

posterior hamule

male genital fossa – *Rialla villosa*

908a lateral view

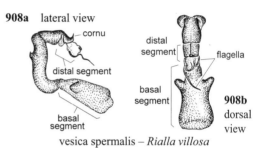

cornu

distal segment

basal segment

distal segment

flagella

basal segment

908b dorsal view

vesica spermalis – *Rialla villosa*

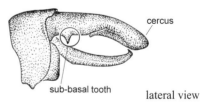

cercus

sub-basal tooth

lateral view

909 male S10 – *Rialla villosa*

Santosia Costa & Santos, 1992: 236.
[♂ pp. 147, couplet 14; ♀ pp. 151, couplet 15]
Type species: *Santosia marshalli* Costa & Santos, 1992 [by original designation]
3 species:

machadoi Costa & Santos, 2000 – **L** [Costa and Santos, 2000]
marshalli Costa & Santos, 1992
newtoni Costa & Santos, 2000 – **L** [Costa and Santos, 2000]

References: Costa and Santos 1992; 2000.
Distribution: SE Brazil.

Medium corduliines (52–57 mm); dark brown with metallic blue or green reflections. Wings (Fig. 910) hyaline; anal loop elongated with well developed midrib and sole. Vertex with prominent tubercles (shared with *Paracordulia*, as in Fig. 900); male tibial keels well developed on all legs; posterior ham-

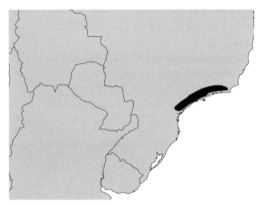

Map 76. Distribution of *Santosia* spp.

ule entire and bent caudally, with pointed tip; distal
segment of vesica spermalis with two apical flagella.
Unique characters: None known.

Status of classification: Good. Female still unknown.
Potential for new species: Likely. Genus extremely
rare in collections, so far known from only 10 specimens.
Habitat: Small marshy areas associated with streams
of clear waters in southeastern Brazilian hills, between 800 and 2400 m (Costa and Santos, 2000).

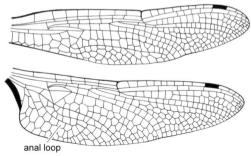

anal loop

910   male wings – *Santosia marshalli*

Somatochlora Selys, 1871a: 279.
[♂ pp. 147, couplet 12; ♀ pp. 150, couplet 11]
Type species: *Libellula metallica* Vander Linden,
1825 [Muttkowski, 1910 by subsequent designation]
43 species; 26 New World species:

albicincta (Burmeister, 1839) [*Epophthalmia*]* – **L**
[Walker, 1925]
    syn *eremita* (Scudder, 1866) [*Cordulia*]
    syn form *massettensis* Whitehouse, 1941
brevicincta Robert, 1954*
calverti Williamson & Gloyd, 1933*
cingulata (Selys, 1871) [*Epitheca*]* – **L** [Walker,
1925]
elongata (Scudder, 1866) [*Cordulia*]* – **L**
[Needham, 1903 as *Somatochlora*? sp. 2]
    syn *saturata (nomen nudum)* (Hagen, 1861)
    [*Cordulia*]
ensigera Martin, 1907* – **L** [Needham, Westfall,
and May, 2000]
    syn *charadraea* Williamson, 1907
filosa (Hagen, 1861) [*Cordulia*]* – **L** [Dunkle,
1977a]
forcipata (Scudder, 1866) [*Cordulia*]* – **L** [Walker,
1925]
    syn *chalybea (nomen nudum)* (Hagen, 1861)
    [*Cordulia*]
franklini (Selys, 1878) [*Epitheca*]* – **L** [Walker,
1925]
    syn *macrotona* Williamson, 1909
georgiana Walker, 1925* – **L** [Daigle, 1994]
hineana Williamson, 1931* – **L** [Cashatt and Vogt,
2001]

hudsonica (Hagen *in* Selys, 1871) [*Epitheca*]* – **L**
[Walker, 1925]
incurvata Walker, 1918*
kennedyi Walker, 1918* – **L** [Walker, 1925]
linearis (Hagen, 1861) [*Cordulia*]* – **L** [Walker,
1925]
    syn *lateralis (lapsus)* Needham, 1901
    syn *procera* (Selys, 1871) [*Epitheca*]
margarita Donnelly, 1962*
minor Calvert, 1898* – **L** [Walker, 1925]
ozarkensis Bird, 1933* – **L** [Pritchard, 1936]
provocans Calvert, 1903* – **L** [Tennessen, 1975]
sahlbergi Trybom, 1889* – **L** [Cannings and
Cannings, 1985]
    syn *walkeri* Kennedy, 1917
semicircularis (Selys, 1871) [*Epitheca*]* – **L**
[Walker, 1925]
    syn *nasalis* (Selys, 1874) [*Epitheca*]
septentrionalis (Hagen, 1861) [*Cordulia*]* – **L**
[Whitehouse, 1941]
    syn *richardsoni (nomen nudum)* (Hagen, 1861)
    [*Cordulia*]
tenebrosa (Say, 1840) [*Libellula*]* – **L** [Walker,
1925]
    syn *tenebrica (nomen nudum)* (Hagen, 1861)
    [*Cordulia*]
walshii (Scudder, 1866) [*Cordulia*]* – **L** [Walker,
1941]
whitehousei Walker, 1925* – **L** [Walker, 1925]
williamsoni Walker, 1907* – **L** [Needham, 1901 as
S. elongata; Walker and Corbet, 1975]

References: Walker, 1925; Needham, Westfall, and

May, 2000.

Distribution: Holarctic; in nearctic region ranging from Canada south to NE United States.

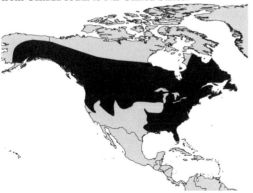

Map 77. Distribution of *Somatochlora* spp.

Medium, slender corduliines (39–68 mm); with bright green eyes in life, dark brown to black with metallic bluc frons and vertex, metallic green reflections on pterothorax and dorsum of basal abdominal segments, and yellow spots on abdomen and sometimes on pterothorax; body hairy, especially pterothorax. Wings (Fig. 912) hyaline or with dark basal spot, in females usually tinged with brown; anal loop with well developed midrib and sole. Male tibial keels on fore and hind legs, on middle legs absent or rarely very short at apical end (in *S. elongata* and *S. linearis*, Fig. 911); sub-basal tooth on male cercus (Figs. 916-918) present or absent. Posterior hamule bent caudally from its base (Figs. 914-915); distal segment of vesica spermalis with two apical flagella. Vulvar lamina spoon or sprout shaped and directed ventrally (Figs. 919, 921), or bifid and not directed ventrally (Figs. 920, 922), varying in length from less than 1/3 to slightly longer than sternum S9. **Unique characters**: None known.

Status of classification: Good species resolution.

Potential for new species: Possible due to secretive habits.

Habitat: Most species rare or uncommon within forests, secretive or high flying, breeding in small bog pools, mossy sedge fens, slow streams, ponds and lakes. Some species apparently gather on hill tops for mating (Dunkle, 2000); mating usually takes place on a tree or bush. One species, *S. hineana*, is included in the Federal Endangered Species List of the United States, but recent studies indicate that the species is more widespread than previously thought (USFWS, 2001).

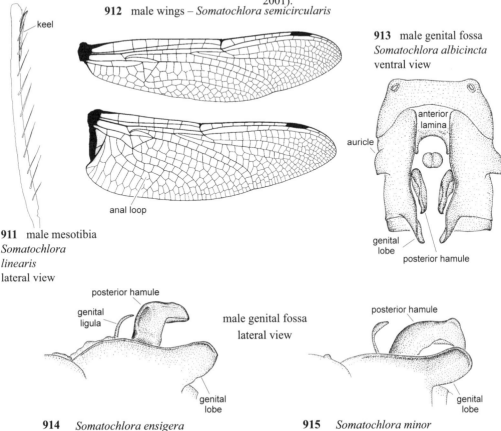

912 male wings – *Somatochlora semicircularis*

keel

anal loop

911 male mesotibia
*Somatochlora
linearis*
lateral view

913 male genital fossa
Somatochlora albicincta
ventral view

anterior lamina

auricle

genital lobe

posterior hamule

posterior hamule

genital ligula

male genital fossa
lateral view

genital lobe

914 *Somatochlora ensigera*

posterior hamule

genital lobe

915 *Somatochlora minor*

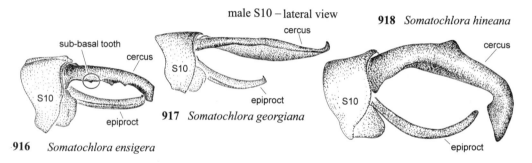

male S10 – lateral view

sub-basal tooth
cercus
S10
epiproct

916 *Somatochlora ensigera*

cercus
S10
epiproct

917 *Somatochlora georgiana*

918 *Somatochlora hineana*

cercus
S10
epiproct

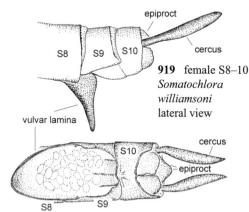

epiproct
S8 S9 S10
cercus

919 female S8–10
Somatochlora williamsoni
lateral view

vulvar lamina

cercus
S10
epiproct
vulvar lamina
S8 S9

921 female S8–10 – *Somatochlora elongata*
ventral view

920 female S8–10 – *Somatochlora linearis*
lateral view

S8 S9 S10
cercus
vulvar lamina

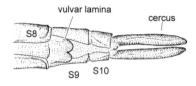

vulvar lamina
cercus
S8
S9 S10

922 female S8–10 – *Somatochlora semicircularis*
ventral view

Williamsonia Davis, 1913: 95.
[♂ pp. 146, couplet 5; ♀ pp. 150-151, couplets 9, 17]
Type species: *Cordulia lintneri* Hagen *in* Selys, 1878
[by original designation]
2 species:

fletcheri Williamson, 1923* – **L** [Charlton and
Cannings, 1993]
  syn *vacua (nomen nudum)* (Hagen, 1867)
  [*Diplax*]
lintneri (Hagen *in* Selys, 1878) [*Cordulia*]* – **L**
[White and Raff, 1970]

References: Needham, Westfall, and May, 2000.
Distribution: SE Canada to NE United States.

Small corduliines (29-35 mm); brown to black with
yellow spots on abdomen, no metallic reflections except on dorsum of head in mature males. Wings (Figs.
923a-b) hyaline, narrowly tinged with orange at base;
anal loop elongate with well-developed midrib, sole
distinct or not. Male tibial keels in fore and hind legs,
absent in middle legs. Posterior hamule bent caudally
and pointed (Figs. 924a-b); distal segment of vesica
spermalis with an apical flagellum (Fig. 925). Vulvar lamina bifid along distal 5/6 and almost as long

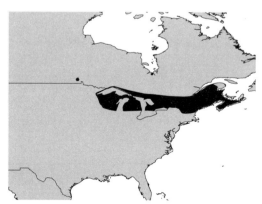

Map 78. Distribution of *Williamsonia* spp.

as sternum S9. **Unique characters**: Distal margin
of HW anal triangle more or less in line with anal
margin of HW (Fig. 923a); medial portion of anterior
lamina free margin costate (Fig. 924b); ventral margin of posterior hamule with a marked medial concavity (Fig. 924a).
Status of classification: Good species resolution.
Potential for new species: Unlikely.
Habitat: Rare and of local distribution in pool bogs
and acid fens, in or near forests. Adults perch on sunny spots on the ground, trunks, moss or dead branches

923b male wings – *Williamsonia litneri*

923a male HW base
Williamsonia litneri

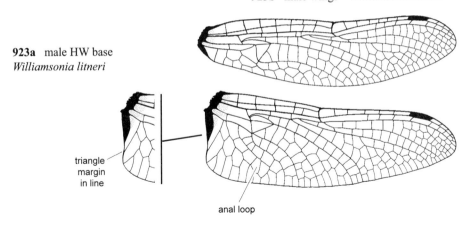

triangle
margin
in line

anal loop

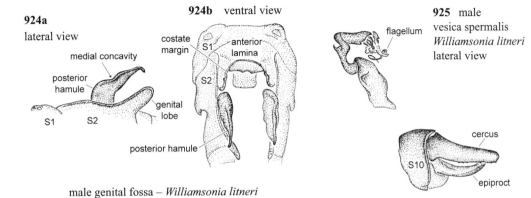

924a
lateral view

medial concavity

posterior
hamule

S1   S2

genital
lobe

924b ventral view

costate
margin

S1   anterior
lamina

S2

posterior hamule

male genital fossa – *Williamsonia litneri*

925 male
vesica spermalis
Williamsonia litneri
lateral view

flagellum

cercus

S10

epiproct

926 male S10 – *Williamsonia litneri*
lateral view

12. Libellulinae

Cosmopolitan: About 945 spp. in 143 genera.

New World: 393 spp. in 46 genera (38 endemic).

Diagnostic characters: Anal loop with well developed midrib and toe (Fraser, 1957) (*i.e.* Fig. 941, shared with most corduliines; indistinct in *Argyrothemis, Nannothemis, Dasythemis esmeralda, D. essequiba*, and some species of *Micrathyria* and *Oligoclada, i.e.* Fig. 940); second crossvein between RP1 and RP2 developed as oblique vein (Figs. 929-930, Bechly, 1996); posterior margin of compound eye evenly curved (Bechly, 1996) (except in *Nothodiplax*, Fig. 1008a); absence of male auricles on S2 (*i.e.* Figs. 1015-1016), and HW anal angle rounded (*i.e.* Figs. 937-941, Needham, Westfall, and May, 2000; shared with *Williamsonia*).

Status of classification: Phylogenetic relationships within this large subfamily are still unresolved (Bechly, 1996). Ris (1909a) was the first to subdivide it into ten groups in his key to genera although he provided no key to these groups. Tillyard (1917) proposed eight tribes (Tetrathemini, Libellulini, Palpopleurini, Brachydiplacini, Sympetrini, Leucorrhiniini, Trithemini, Trameini), all of which were later (Tillyard and Fraser, 1940) raised to subfamily (with Diastatopinae proposed as a replacement name for Palpopleurini and Zyxommatinae for Trameini), and included two more (Onychotheminae and Rhyotheminae) making a total of ten taxa. Two more (Zygonyctinae and Pantaliinae [replacement name for Trameini]) were added by Fraser (1957). These names have been loosely used throughout the literature either as tribes (*e.g.* Needham and Broughton, 1927) or as subfamilies (*e.g.* Davies and Tobin, 1985; Bridges, 1994; Steinmann, 1997). Bechly (1996) used 19 names to include all members of what we consider to constitute this subfamily. Pending a phylogenetic assessment for this group, we believe it unwise to formally use any of these names. Some genera (*Macrodiplax* and Old World genera *Aethriamanta, Selysiothemis*, and *Urothemis*) have been assigned to their own family, the Macrodiplactidae (Fraser, 1957), based on presence of strong oblique vein, open wing venation, and very few antenodal crossveins (Bechly, 1996), though most odonatologists do not recognize this taxon as a distinct family (these character states are observed in other libelluline genera as well, *i.e. Edonis* and *Nephepeltia*).

Key to males

1. Sectors of arculus in FW stalked (Fig. 927) 2

1'. Sectors of arculus in FW separated (Fig. 928) 85

927

sectors stalked

FW base – *Orthemis attenuata*

928

sectors separated

FW base – *Libellula herculea*

2(1). Last antenodal in FW complete (Fig. 929) 3

2'. Last antenodal in FW incomplete (Fig. 930) 33

929

last antenodal complete   oblique vein   **no** crossveins

RP1
RP2

FW – *Pachydiplax longipennis*   bridge crossvein

930

last antenodal incomplete   oblique vein   crossvein

RP1
RP2

bridge crossveins   FW – *Micrathyria aequalis*

3(2). Crossveins under proximal 1/2 of pterostigma in FW absent (Fig. 929) .... ***Pachydiplax*** (Page 265)

3'. Crossveins under proximal 1/2 of pterostigma in FW present (as in Fig. 930) 4

4(3). Cerci narrowly cylindrical throughout (Figs. 931a-b); distal segment of vesica spermalis with 2 laterodistal cornua ending in a recurved hook (Figs. 934a-b) .... ***Elga* in part** (Page 239)

4'. Cerci not narrowly cylindrical throughout (Figs. 932a-b-933); distal segment of vesica spermalis lacking 2 laterodistal cornua ending in a recurved hook (Figs. 935a-b, 936) **5**

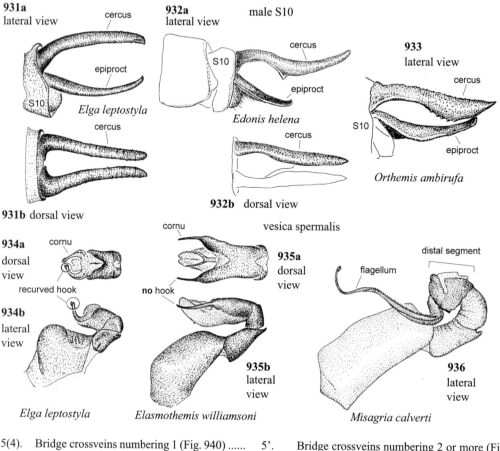

931a lateral view — cercus — epiproct — S10 — *Elga leptostyla*

931b dorsal view — cercus

932a lateral view — male S10 — S10 — cercus — epiproct — *Edonis helena* — cercus

932b dorsal view

933 lateral view — cercus — S10 — epiproct — *Orthemis ambirufa*

vesica spermalis

934a dorsal view — cornu — recurved hook

934b lateral view

Elga leptostyla

935a dorsal view — cornu — **no** hook

935b lateral view

Elasmothemis williamsoni

936 lateral view — distal segment — flagellum

Misagria calverti

5(4). Bridge crossveins numbering 1 (Fig. 940) **6**

5'. Bridge crossveins numbering 2 or more (Fig. 941) **25**

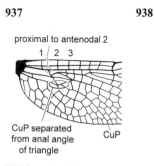

937 — proximal to antenodal 2 — 1 / 2 3 — CuP separated from anal angle of triangle — CuP

HW base – *Oligoclada rhea*

938 — closer to antenodal 2 than to 3 — 1 2 / 3

HW base – *Orthemis attenuata*

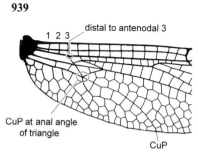

939 — distal to antenodal 3 — 1 2 3 — CuP at anal angle of triangle — CuP

HW base – *Misagria parana*

6(5). Arculus in HW proximal to antenodal 2 (Fig. 937) **7**

6'. Arculus in HW opposite to antenodal 2, or distal to 2 but closer to antenodal 2 than to 3 (Fig. 938) **18**

6". Arculus in HW from slightly proximal to distal to antenodal 3 (Fig. 939) .... ***Misagria* in part** (Page 258)

7(6). Costal side of FW triangle (**A**) less than 1/2 as long as proximal side (**B**) (Fig. 943) **8**

7'. Costal side of FW triangle (**A**) at least 1/2 as long as or a little longer than 1/2 of proximal side (**B**) (Fig. 942) **15**

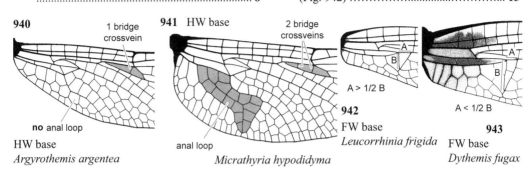

940

1 bridge crossvein

no anal loop

HW base
Argyrothemis argentea

941 HW base

2 bridge crossveins

anal loop

Micrathyria hypodidyma

A > 1/2 B

942
FW base
Leucorrhinia frigida

A < 1/2 B

943
FW base
Dythemis fugax

8(7). Anterior lamina in lateral view as high as 4/5 of hamule or higher (Fig. 944) **9**

8'. Anterior lamina in lateral view shorter than 4/5 of hamule (Fig. 945) **10**

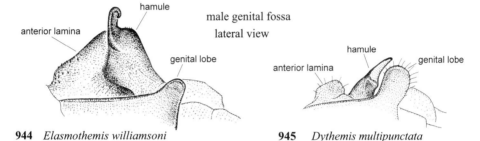

hamule

anterior lamina

genital lobe

male genital fossa
lateral view

hamule

anterior lamina

genital lobe

944 *Elasmothemis williamsoni*

945 *Dythemis multipunctata*

9(8). CuP in HW arising at or near anal angle of triangle (as in Fig. 939); posterior lobe of prothorax in dorsal view widest at base (as in Fig. 946), in lateral view bent caudally (as in Fig. 947b) .... ***Elasmothemis*** **in part** (Page 238)

9'. CuP in HW distinctly separated from anal angle of triangle (Fig. 937); posterior lobe of prothorax in dorsal view constricted at base (as in Fig. 947a), in lateral view upright (as in Fig. 948) .... ***Oligoclada*** **in part** (Page 262)

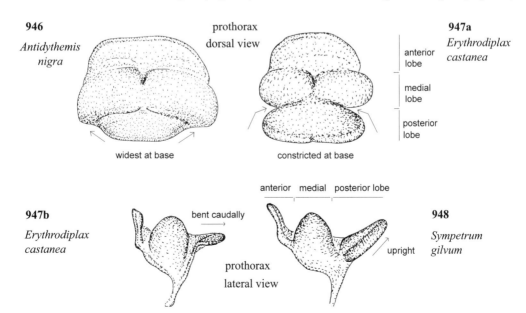

946
*Antidythemis
nigra*

prothorax
dorsal view

anterior
lobe

medial
lobe

posterior
lobe

947a
*Erythrodiplax
castanea*

widest at base

constricted at base

anterior   medial   posterior lobe

947b
*Erythrodiplax
castanea*

bent caudally

upright

948
*Sympetrum
gilvum*

prothorax
lateral view

10(8). Inner branch of hamule as large as or larger than outer branch (Fig. 949) **11**

10'. Inner branch of hamule smaller than outer branch (Fig. 950) **12**

male genital fossa

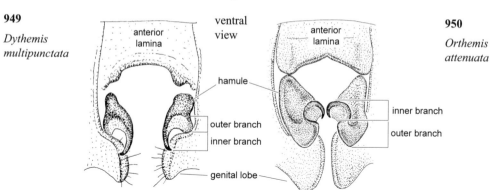

949

Dythemis multipunctata

950

Orthemis attenuata

11(10). Mspl in FW indistinct (Fig. 951); CuP in HW distinctly separated from anal angle of triangle (Fig. 953); radial planate consisting of 1 row of cells throughout (Fig. 951) ....
.... ***Oligoclada*** **in part** (Page 262)

11'. Mspl in FW distinct (Fig. 952); CuP in HW arising at or near anal angle of triangle (as in Fig. 954); radial planate consisting of 2 or more rows of cells (Fig. 952) ....
.... ***Dythemis*** **in part** (Page 235)

951

952

FW – *Oligoclada abbreviata*

FW – *Dythemis fugax*

12(10). CuP in HW arising at or near anal angle of triangle (as in Fig. 954) **13**

12'. CuP in HW distinctly separated from anal angle of triangle (as in Fig. 953) **14**

953

954

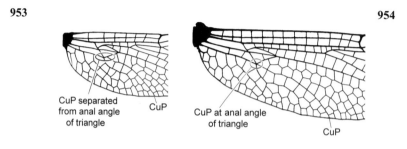

HW base – *Oligoclada rhea*

HW base – *Misagria parana*

13(12). Distal segment of vesica spermalis long and cylindrical (Figs. 955-957) ....
.... ***Erythrodiplax*** **in part** (Page 241)

13'. Distal segment of vesica spermalis not long and cylindrical (as in Figs. 958-959) ....
.... ***Orthemis*** **in part** (Page 264)

14(12).   Distal segment of vesica spermalis long and cylindrical (Figs. 955-957) ....
.... ***Erythrodiplax*** **in part** (Page 241)

14'.   Distal segment of vesica spermalis not long and cylindrical (as in Figs. 958-959) ....
.... ***Cannaphila*** **in part** (Page 230)

vesica spermalis
lateral view

distal segment

955

Erythrodiplax juliana

956   *Crocothemis servilia*

957

Erythrodiplax basifusca

distal segment

958

Sympetrum villosum

959

Sympetrum signiferum

15(7).   Inner branch of hamule as large as or larger than outer branch (as in Fig. 949); distal segment of vesica spermalis not long and cylindrical (as in Figs. 958-959) **16**

15'.   Inner branch of hamule smaller than outer branch (as in Fig. 950); distal segment of vesica spermalis long and cylindrical (Figs. 955-957) .... ***Erythrodiplax*** **in part** (Page 241)

16(15).   Posterior lobe of prothorax in dorsal view widest at base and narrower than other prothoracic lobes (as in Fig. 960); in lateral view bent caudally (as in Fig. 961b) ....
.... ***Dasythemis*** **in part** (Page 233)

16'.   Posterior lobe of prothorax in dorsal view constricted at base (as in Figs. 961a, 962a) and wider than other prothoracic lobes (as in Fig. 962a); in lateral view upright (as in Fig. 962b)
.... **17**

prothorax   dorsal view

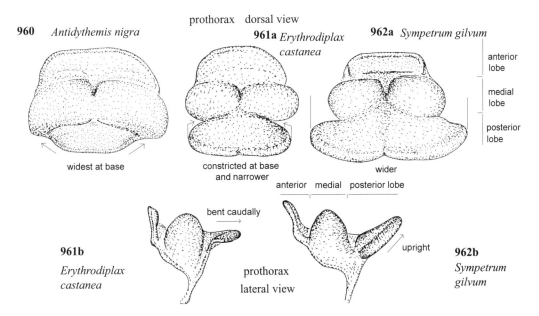

960   *Antidythemis nigra*

961a *Erythrodiplax castanea*

962a *Sympetrum gilvum*

anterior lobe

medial lobe

posterior lobe

widest at base

constricted at base and narrower

wider

anterior   medial   posterior lobe

bent caudally

961b
Erythrodiplax castanea

prothorax
lateral view

upright

962b
Sympetrum gilvum

17(16). Wings with basal dark spot not reaching no-
dus (Fig. 963) ....
.... *Leucorrhinia* in part (Page 248)

17'. Wings lacking basal dark spot (Fig. 964) ....
.... *Oligoclada* in part (Page 262)

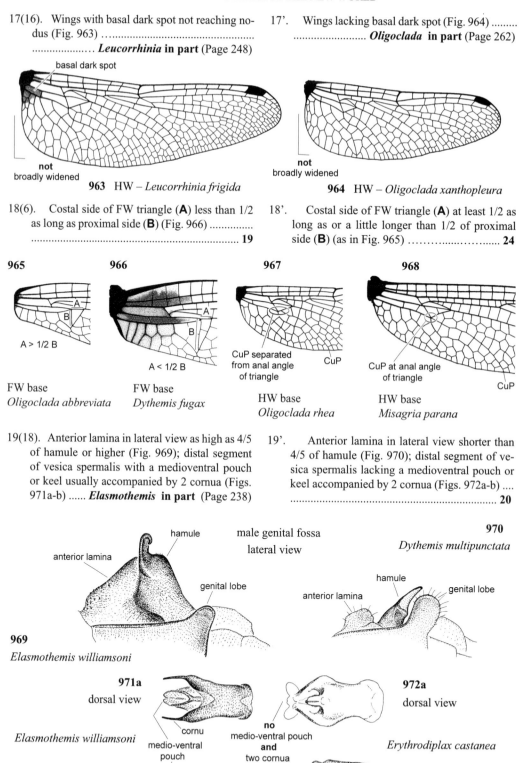

963 HW – *Leucorrhinia frigida*

964 HW – *Oligoclada xanthopleura*

18(6). Costal side of FW triangle (**A**) less than 1/2
as long as proximal side (**B**) (Fig. 966) ....
.... **19**

18'. Costal side of FW triangle (**A**) at least 1/2 as
long as or a little longer than 1/2 of proximal
side (**B**) (as in Fig. 965) **24**

965   **966**   **967**   **968**

A > 1/2 B

FW base
Oligoclada abbreviata

A < 1/2 B

FW base
Dythemis fugax

CuP separated
from anal angle
of triangle

CuP

HW base
Oligoclada rhea

CuP at anal angle
of triangle

HW base
Misagria parana

CuP

19(18). Anterior lamina in lateral view as high as 4/5
of hamule or higher (Fig. 969); distal segment
of vesica spermalis with a medioventral pouch
or keel usually accompanied by 2 cornua (Figs.
971a-b) ...... *Elasmothemis* in part (Page 238)

19'. Anterior lamina in lateral view shorter than
4/5 of hamule (Fig. 970); distal segment of vesica spermalis lacking a medioventral pouch or
keel accompanied by 2 cornua (Figs. 972a-b) ....
.... **20**

hamule

male genital fossa
lateral view

970
Dythemis multipunctata

anterior lamina

genital lobe

hamule

anterior lamina

genital lobe

969
Elasmothemis williamsoni

971a
dorsal view

972a
dorsal view

Elasmothemis williamsoni

cornu

medio-ventral
pouch

no
medio-ventral pouch
and
two cornua

Erythrodiplax castanea

971b
lateral view

cornu

972b
lateral view

vesica spermalis distal segment

20(19). Inner branch of hamule as large as or larger than outer branch (Fig. 973) .... ***Dythemis*** **in part** (Page 235)

20'. Inner branch of hamule smaller than outer branch (Fig. 974) **21**

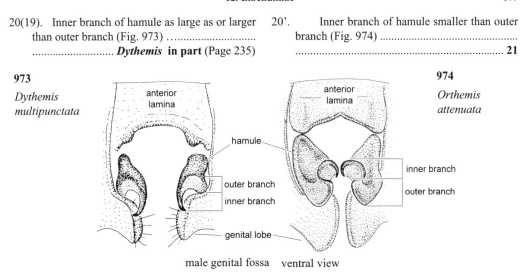

973

Dythemis multipunctata

anterior lamina

hamule

outer branch

inner branch

genital lobe

male genital fossa   ventral view

974

Orthemis attenuata

anterior lamina

inner branch

outer branch

21(20). CuP in HW arising at or near anal angle of triangle (Fig. 968) **22**

21'. CuP in HW distinctly separated from anal angle of triangle (Fig. 967) **23**

22(21). Distal segment of vesica spermalis long and cylindrical (Figs. 975-977) .... ***Erythrodiplax*** **in part** (Page 241)

22' Distal segment of vesica spermalis not long and cylindrical (as in Figs. 978-979) .... ***Orthemis*** **in part** (Page 264)

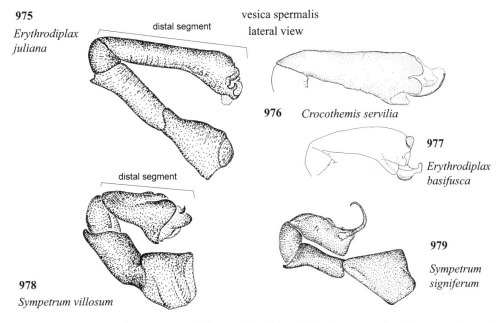

975

Erythrodiplax juliana

distal segment

vesica spermalis lateral view

976   *Crocothemis servilia*

977

Erythrodiplax basifusca

distal segment

978

Sympetrum villosum

979

Sympetrum signiferum

23(21). Distal segment of vesica spermalis long and cylindrical (Figs. 975-977) .... ***Erythrodiplax*** **in part** (Page 241)

23'. Distal segment of vesica spermalis not long and cylindrical (as in Figs. 978-979) .... ***Cannaphila*** **in part** (Page 230)

24(18). Inner branch of hamule as large or larger than outer branch (as in Fig. 973); distal segment of vesica spermalis not long and cylindrical (as in Figs. 978-979) .... ***Dasythemis*** **in part** (Page 233)

24'. Inner branch of hamule smaller than outer branch (as in Fig. 974); distal segment of vesica spermalis long and cylindrical (as in Figs. 975-977) .... ***Erythrodiplax*** **in part** (Page 241)

25(5).   HW anal loop with toe (Fig. 980) .... **26**   25'.   HW anal loop without toe (as in Fig. 981)
.... **32**

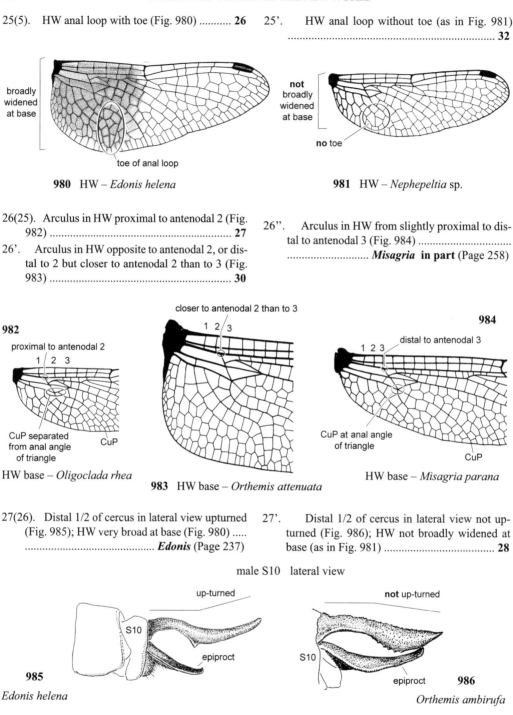

980   HW – *Edonis helena*   **981**   HW – *Nephepeltia* sp.

26(25).   Arculus in HW proximal to antenodal 2 (Fig.
982) .... **27**   26".   Arculus in HW from slightly proximal to dis-
                                                            tal to antenodal 3 (Fig. 984) ....
26'.   Arculus in HW opposite to antenodal 2, or dis-   .... *Misagria* **in part** (Page 258)
tal to 2 but closer to antenodal 2 than to 3 (Fig.
983) **30**

HW base – *Oligoclada rhea*   HW base – *Misagria parana*

983   HW base – *Orthemis attenuata*

27(26).   Distal 1/2 of cercus in lateral view upturned   27'.   Distal 1/2 of cercus in lateral view not up-
(Fig. 985); HW very broad at base (Fig. 980) .....   turned (Fig. 986); HW not broadly widened at
.... *Edonis* (Page 237) base (as in Fig. 981) **28**

Edonis helena   *Orthemis ambirufa*

28(27).   Anterior lamina in lateral view as high as   28'.   Anterior lamina in lateral view shorter than 4/5
4/5 of hamule or higher (Fig. 987); hamule not   of hamule (as in Fig. 988); hamule bifid (Figs.
bifid (Fig. 989); distal segment of vesica sper-   990-991); distal segment of vesica spermalis
malis with a medioventral pouch or keel usually   lacking a medioventral pouch or keel accompa-
accompanied by 2 flagella (Figs. 992a-b) ....   nied by 2 flagella (Figs. 993a-b) **29**
.... *Elasmothemis* **in part** (Page 238)

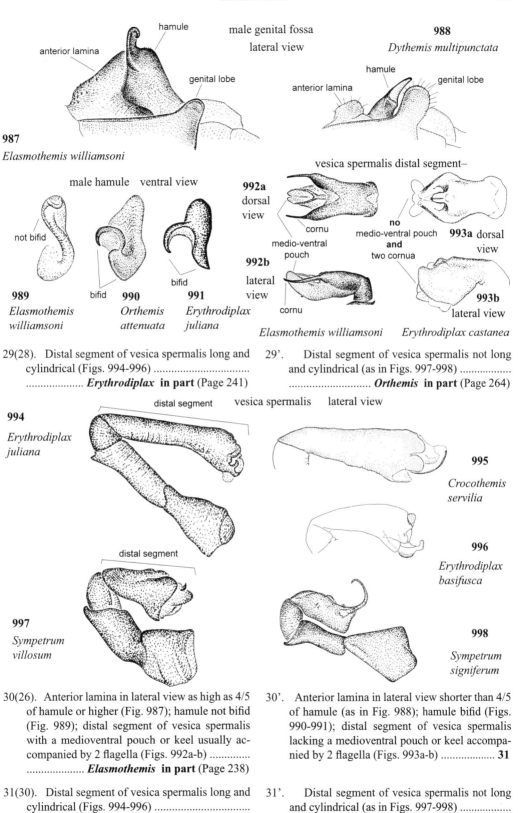

987
Elasmothemis williamsoni

male genital fossa
lateral view

988
Dythemis multipunctata

male hamule   ventral view

989
Elasmothemis williamsoni

990
Orthemis attenuata

991
Erythrodiplax juliana

vesica spermalis distal segment–

992a dorsal view

992b lateral view

Elasmothemis williamsoni

993a dorsal view

993b lateral view

Erythrodiplax castanea

29(28). Distal segment of vesica spermalis long and cylindrical (Figs. 994-996) .... ***Erythrodiplax*** **in part** (Page 241)

29'. Distal segment of vesica spermalis not long and cylindrical (as in Figs. 997-998) .... ***Orthemis*** **in part** (Page 264)

vesica spermalis   lateral view

994
Erythrodiplax juliana

distal segment

995
Crocothemis servilia

996
Erythrodiplax basifusca

997
Sympetrum villosum

distal segment

998
Sympetrum signiferum

30(26). Anterior lamina in lateral view as high as 4/5 of hamule or higher (Fig. 987); hamule not bifid (Fig. 989); distal segment of vesica spermalis with a medioventral pouch or keel usually accompanied by 2 flagella (Figs. 992a-b) .... ***Elasmothemis*** **in part** (Page 238)

30'. Anterior lamina in lateral view shorter than 4/5 of hamule (as in Fig. 988); hamule bifid (Figs. 990-991); distal segment of vesica spermalis lacking a medioventral pouch or keel accompanied by 2 flagella (Figs. 993a-b) **31**

31(30). Distal segment of vesica spermalis long and cylindrical (Figs. 994-996) .... ***Erythrodiplax*** **in part** (Page 241)

31'. Distal segment of vesica spermalis not long and cylindrical (as in Figs. 997-998) .... ***Orthemis*** **in part** (Page 264)

32(25). Arculus in HW proximal to antenodal 2 (as in Fig. 1001); crossveins in FW triangle absent (as in Fig. 999); S7-9 widened and flattened, widest point less than twice as wide as base of S7 .... *Nephepeltia* (Page 260)

32'. Arculus in HW from slightly proximal to distal to antenodal 3 (Fig. 1002); crossveins in FW triangle present (Fig. 1000); S7-9 not widened .... ***Misagria* in part** (Page 258)

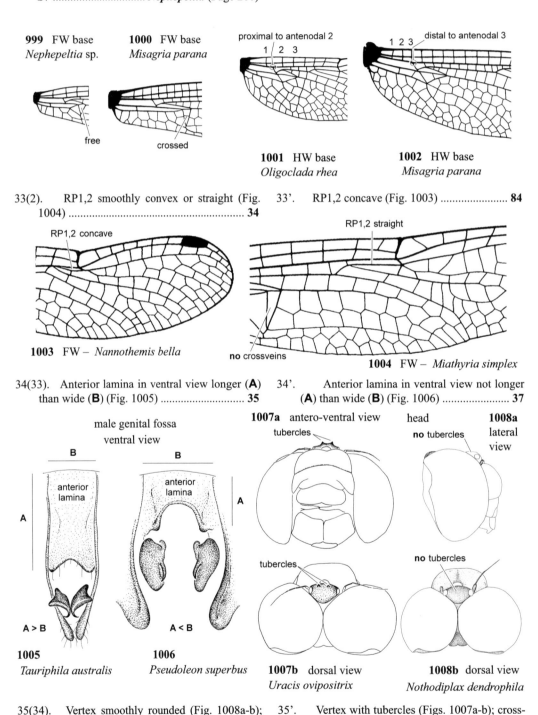

999 FW base
Nephepeltia sp.

1000 FW base
Misagria parana

free

crossed

proximal to antenodal 2
1 2 3

distal to antenodal 3
1 2 3

1001 HW base
Oligoclada rhea

1002 HW base
Misagria parana

33(2). RP1,2 smoothly convex or straight (Fig. 1004) **34**

33'. RP1,2 concave (Fig. 1003) **84**

RP1,2 concave

RP1,2 straight

1003 FW – *Nannothemis bella*

no crossveins

1004 FW – *Miathyria simplex*

34(33). Anterior lamina in ventral view longer (**A**) than wide (**B**) (Fig. 1005) **35**

34'. Anterior lamina in ventral view not longer (**A**) than wide (**B**) (Fig. 1006) **37**

male genital fossa
ventral view

B

B

anterior
lamina

anterior
lamina

A

A

A

A > B

A < B

1005

Tauriphila australis

1006

Pseudoleon superbus

1007a antero-ventral view
tubercles

head

no tubercles

1008a
lateral
view

tubercles

no tubercles

1007b dorsal view
Uracis ovipositrix

1008b dorsal view
Nothodiplax dendrophila

35(34). Vertex smoothly rounded (Fig. 1008a-b); crossveins in FW triangle absent (Figs. 999, 1004) .... ***Miathyria* in part** (Page 255)

35'. Vertex with tubercles (Figs. 1007a-b); crossveins in FW triangle present (Fig. 1000) ....
.... **36**

36(35).   Median planate consisting of 1 row of cells throughout (Fig. 1009); cell rows at base of FW discoidal field numbering 3 (Fig. 1009) ....………. ***Tauriphila*** (Page 275)

36'.   Median planate consisting of 2 or more rows of cells (Fig. 1010); cell rows at base of FW discoidal field numbering 4 or more (Fig. 1010) .... ***Tramea*** **in part** (Page 277)

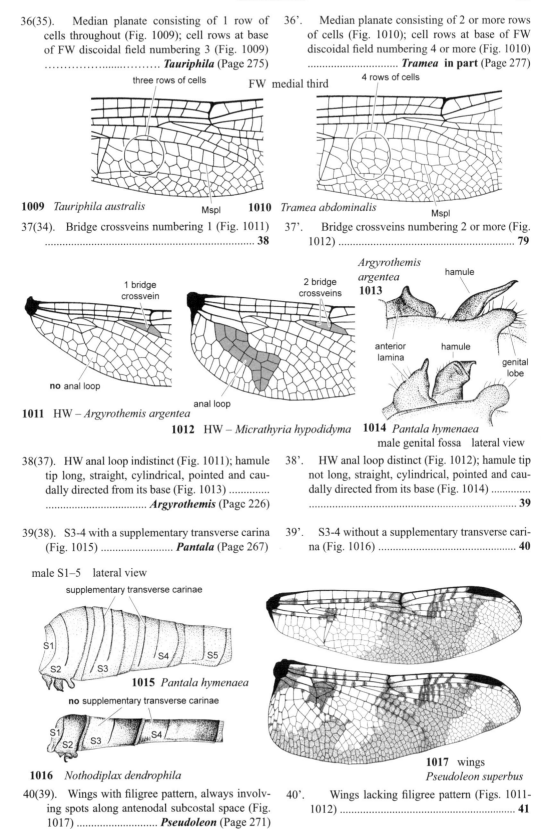

1009 *Tauriphila australis*

1010 *Tramea abdominalis*

37(34).   Bridge crossveins numbering 1 (Fig. 1011) **38**

37'.   Bridge crossveins numbering 2 or more (Fig. 1012) **79**

1011 HW – *Argyrothemis argentea*

1012 HW – *Micrathyria hypodidyma*

1013 *Argyrothemis argentea*

1014 *Pantala hymenaea*
male genital fossa   lateral view

38(37).   HW anal loop indistinct (Fig. 1011); hamule tip long, straight, cylindrical, pointed and caudally directed from its base (Fig. 1013) .... ***Argyrothemis*** (Page 226)

38'.   HW anal loop distinct (Fig. 1012); hamule tip not long, straight, cylindrical, pointed and caudally directed from its base (Fig. 1014) **39**

39(38).   S3-4 with a supplementary transverse carina (Fig. 1015) .... ***Pantala*** (Page 267)

39'.   S3-4 without a supplementary transverse carina (Fig. 1016) **40**

1015 *Pantala hymenaea*

1016 *Nothodiplax dendrophila*

1017 wings
Pseudoleon superbus

40(39).   Wings with filigree pattern, always involving spots along antenodal subcostal space (Fig. 1017) .... ***Pseudoleon*** (Page 271)

40'.   Wings lacking filigree pattern (Figs. 1011-1012) **41**

41(40). Hind femur (CAUTION: check to see that spines are not broken) lacking 3-4 long and strong spines on distal 1/2 and numerous short spines distally directed on basal 1/2 (Figs. 1018-1019, 1021) **42**

41'. Hind femur (CAUTION: check to see that spines are not broken) with 3-4 long and strong spines on distal 1/2, numerous short and distally directed spines on basal 1/2 (Fig. 1020) .... ***Erythemis* in part** (Page 240)

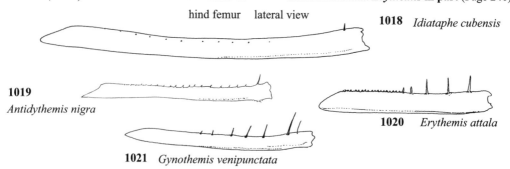

hind femur   lateral view

1018   *Idiataphe cubensis*

1019

Antidythemis nigra

1020   *Erythemis attala*

1021   *Gynothemis venipunctata*

42(41). Cerci narrowly cylindrical throughout (Figs. 1022a-b); distal segment of vesica spermalis with 2 laterodistal cornua ending in a recurved hook (Figs. 1025a-b) .. ***Elga* in part** (Page 239)

42'. Cerci not narrowly cylindrical throughout (Figs. 1023a-b, 1024); distal segment of vesica spermalis lacking 2 laterodistal cornua ending in a recurved hook (Figs. 1026a-b, 1027) **43**

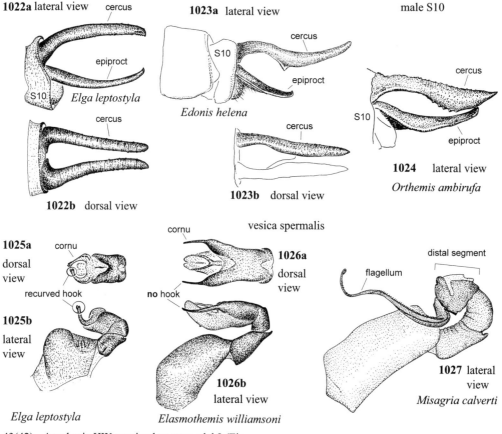

1022a lateral view   cercus

epiproct

S10   *Elga leptostyla*

cercus

1022b   dorsal view

1023a   lateral view

cercus

S10

epiproct

Edonis helena

cercus

1023b   dorsal view

male S10

cercus

S10

epiproct

1024   lateral view

Orthemis ambirufa

vesica spermalis

cornu

1025a
dorsal
view

cornu

recurved hook

1025b
lateral
view

Elga leptostyla

no hook

1026b
lateral view

Elasmothemis williamsoni

1026a
dorsal
view

distal segment

flagellum

1027 lateral
view

Misagria calverti

43(42). Arculus in HW proximal to antenodal 2 (Fig. 1028) **44**

43'. Arculus in HW opposite to antenodal 2, or distal to 2 but closer to antenodal 2 than to 3 (Fig. 1029) **65**

43". Arculus in HW from slightly proximal to distal to antenodal 3 (Fig. 1030) .... ***Uracis* in part** (Page 279)

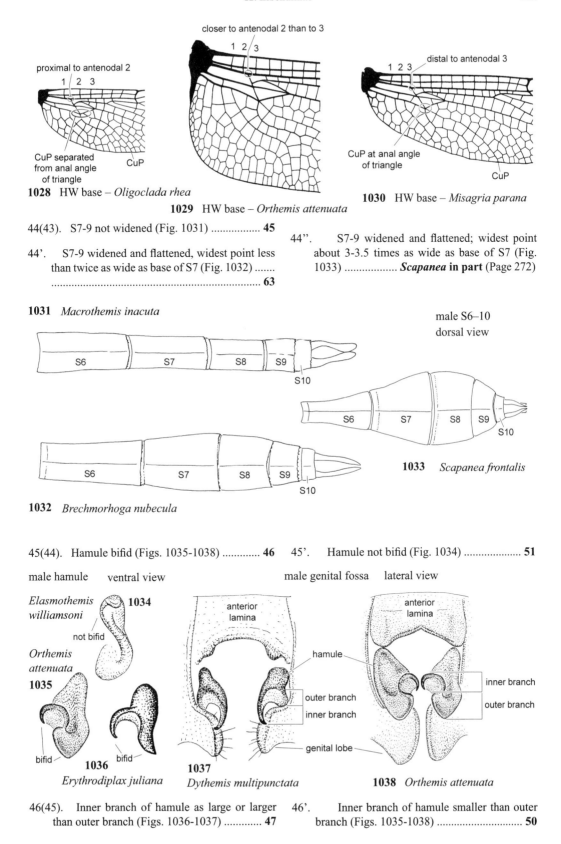

closer to antenodal 2 than to 3

1 2 3

proximal to antenodal 2

1 2 3

distal to antenodal 3

1 2 3

CuP separated
from anal angle
of triangle

CuP

CuP at anal angle
of triangle

CuP

1028 HW base – *Oligoclada rhea*

1029 HW base – *Orthemis attenuata*

1030 HW base – *Misagria parana*

44(43). S7-9 not widened (Fig. 1031) **45**

44'. S7-9 widened and flattened, widest point less than twice as wide as base of S7 (Fig. 1032) **63**

44". S7-9 widened and flattened; widest point about 3-3.5 times as wide as base of S7 (Fig. 1033) .... ***Scapanea* in part** (Page 272)

1031 *Macrothemis inacuta*

male S6–10
dorsal view

S6　S7　S8　S9
S10

S6　S7　S8　S9
S10

1033 *Scapanea frontalis*

S6　S7　S8　S9
S10

1032 *Brechmorhoga nubecula*

45(44). Hamule bifid (Figs. 1035-1038) **46**

45'. Hamule not bifid (Fig. 1034) **51**

male hamule ventral view

male genital fossa lateral view

Elasmothemis williamsoni **1034**

not bifid

anterior lamina

anterior lamina

Orthemis attenuata **1035**

hamule

inner branch

outer branch

outer branch
inner branch

bifid **1036** bifid

genital lobe

1037

Erythrodiplax juliana

Dythemis multipunctata

1038 *Orthemis attenuata*

46(45). Inner branch of hamule as large or larger than outer branch (Figs. 1036-1037) **47**

46'. Inner branch of hamule smaller than outer branch (Figs. 1035-1038) **50**

47(46).   Crossveins in FW triangle present (as in Fig. 1040) **48**

47'.   Crossveins in FW triangle absent (as in Fig. 1039) **49**

1041   HW base – *Oligoclada rhea*   **1042**   HW base – *Misagria parana*

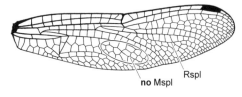

1039   FW base
Nephepeltia sp.

free

1040   FW base
Misagria parana

crossed

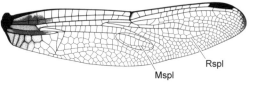

CuP separated
from anal angle
of triangle

CuP

CuP at anal angle
of triangle

CuP

48(47).   Mspl in FW indistinct (Fig. 1043); CuP in HW distinctly separated from anal angle of triangle (Fig. 1041); radial planate consisting of 1 row of cells throughout (Fig. 1043) ....
.... ***Oligoclada*** **in part** (Page 262)

48'.   Mspl in FW distinct (Fig. 1044); CuP in HW arising at or near anal angle of triangle (Fig. 1042); radial planate consisting of 2 or more rows of cells (Fig. 1044) ....
.... ***Dythemis*** **in part** (Page 235)

1043   FW – *Oligoclada abbreviata*

Rspl

no Mspl

1044   FW – *Dythemis fugax*

Rspl

Mspl

49(47).   S9 with lateral carina (Fig. 1045); spines of hind femur (CAUTION: check to see that spines are not broken) gradually increasing in length distally (Fig. 1046) ..... ***Nothodiplax*** (Page 261)

49'.   S9 without lateral carina; spines of hind femur (CAUTION: check to see that spines are not broken) dimorphic; with long and short series (Fig. 1047) .... ***Oligoclada*** **in part** (Page 262)

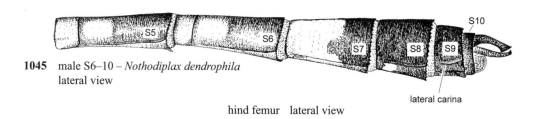

S10

S5

S6

S7

S8

S9

1045   male S6–10 – *Nothodiplax dendrophila*
lateral view

lateral carina

hind femur   lateral view

1046   *Nothodiplax dendrophila*

1047   *Oligoclada crocogaster*

50(46).   Posterior lobe of prothorax in lateral view bent caudally (Fig. 1048a), in dorsal view narrower than other prothoracic lobes (Fig. 1048b); distal segment of vesica spermalis long and cylindrical (Figs. 1050-1052) ....
.... ***Erythrodiplax*** **in part** / ***Crocothemis***

50'.   Posterior lobe of prothorax in lateral view upright (Fig. 1049a), in dorsal view wider than other prothoracic lobes (Fig. 1049b); distal segment of vesica spermalis not long and cylindrical (Figs. 1053-1054) ....
.... ***Sympetrum*** **in part** (Page 274)

Erythrodiplax (Page 241): *If* from Florida in US or Cuba *and* small size (total length 36 mm or less, except *E. umbrata* and *E. funerea*) *and* 2 rows of cells in radial planate (as in Fig. 1044)

Crocothemis (Page 232): Occurring *only* in Florida in US and Cuba (as of 2005) *and* large size (total length more than 40 mm) *and* 1 row of cells in the radial planate (as in Fig. 1043)

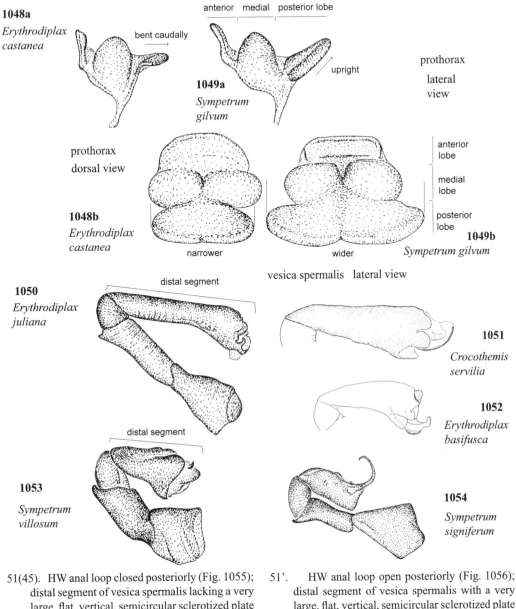

1048a *Erythrodiplax castanea*

bent caudally

anterior   medial   posterior lobe

upright

1049a *Sympetrum gilvum*

prothorax lateral view

prothorax dorsal view

anterior lobe

medial lobe

posterior lobe

1048b *Erythrodiplax castanea*

narrower

wider

1049b *Sympetrum gilvum*

vesica spermalis   lateral view

distal segment

1050 *Erythrodiplax juliana*

1051 *Crocothemis servilia*

1052 *Erythrodiplax basifusca*

distal segment

1053 *Sympetrum villosum*

1054 *Sympetrum signiferum*

51(45).   HW anal loop closed posteriorly (Fig. 1055); distal segment of vesica spermalis lacking a very large, flat, vertical, semicircular sclerotized plate (Fig. 1057) **52**

51'.   HW anal loop open posteriorly (Fig. 1056); distal segment of vesica spermalis with a very large, flat, vertical, semicircular sclerotized plate (Fig. 1058) .... ***Tholymis* in part** (Page 276)

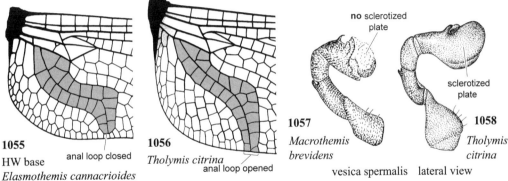

1055 HW base *Elasmothemis cannacrioides*   anal loop closed

1056 *Tholymis citrina*   anal loop opened

no sclerotized plate

sclerotized plate

1057 *Macrothemis brevidens*

1058 *Tholymis citrina*

vesica spermalis   lateral view

52(51). Crossveins in FW triangle present (as in Fig. 1060) **53**

52'.   Crossveins in FW triangle absent (as in Fig. 1059) **59**

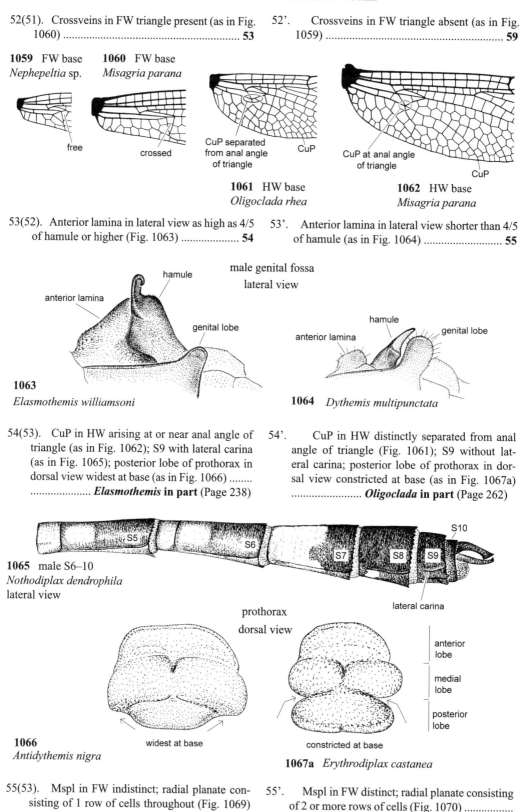

1059 FW base
Nephepeltia sp.

1060 FW base
Misagria parana

free

crossed

CuP separated
from anal angle
of triangle

CuP

CuP at anal angle
of triangle

CuP

1061 HW base
Oligoclada rhea

1062 HW base
Misagria parana

53(52). Anterior lamina in lateral view as high as 4/5 of hamule or higher (Fig. 1063) **54**

53'.   Anterior lamina in lateral view shorter than 4/5 of hamule (as in Fig. 1064) **55**

hamule

anterior lamina

genital lobe

male genital fossa
lateral view

hamule

anterior lamina

genital lobe

1063

Elasmothemis williamsoni

1064   *Dythemis multipunctata*

54(53). CuP in HW arising at or near anal angle of triangle (as in Fig. 1062); S9 with lateral carina (as in Fig. 1065); posterior lobe of prothorax in dorsal view widest at base (as in Fig. 1066) .... *Elasmothemis* **in part** (Page 238)

54'.   CuP in HW distinctly separated from anal angle of triangle (Fig. 1061); S9 without lateral carina; posterior lobe of prothorax in dorsal view constricted at base (as in Fig. 1067a) .... *Oligoclada* **in part** (Page 262)

S5

S6

S7

S8

S9

S10

1065 male S6–10
Nothodiplax dendrophila
lateral view

lateral carina

prothorax
dorsal view

anterior
lobe

medial
lobe

posterior
lobe

1066

Antidythemis nigra

widest at base

constricted at base

1067a *Erythrodiplax castanea*

55(53). Mspl in FW indistinct; radial planate consisting of 1 row of cells throughout (Fig. 1069) **56**

55'.   Mspl in FW distinct; radial planate consisting of 2 or more rows of cells (Fig. 1070) **57**

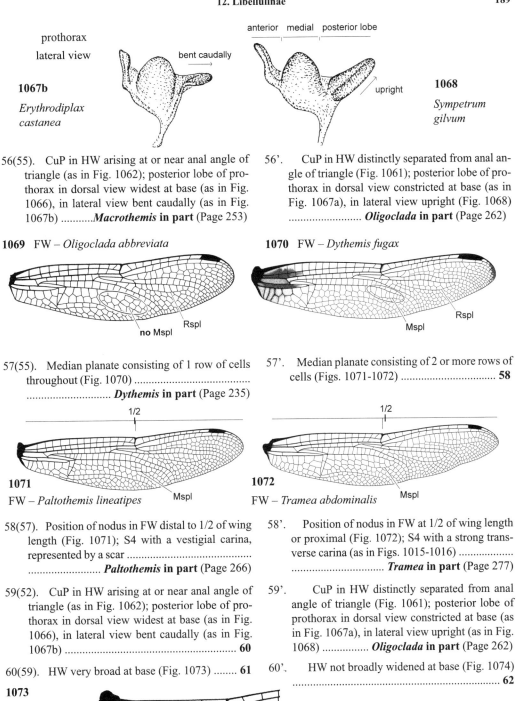

prothorax

lateral view

anterior | medial | posterior lobe

bent caudally →

upright ↗

1067b

Erythrodiplax castanea

1068

Sympetrum gilvum

56(55). CuP in HW arising at or near anal angle of triangle (as in Fig. 1062); posterior lobe of prothorax in dorsal view widest at base (as in Fig. 1066), in lateral view bent caudally (as in Fig. 1067b) ....*Macrothemis* **in part** (Page 253)

56'. CuP in HW distinctly separated from anal angle of triangle (Fig. 1061); posterior lobe of prothorax in dorsal view constricted at base (as in Fig. 1067a), in lateral view upright (Fig. 1068) .... *Oligoclada* **in part** (Page 262)

1069 FW – *Oligoclada abbreviata*

Rspl

no Mspl

1070 FW – *Dythemis fugax*

Rspl

Mspl

57(55). Median planate consisting of 1 row of cells throughout (Fig. 1070) ....
.... *Dythemis* **in part** (Page 235)

57'. Median planate consisting of 2 or more rows of cells (Figs. 1071-1072) **58**

1/2

1/2

1071

FW – *Paltothemis lineatipes* Mspl

1072

FW – *Tramea abdominalis* Mspl

58(57). Position of nodus in FW distal to 1/2 of wing length (Fig. 1071); S4 with a vestigial carina, represented by a scar ....
.... *Paltothemis* **in part** (Page 266)

58'. Position of nodus in FW at 1/2 of wing length or proximal (Fig. 1072); S4 with a strong transverse carina (as in Figs. 1015-1016) ....
.... *Tramea* **in part** (Page 277)

59(52). CuP in HW arising at or near anal angle of triangle (as in Fig. 1062); posterior lobe of prothorax in dorsal view widest at base (as in Fig. 1066), in lateral view bent caudally (as in Fig. 1067b) **60**

59'. CuP in HW distinctly separated from anal angle of triangle (Fig. 1061); posterior lobe of prothorax in dorsal view constricted at base (as in Fig. 1067a), in lateral view upright (as in Fig. 1068) .... *Oligoclada* **in part** (Page 262)

60(59). HW very broad at base (Fig. 1073) **61**

60'. HW not broadly widened at base (Fig. 1074) **62**

1073

HW base
Miathryria simplex

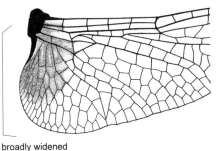

not broadly widened

broadly widened

1074

HW base
Gynothemis venipunctata

61(60).   Spines of hind femur (CAUTION: check to see that spines are not broken) all short, stout, and directed proximally (Figs. 1075-1076) .... *Macrothemis* **in part** (Page 253)

61'.   Spines of hind femur (CAUTION: check to see that spines are not broken) dimorphic; with long and short series (as in Figs. 1077-1078) .... *Miathyria* **in part** (Page 255)

1075

Macrothemis heteronycha

hind femur   lateral view

1077 *Antidythemis nigra*

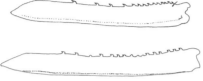

1076

Macrothemis celeno

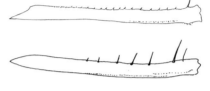

1078   *Gynothemis venipunctata*

62(60).   Spines of hind femur (CAUTION: check to see that spines are not broken) all short, stout, and directed proximally (Figs. 1075-1076) .... *Macrothemis* **in part** (Page 253)

62'.   Spines of hind femur (CAUTION: check to see that spines are not broken) gradually increasing in length distally (Fig. 1078) or dimorphic, with long and short series (as in Fig. 1077) .... *Gynothemis* **in part** (Page 245)

63(44).   Hamule bifid (as in Figs. 1087-1088, 1091-1092); posterior lobe of prothorax in dorsal view constricted at base (as in Fig. 1080a), in lateral view upright (Fig. 1081) .... *Sympetrum* **in part** (Page 274)

63'.   Hamule not bifid (as in Fig. 1086); posterior lobe of prothorax in dorsal view widest at base (as in Fig. 1079), in lateral view bent caudally (as in Fig. 1080b) **64**

prothorax

1079

Antidythemis nigra

dorsal view

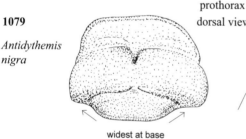

widest at base

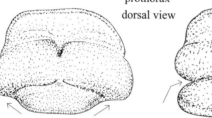

constricted at base

1080a

anterior lobe

Erythrodiplax castanea

medial lobe

posterior lobe

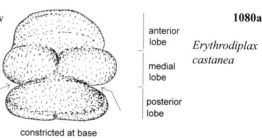

1080b

Erythrodiplax castanea

bent caudally

anterior medial posterior lobe

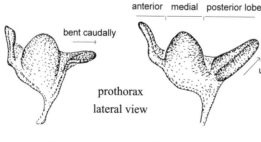

upright

1081

Sympetrum gilvum

prothorax

lateral view

64(63).   Mspl in FW distinct (as in Fig. 1082) .... *Brechmorhoga* **in part** (Page 228)

64'.   Mspl in FW indistinct (Fig. 1083) (except in *M. griseofrons*)* .... *Macrothemis* **in part** (Page 253)

Brechmorhoga: Short inner tarsal claw (as in Fig. 1084); relatively wide discoidal field in FW, with discoidal index approximately 1.4-1.8 [index consisting of ratio of distance between MA and MP at wing margin divided by distance at proximal portion] (as in Fig. 1082); longer body (40-62 mm)

Macrothemis: Long inner tarsal claw (Fig. 1085) (except for members of *M. tessellata* group of species); relatively narrow discoidal field in FW, with discoidal index <1.0-1.3 (Fig. 1083); shorter body (21-42 mm)

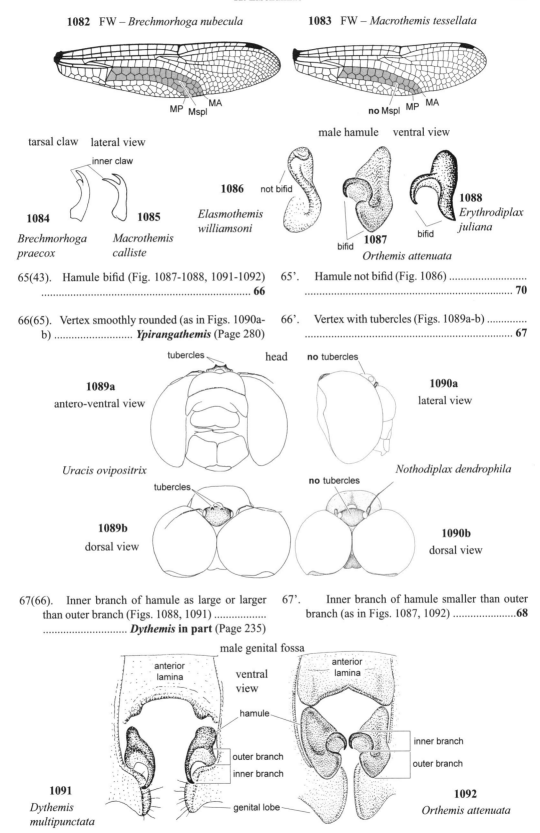

1082 FW – *Brechmorhoga nubecula*

MP Mspl MA

1083 FW – *Macrothemis tessellata*

no Mspl MP MA

tarsal claw   lateral view

inner claw

1084   **1085**
Brechmorhoga   *Macrothemis*
praecox   *calliste*

male hamule   ventral view

1086 not bifid
Elasmothemis
williamsoni

bifid   **1087**   bifid
Orthemis attenuata

1088
Erythrodiplax
juliana

65(43). Hamule bifid (Fig. 1087-1088, 1091-1092) **66**

65'. Hamule not bifid (Fig. 1086) **70**

66(65). Vertex smoothly rounded (as in Figs. 1090a-b) .... ***Ypirangathemis*** (Page 280)

66'. Vertex with tubercles (Figs. 1089a-b) **67**

tubercles   head

no tubercles

1089a
antero-ventral view

1090a
lateral view

Uracis ovipositrix

tubercles

no tubercles

Nothodiplax dendrophila

1089b
dorsal view

1090b
dorsal view

67(66). Inner branch of hamule as large or larger than outer branch (Figs. 1088, 1091) .... ***Dythemis* in part** (Page 235)

67'. Inner branch of hamule smaller than outer branch (as in Figs. 1087, 1092) **68**

male genital fossa

anterior
lamina

ventral
view

anterior
lamina

hamule

outer branch
inner branch

inner branch

outer branch

genital lobe

1091
Dythemis
multipunctata

1092
Orthemis attenuata

68(67). S9 with lateral carina (as in Fig. 1065) .. **69**

Erythrodiplax (Page 241): Transverse carina on S2 rounded, or *if* roundly angled (Figs. 1096-1098, in *E. amazonica* and *E. castanea*), *then* posterior lobe of prothorax narrowed at base (Fig. 1100), *and* 1 or 2 crossveins in HW CuA space (Fig. 1101).

68'. S9 without lateral carina ....
.... ***Erythrodiplax*** in part /***Uracis*** in part

Uracis (Page 279): Transverse carina on S2 forming a near right angle (Figs. 1093-1095; posterior lobe of prothorax widest at base (Fig. 1099); 3 or more crossveins in HW CuA space (Fig. 1102).

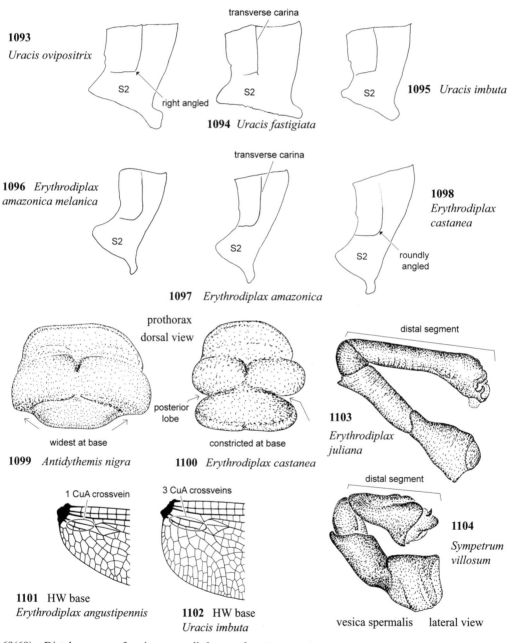

male S2   lateral view

transverse carina

1093
Uracis ovipositrix

S2

right angled

1094 *Uracis fastigiata*

S2

S2

1095 *Uracis imbuta*

transverse carina

1096 *Erythrodiplax amazonica melanica*

S2

S2

1098
Erythrodiplax castanea

S2

roundly angled

1097 *Erythrodiplax amazonica*

prothorax dorsal view

posterior lobe

distal segment

1103
Erythrodiplax juliana

widest at base

1099 *Antidythemis nigra*

constricted at base

1100 *Erythrodiplax castanea*

1 CuA crossvein

3 CuA crossveins

distal segment

1104
Sympetrum villosum

1101 HW base
Erythrodiplax angustipennis

1102 HW base
Uracis imbuta

vesica spermalis   lateral view

69(68). Distal segment of vesica spermalis long and cylindrical (Fig. 1103) ....
.... ***Erythrodiplax*** in part (Page 241)

69'. Distal segment of vesica spermalis not long and cylindrical (as in Fig. 1104) ....
.... ***Rhodopygia*** (Page 272)

70(65). S7-9 not widened (Fig. 1105) **71**

70'.   S7-9 widened and flattened, widest point less than twice as wide as base of S7 (Fig. 1106) **78**

70".   S7-9 widened and flattened; widest point about 3-3.5 times as wide as base of S7 (Fig. 1107) .... ***Scapanea*** (Page 272)

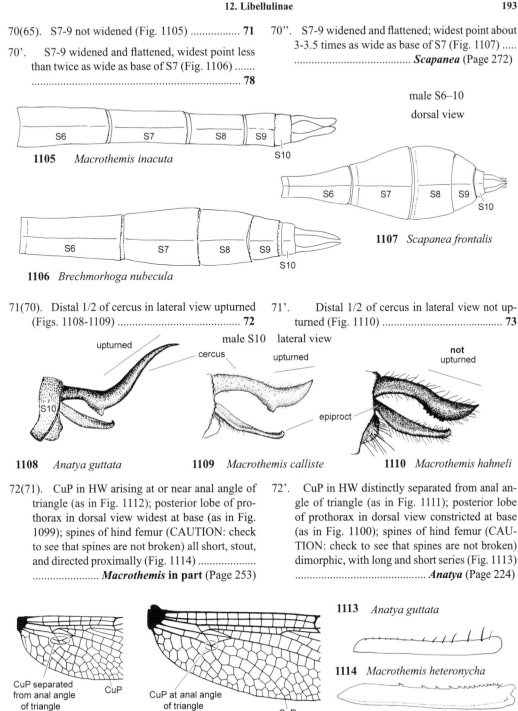

1105   *Macrothemis inacuta*

male S6–10
dorsal view

1107   *Scapanea frontalis*

1106   *Brechmorhoga nubecula*

71(70). Distal 1/2 of cercus in lateral view upturned (Figs. 1108-1109) **72**

71'.   Distal 1/2 of cercus in lateral view not upturned (Fig. 1110) **73**

male S10   lateral view

upturned

cercus

upturned

not upturned

S10

epiproct

1108   *Anatya guttata*

1109   *Macrothemis calliste*

1110   *Macrothemis hahneli*

72(71). CuP in HW arising at or near anal angle of triangle (as in Fig. 1112); posterior lobe of prothorax in dorsal view widest at base (as in Fig. 1099); spines of hind femur (CAUTION: check to see that spines are not broken) all short, stout, and directed proximally (Fig. 1114) .... ***Macrothemis* in part** (Page 253)

72'.   CuP in HW distinctly separated from anal angle of triangle (as in Fig. 1111); posterior lobe of prothorax in dorsal view constricted at base (as in Fig. 1100); spines of hind femur (CAUTION: check to see that spines are not broken) dimorphic, with long and short series (Fig. 1113) .... ***Anatya*** (Page 224)

1113   *Anatya guttata*

1114   *Macrothemis heteronycha*

CuP separated from anal angle of triangle

CuP

CuP at anal angle of triangle

CuP

1111 HW base *Oligoclada rhea*

1112 HW base *Misagria parana*

hind femur   lateral view

73(71). Anterior lamina in lateral view as high as 4/5 of hamule or higher (Fig. 1115); distal segment of vesica spermalis with a medioventral pouch or keel usually accompanied by 2 flagella (Figs. 1117a-b) ...... ***Elasmothemis* in part** (Page 238)

73'.   Anterior lamina in lateral view shorter than 4/5 of hamule (Fig. 1116); distal segment of vesica spermalis lacking a medioventral pouch or keel accompanied by 2 flagella (Figs. 1118a-b) **74**

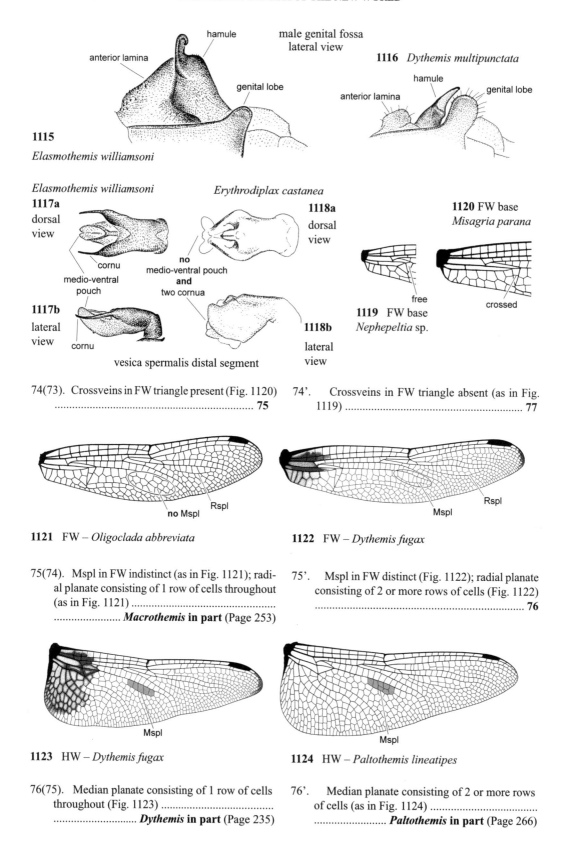

male genital fossa
lateral view

1116 *Dythemis multipunctata*

1115

Elasmothemis williamsoni

Elasmothemis williamsoni

1117a
dorsal
view

Erythrodiplax castanea

1118a
dorsal
view

1120 FW base
Misagria parana

no
medio-ventral pouch
and
two cornua

1117b
lateral
view

1119 FW base
Nephepeltia sp.

vesica spermalis distal segment

1118b
lateral
view

74(73). Crossveins in FW triangle present (Fig. 1120) **75**

74'. Crossveins in FW triangle absent (as in Fig. 1119) **77**

1121 FW – *Oligoclada abbreviata*

1122 FW – *Dythemis fugax*

75(74). Mspl in FW indistinct (as in Fig. 1121); radial planate consisting of 1 row of cells throughout (as in Fig. 1121) .... **Macrothemis in part** (Page 253)

75'. Mspl in FW distinct (Fig. 1122); radial planate consisting of 2 or more rows of cells (Fig. 1122) **76**

1123 HW – *Dythemis fugax*

1124 HW – *Paltothemis lineatipes*

76(75). Median planate consisting of 1 row of cells throughout (Fig. 1123) .... **Dythemis in part** (Page 235)

76'. Median planate consisting of 2 or more rows of cells (as in Fig. 1124) .... **Paltothemis in part** (Page 266)

77(74). Spines of hind femur (CAUTION: check to see that spines are not broken) all short, stout, and directed proximally (Figs. 1125-1126) .... *Macrothemis* **in part** (Page 253)

77'. Spines of hind femur (CAUTION: check to see that spines are not broken) gradually increasing in length distally (Fig. 1128) or dimorphic, with long and short series (as in Fig. 1127) .... *Gynothemis* **in part** (Page 245)

1125
Macrothemis heteronycha

hind femur   lateral view

1127 *Antidythemis nigra*

1126
Macrothemis celeno

1128 *Gynothemis venipunctata*

78(70). Mspl in FW distinct (as in Fig. 1129) .... *Brechmorhoga* **in part**

78'. Mspl in FW indistinct (Fig. 1130) (except in *M. griseofrons*)* .... *Macrothemis* **in part**

1129 FW – *Brechmorhoga nubecula*

MP Mspl MA

1130 FW – *Macrothemis tesselata*

no Mspl MP MA

inner claw

1131
Brechmorhoga praecox

1132
Macrothemis calliste

tarsal claw   lateral view

Brechmorhoga (Page 228): Short inner tarsal claw (as in Fig. 1131); relatively wide discoidal field in FW, with discoidal index approximately 1.4-1.8 (as in Fig. 1129) [index consisting of ratio of distance between MA and MP at wing margin divided by distance at proximal portion]; longer body (40-62 mm)

****Macrothemis*** (Page 253): Long inner tarsal claw (Fig. 1132) (except for members of *M. tessellata* group of species); relatively narrow discoidal field in FW, with discoidal index <1.0-1.3 (Fig. 1130); shorter body (21-42 mm)

79(37). Aspl in HW more or less straight (Fig. 1133) .... *Micrathyria* **in part** (Page 256)

79'. Aspl in HW bent at heel level of anal loop at an obtuse angle (Fig. 1134) **80**

79". Aspl in HW bent at heel level of anal loop at a right (90 degree) to acute angle (Fig. 1135) .... *Antidythemis* (Page 225)

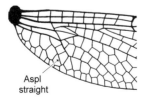
Aspl straight

1133 HW base
Micrathyria dido

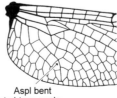
Aspl bent at obtuse angle

1134 HW base
Micrathyria aequalis

Aspl bent at acute angle

1135 HW base
Antidythemis nigra

80(79). HW anal loop closed posteriorly (Fig. 1136); distal segment of vesica spermalis lacking a very large, flat, vertical, semicircular sclerotized plate (as in Fig. 1138) **81**

80'. HW anal loop open posteriorly (Fig. 1137); distal segment of vesica spermalis with a very large, flat, vertical, semicircular sclerotized plate (Fig. 1139) ....***Tholymis* in part** (Page 276)

vesica spermalis   lateral view

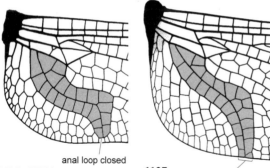

no sclerotized plate

sclerotized plate

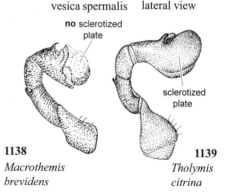

1136 HW base
Elasmothemis cannacrioides

anal loop closed

1137
Tholymis citrina

anal loop opened

1138
Macrothemis brevidens

1139
Tholymis citrina

81(80). Anterior lamina in lateral view as high as 4/5 of hamule or higher (Fig. 1140) **82**

81'. Anterior lamina in lateral view shorter than 4/5 of hamule (as in Fig. 1141) **83**

male genital fossa
lateral view

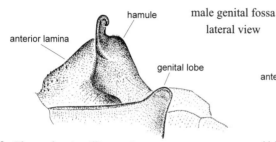

hamule

anterior lamina

genital lobe

hamule

anterior lamina

genital lobe

1140 *Elasmothemis williamsoni*

1141 *Dythemis multipunctata*

82(81). CuP in HW arising at or near anal angle of triangle (as in Fig. 1143); posterior lobe of prothorax in dorsal view widest at base (as in Fig. 1145); S9 with lateral carina (as in Fig. 1144) .... ***Elasmothemis* in part** (Page 238)

82'. CuP in HW distinctly separated from anal angle of triangle (as in Fig. 1142); posterior lobe of prothorax in dorsal view constricted at base (as in Fig. 1146); S9 without lateral carina .... ***Micrathyria* in part** (Page 256)

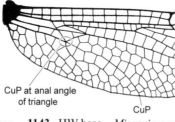

1144 male S8–10
Nothodiplax dendrophila

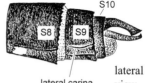

S10

S8   S9

CuP separated from anal angle of triangle

CuP

CuP at anal angle of triangle

CuP

lateral carina

lateral view

1142 HW base – *Oligoclada rhea*   **1143** HW base – *Misagria parana*

prothorax
dorsal view

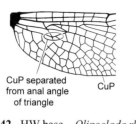

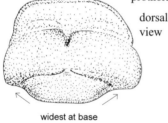

anterior lobe

medial lobe

posterior lobe

1146
Erythrodiplax castanea

1145
Antidythemis nigra

widest at base

constricted at base

83(81). Distal segment of vesica spermalis long and cylindrical (Fig. 1147) ....
.... ***Erythrodiplax*** **in part** (Page 241)

83'. Distal segment of vesica spermalis not long and cylindrical (Figs. 1148a-b) ....
.... ***Micrathyria*** **in part** (Page 256)

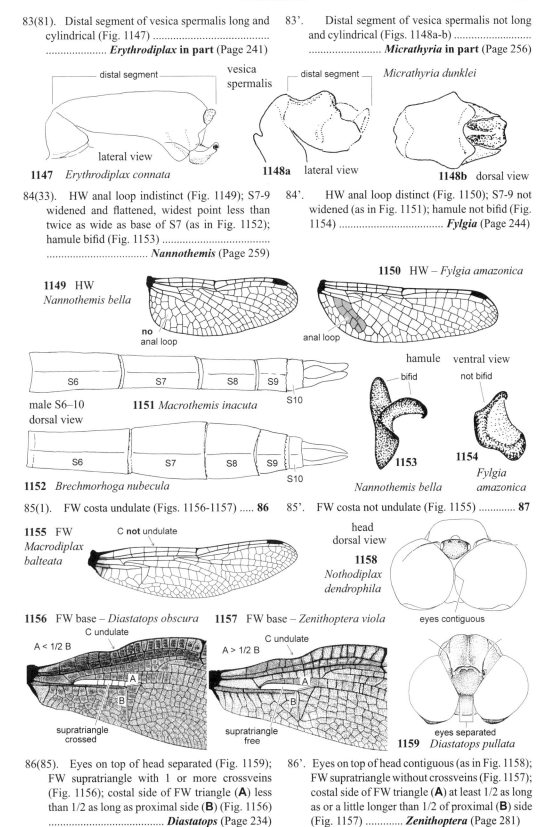

vesica spermalis

distal segment

lateral view

1147 *Erythrodiplax connata*

Micrathyria dunklei

distal segment

1148a lateral view

1148b dorsal view

84(33). HW anal loop indistinct (Fig. 1149); S7-9 widened and flattened, widest point less than twice as wide as base of S7 (as in Fig. 1152); hamule bifid (Fig. 1153) ....
.... ***Nannothemis*** (Page 259)

84'. HW anal loop distinct (Fig. 1150); S7-9 not widened (as in Fig. 1151); hamule not bifid (Fig. 1154) .... ***Fylgia*** (Page 244)

1150 HW – *Fylgia amazonica*

1149 HW
Nannothemis bella

no
anal loop

anal loop

hamule   ventral view

bifid   not bifid

male S6–10
dorsal view

1151 *Macrothemis inacuta*

S6   S7   S8   S9   S10

1153
Nannothemis bella

1154
Fylgia amazonica

1152 *Brechmorhoga nubecula*

S6   S7   S8   S9   S10

85(1). FW costa undulate (Figs. 1156-1157) **86**

85'. FW costa not undulate (Fig. 1155) **87**

1155 FW
Macrodiplax balteata

C **not** undulate

head
dorsal view

1158
Nothodiplax dendrophila

eyes contiguous

1156 FW base – *Diastatops obscura*

C undulate

A < 1/2 B

A

B

supratriangle crossed

1157 FW base – *Zenithoptera viola*

C undulate

A > 1/2 B

A

B

supratriangle free

eyes separated

1159 *Diastatops pullata*

86(85). Eyes on top of head separated (Fig. 1159); FW supratriangle with 1 or more crossveins (Fig. 1156); costal side of FW triangle (**A**) less than 1/2 as long as proximal side (**B**) (Fig. 1156) .... ***Diastatops*** (Page 234)

86'. Eyes on top of head contiguous (as in Fig. 1158); FW supratriangle without crossveins (Fig. 1157); costal side of FW triangle (**A**) at least 1/2 as long as or a little longer than 1/2 of proximal (**B**) side (Fig. 1157) .... ***Zenithoptera*** (Page 281)

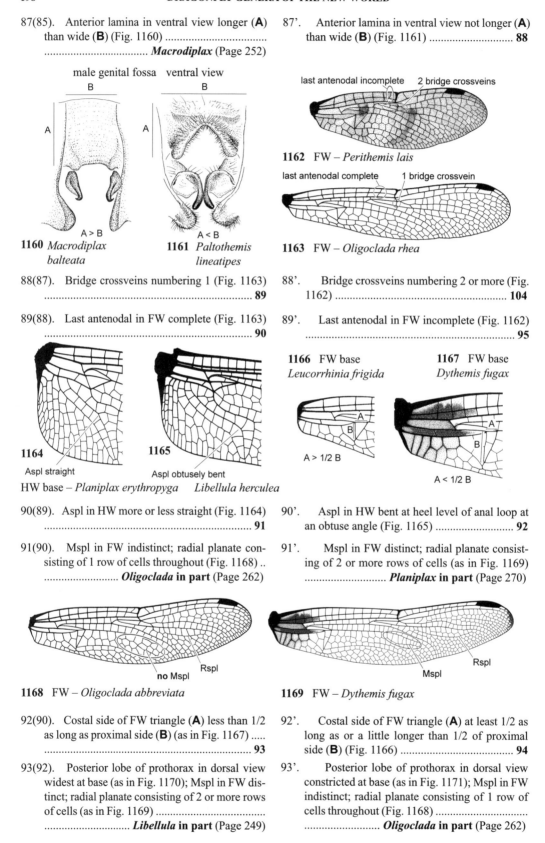

87(85). Anterior lamina in ventral view longer (**A**) than wide (**B**) (Fig. 1160) .... *Macrodiplax* (Page 252)

87'. Anterior lamina in ventral view not longer (**A**) than wide (**B**) (Fig. 1161) **88**

male genital fossa   ventral view

B   B

A   A

A > B   A < B

1160 *Macrodiplax balteata*

1161 *Paltothemis lineatipes*

last antenodal incomplete   2 bridge crossveins

1162 FW – *Perithemis lais*

last antenodal complete   1 bridge crossvein

1163 FW – *Oligoclada rhea*

88(87). Bridge crossveins numbering 1 (Fig. 1163) **89**

88'. Bridge crossveins numbering 2 or more (Fig. 1162) **104**

89(88). Last antenodal in FW complete (Fig. 1163) **90**

89'. Last antenodal in FW incomplete (Fig. 1162) **95**

1166 FW base *Leucorrhinia frigida*

1167 FW base *Dythemis fugax*

1164

1165

Aspl straight

Aspl obtusely bent

HW base – *Planiplax erythropyga*   *Libellula herculea*

A > 1/2 B

A < 1/2 B

90(89). Aspl in HW more or less straight (Fig. 1164) **91**

90'. Aspl in HW bent at heel level of anal loop at an obtuse angle (Fig. 1165) **92**

91(90). Mspl in FW indistinct; radial planate consisting of 1 row of cells throughout (Fig. 1168) .. .. *Oligoclada* **in part** (Page 262)

91'. Mspl in FW distinct; radial planate consisting of 2 or more rows of cells (as in Fig. 1169) .... *Planiplax* **in part** (Page 270)

Rspl

no Mspl

1168 FW – *Oligoclada abbreviata*

Rspl

Mspl

1169 FW – *Dythemis fugax*

92(90). Costal side of FW triangle (**A**) less than 1/2 as long as proximal side (**B**) (as in Fig. 1167) **93**

92'. Costal side of FW triangle (**A**) at least 1/2 as long as or a little longer than 1/2 of proximal side (**B**) (Fig. 1166) **94**

93(92). Posterior lobe of prothorax in dorsal view widest at base (as in Fig. 1170); Mspl in FW distinct; radial planate consisting of 2 or more rows of cells (as in Fig. 1169) .... *Libellula* **in part** (Page 249)

93'. Posterior lobe of prothorax in dorsal view constricted at base (as in Fig. 1171); Mspl in FW indistinct; radial planate consisting of 1 row of cells throughout (Fig. 1168) .... *Oligoclada* **in part** (Page 262)

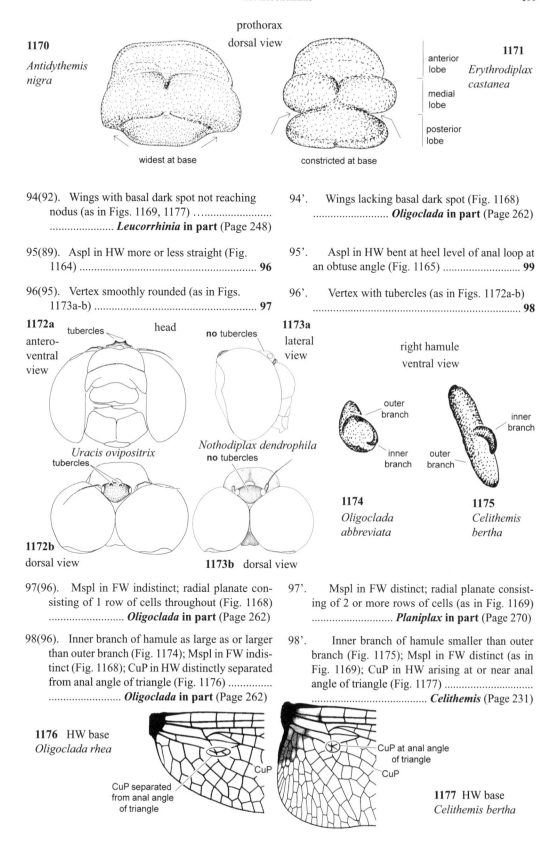

1170
Antidythemis nigra

prothorax
dorsal view

anterior lobe
medial lobe
posterior lobe

1171
Erythrodiplax castanea

widest at base

constricted at base

94(92). Wings with basal dark spot not reaching nodus (as in Figs. 1169, 1177) …..................... .... ***Leucorrhinia* in part** (Page 248)

94'. Wings lacking basal dark spot (Fig. 1168) .... ***Oligoclada* in part** (Page 262)

95(89). Aspl in HW more or less straight (Fig. 1164) **96**

95'. Aspl in HW bent at heel level of anal loop at an obtuse angle (Fig. 1165) **99**

96(95). Vertex smoothly rounded (as in Figs. 1173a-b) **97**

96'. Vertex with tubercles (as in Figs. 1172a-b) **98**

1172a
antero-ventral view

tubercles   head

Uracis ovipositrix
tubercles

1172b
dorsal view

no tubercles

1173a
lateral view

Nothodiplax dendrophila
no tubercles

1173b dorsal view

right hamule
ventral view

outer branch
inner branch
inner branch
outer branch

1174
Oligoclada abbreviata

1175
Celithemis bertha

97(96). Mspl in FW indistinct; radial planate consisting of 1 row of cells throughout (Fig. 1168) .... ***Oligoclada* in part** (Page 262)

97'. Mspl in FW distinct; radial planate consisting of 2 or more rows of cells (as in Fig. 1169) .... ***Planiplax* in part** (Page 270)

98(96). Inner branch of hamule as large as or larger than outer branch (Fig. 1174); Mspl in FW indistinct (Fig. 1168); CuP in HW distinctly separated from anal angle of triangle (Fig. 1176) .... ***Oligoclada* in part** (Page 262)

98'. Inner branch of hamule smaller than outer branch (Fig. 1175); Mspl in FW distinct (as in Fig. 1169); CuP in HW arising at or near anal angle of triangle (Fig. 1177) .... ***Celithemis*** (Page 231)

1176 HW base
Oligoclada rhea

CuP

CuP separated from anal angle of triangle

CuP at anal angle of triangle

CuP

1177 HW base
Celithemis bertha

99(95). Costal side (**A**) of FW triangle less than 1/2 as long as proximal side (**B**) (as in Fig. 1179) **100**

99'. Costal side of FW triangle (**A**) at least 1/2 as long as or a little longer than 1/2 of proximal side (**B**) (as in Fig. 1178) **103**

1178 FW base
Leucorhinia frigida

A > 1/2 B

1179 FW base
Dythemis fugax

A < 1/2 B

male genital fossa
lateral view

anterior lamina — hamule — genital lobe

1180 *Brachymesia furcata*

anterior lamina — hamule — genital lobe

1181 *Idiataphe cubensis*

100(99). Anterior lamina in lateral view as high as 4/5 of hamule or higher (Fig. 1180) **101**

100'. Anterior lamina in lateral view shorter than 4/5 of hamule (Fig. 1181) **102**

101(100). Mspl in FW indistinct (Fig. 1182); CuP in HW distinctly separated from anal angle of triangle (Fig. 1184) ....
.... ***Oligoclada* in part** (Page 262)

101'. Mspl in FW distinct (as in Fig. 1183); CuP in HW arising at or near anal angle of triangle (as in Fig. 1185) ....
.... ***Brachymesia*** (Page 227)

1182 FW – *Oligoclada abbreviata*

no Mspl — Rspl

1183 FW – *Dythemis fugax*

Mspl — Rspl

1184 HW base
Oligoclada rhea

CuP separated from anal angle of triangle — CuP

CuP at anal angle of triangle — CuP

1185 HW base
Celithemis bertha

102(100). Posterior lobe of prothorax in dorsal view widest at base (as in Fig. 1186); Mspl in FW distinct (as in Fig. 1183); CuP in HW arising at or near anal angle of triangle (as in Fig. 1185) .... ***Libellula* in part** (Page 249)

102'. Posterior lobe of prothorax in dorsal view constricted at base (as in Fig. 1187); Mspl in FW indistinct (Fig. 1182); CuP in HW distinctly separated from anal angle of triangle (Fig. 1184) .... ***Oligoclada* in part** (Page 262)

1186
Antidythemis nigra

prothorax dorsal view

widest at base

anterior lobe — medial lobe — posterior lobe

1187
Erythrodiplax castanea

constricted at base

103(99). Inner branch of hamule as large or larger than outer branch (Fig. 1188); posterior lobe of prothorax in dorsal view constricted at base (as in Fig. 1187); Mspl in FW indistinct (Fig. 1182) .... *Oligoclada* **in part** (Page 262)

103'. Inner branch of hamule smaller than outer branch (Fig. 1189); posterior lobe of prothorax in dorsal view widest at base (as in Fig. 1186); Mspl in FW distinct (as in Fig. 1183) .... *Idiataphe* (Page 247)

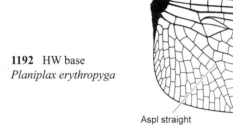

right hamule
ventral view

outer branch

inner branch

inner branch

outer branch

1188
Oligoclada abbreviata

1189
Idiataphe cubensis

prothorax   lateral view

anterior medial posterior lobe

bent caudally

1190
Erythrodiplax castanea

upright

1191
Sympetrum gilvum

104(88). Aspl in HW more or less straight (Fig. 1192); posterior lobe of prothorax in dorsal view constricted at base (as in Fig. 1187), in lateral view upright (as in Fig. 1191) **105**

104'. Aspl in HW bent at heel level of anal loop at an obtuse angle (Fig. 1193); posterior lobe of prothorax in dorsal view widest at base (as in Fig. 1186), in lateral view bent caudally (as in Fig. 1190) .... *Libellula* **in part** (Page 249)

1192 HW base
Planiplax erythropyga

Aspl straight

1193 HW base
Libellula herculea

Aspl obtusely bent

105(104). Vertex smoothly rounded (as in Figs. 1195a-b) .... *Planiplax* **in part** (Page 270)

105'. Vertex with tubercles (as in Figs. 1194a-b) .... *Perithemis* (Page 268)

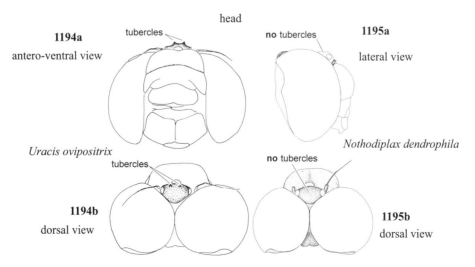

head

1194a
antero-ventral view

tubercles

no tubercles

1195a

lateral view

Uracis ovipositrix

Nothodiplax dendrophila

tubercles

no tubercles

1194b
dorsal view

1195b
dorsal view

Key to females

1. Sectors of arculus in FW stalked (as in Fig. 1196) **2**

1'. Sectors of arculus in FW not stalked (as in Fig. 1197) **71**

sectors stalked

sectors separated

1196 FW base – *Orthemis attenuata*

1197 FW base – *Libellula herculea*

2(1). Sternum of S9 not projected distally or only slightly so (Figs. 1198-1200) **3**

2'. Sternum of S9 projected distally (Figs. 1201-1203) **60**

female S8–10

1198 *Erythrodiplax basalis*

1199 *Brechmorhoga nubecula*

1200   *Argyrothemis argentea*

S10

S9

S8

vulvar lamina

sternum of S9

lateral view

S10

S9

lateral view   sternum of S9

vulvar lamina

S8   S9   S10

vulvar lamina

lateral view

S8   S9   S10

vulvar lamina   sternum of S9

S10

sternum of S9

S8   S9   S10

S8   S9

vulvar lamina   sternum of S9

1201   *Dasythemis esmeralda*

vulvar lamina   S10

S8   S9   sternum of S9

1202   *Fylgia amazonica*
ventral view

S10   sternum
of S9

S8   S9

vulvar lamina

1203   *Ypirangathemis calverti*

3(2). S8 with lateral flanges (Fig. 1204) **4**

3'. S8 without lateral flanges (Fig. 1205) **9**

female S7–10
lateral view

S10

S7   S8   S9

lateral flange

1204   *Orthemis attenuata*

S10

S7   S8   S9

no lateral flange

1205   *Micrathyria aequalis*

4(3). Arculus in HW proximal to antenodal 2 (as in Fig. 1206) **5**

4'. Arculus in HW opposite to antenodal 2, or distal to 2 but closer to antenodal 2 than to 3 (as in Fig. 1207) **8**

4''. Arculus in HW from slightly proximal to distal to antenodal 3 (as in Fig. 1208) .... ***Misagria*** (Page 258)

1207 HW base – *Orthemis attenuata*

1206 HW base – *Oligoclada rhea*   closer to antenodal 2 than to 3   **1208** HW base – *Misagria parana*

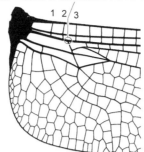

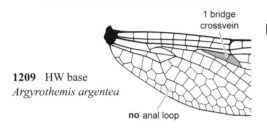

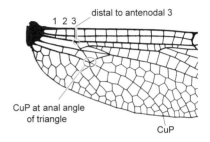

proximal to antenodal 2
1 2 3

CuP separated
from anal angle   CuP
of triangle

1 2/3

1 2 3

distal to antenodal 3

CuP at anal angle
of triangle

CuP

5(4). Bridge crossveins numbering 1 (as in Fig. 1209) **6**

5'.   Bridge crossveins numbering 2 or more (as in Fig. 1210) **7**

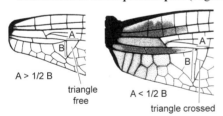

1 bridge
crossvein

1209 HW base
Argyrothemis argentea

no anal loop

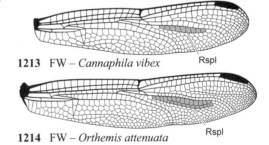

2 bridge
crossveins

1210 HW base
Micrathyria hypodidyma

anal loop

6(5). Vulvar lamina not projected ventrally (as in Figs. 1199-1203); CuP in HW distinctly separated from anal angle of triangle (as in Fig. 1206); radial planate consisting of 1 row of cells throughout (as in Fig. 1213) .... *Cannaphila* **in part** (Page 230)

6'.   Vulvar lamina projected ventrally (as in Fig. 1198); CuP in HW arising at or near anal angle of triangle (as in Fig. 1208); radial planate consisting of 2 or more rows of cells (as in Fig. 1214) .... *Orthemis* **in part** (Page 264)

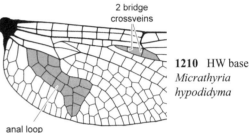

A
B

A > 1/2 B

triangle
free

A
B

A < 1/2 B
triangle crossed

1211 FW base
Oligoclada abbreviata

1212 FW base
Dythemis fugax

1213 FW – *Cannaphila vibex*   Rspl

1214 FW – *Orthemis attenuata*   Rspl

7(5). Costal side of FW triangle (**A**) less than 1/2 as long as proximal side (**B**) (as in Fig. 1212); crossveins in FW triangle present (as in Fig. 1212); radial planate consisting of 2 or more rows of cells (as in Fig. 1214) .... *Orthemis* **in part** (Page 264)

7'. Costal side of FW triangle (**A**) at least 1/2 as long as or a little longer than 1/2 of proximal side (**B**) (as in Fig. 1211); crossveins in FW triangle absent (as in Fig. 1211); radial planate consisting of 1 row of cells throughout (as in Fig. 1213) .... *Edonis* **in part** (Page 237)

8(4). Vulvar lamina not projected ventrally (Fig. 1199-1203); CuP in HW distinctly separated from anal angle of triangle (as in Fig. 1206); radial planate consisting of 1 row of cells throughout (Fig. 1213) *Cannaphila* **in part** (Page 230)

8'. Vulvar lamina projected ventrally (Fig. 1198); CuP in HW arising at or near anal angle of triangle (as in Fig. 1208); radial planate consisting of 2 or more rows of cells (as in Fig. 1214) .... *Orthemis* **in part** (Page 264)

9(3). HW anal loop indistinct (as in Fig. 1209) ...**10**

9'.   HW anal loop distinct (as in Fig. 1210) **11**

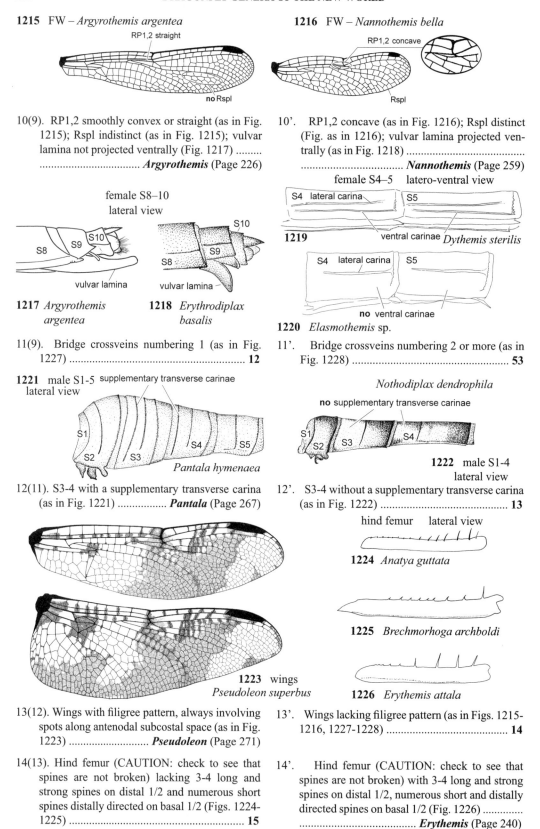

1215 FW – *Argyrothemis argentea*

RP1,2 straight

no Rspl

1216 FW – *Nannothemis bella*

RP1,2 concave

Rspl

10(9). RP1,2 smoothly convex or straight (as in Fig. 1215); Rspl indistinct (as in Fig. 1215); vulvar lamina not projected ventrally (Fig. 1217) .... ***Argyrothemis*** (Page 226)

10'. RP1,2 concave (as in Fig. 1216); Rspl distinct (Fig. as in 1216); vulvar lamina projected ventrally (as in Fig. 1218) .... ***Nannothemis*** (Page 259)

female S4–5   latero-ventral view

S4 lateral carina   S5

1219   ventral carinae *Dythemis sterilis*

S4 lateral carina   S5

no ventral carinae

1220 *Elasmothemis* sp.

female S8–10
lateral view

S8   S9 S10

vulvar lamina

S10 S9 S8

vulvar lamina

1217 *Argyrothemis argentea*

1218 *Erythrodiplax basalis*

11(9). Bridge crossveins numbering 1 (as in Fig. 1227) **12**

11'. Bridge crossveins numbering 2 or more (as in Fig. 1228) **53**

1221 male S1-5   supplementary transverse carinae
lateral view

S1 S2 S3 S4 S5

Pantala hymenaea

Nothodiplax dendrophila

no supplementary transverse carinae

S1 S2 S3 S4

1222   male S1-4
lateral view

12(11). S3-4 with a supplementary transverse carina (as in Fig. 1221) .... ***Pantala*** (Page 267)

12'. S3-4 without a supplementary transverse carina (as in Fig. 1222) **13**

hind femur   lateral view

1224 *Anatya guttata*

1225 *Brechmorhoga archboldi*

1223 wings
Pseudoleon superbus

1226 *Erythemis attala*

13(12). Wings with filigree pattern, always involving spots along antenodal subcostal space (as in Fig. 1223) .... ***Pseudoleon*** (Page 271)

13'. Wings lacking filigree pattern (as in Figs. 1215-1216, 1227-1228) **14**

14(13). Hind femur (CAUTION: check to see that spines are not broken) lacking 3-4 long and strong spines on distal 1/2 and numerous short spines distally directed on basal 1/2 (Figs. 1224-1225) **15**

14'. Hind femur (CAUTION: check to see that spines are not broken) with 3-4 long and strong spines on distal 1/2, numerous short and distally directed spines on basal 1/2 (Fig. 1226) .... ***Erythemis*** (Page 240)

15(14). Crossveins under proximal 1/2 of pterostigma in FW absent (as in Fig. 1227) .... ***Pachydiplax* in part** (Page 265)

15'. Crossveins under proximal 1/2 of pterostigma in FW present (as in Fig. 1228) …..………... **16**

1227 FW – *Pachydiplax longipennis*

1228 FW – *Micrathyria aequalis*

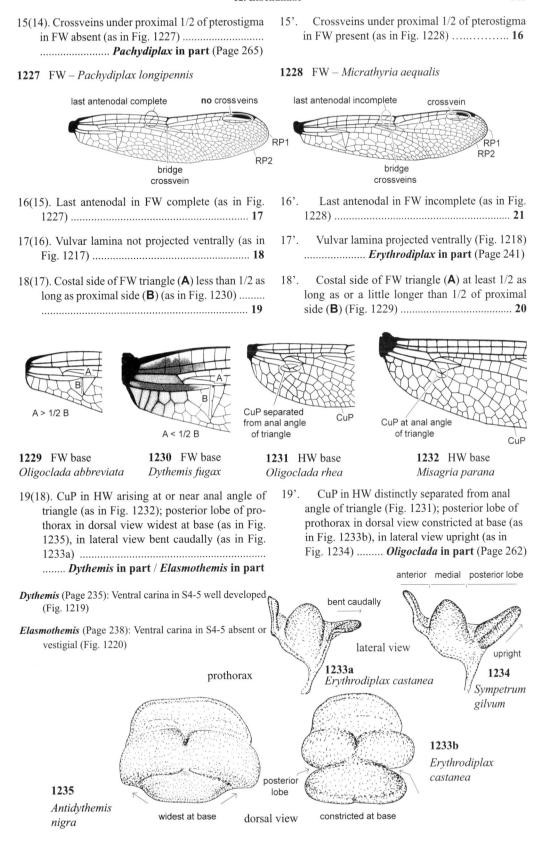

16(15). Last antenodal in FW complete (as in Fig. 1227) **17**

16'. Last antenodal in FW incomplete (as in Fig. 1228) **21**

17(16). Vulvar lamina not projected ventrally (as in Fig. 1217) **18**

17'. Vulvar lamina projected ventrally (Fig. 1218) .... ***Erythrodiplax* in part** (Page 241)

18(17). Costal side of FW triangle (**A**) less than 1/2 as long as proximal side (**B**) (as in Fig. 1230) **19**

18'. Costal side of FW triangle (**A**) at least 1/2 as long as or a little longer than 1/2 of proximal side (**B**) (Fig. 1229) **20**

1229 FW base
Oligoclada abbreviata

1230 FW base
Dythemis fugax

1231 HW base
Oligoclada rhea

1232 HW base
Misagria parana

19(18). CuP in HW arising at or near anal angle of triangle (as in Fig. 1232); posterior lobe of prothorax in dorsal view widest at base (as in Fig. 1235), in lateral view bent caudally (as in Fig. 1233a) .... ***Dythemis* in part** / ***Elasmothemis* in part**

19'. CuP in HW distinctly separated from anal angle of triangle (Fig. 1231); posterior lobe of prothorax in dorsal view constricted at base (as in Fig. 1233b), in lateral view upright (as in Fig. 1234) .... ***Oligoclada* in part** (Page 262)

Dythemis (Page 235): Ventral carina in S4-5 well developed (Fig. 1219)

Elasmothemis (Page 238): Ventral carina in S4-5 absent or vestigial (Fig. 1220)

1233a
Erythrodiplax castanea

1234
Sympetrum gilvum

1233b
Erythrodiplax castanea

1235
Antidythemis nigra

20(18). Wings with basal dark spot not reaching nodus (as in Fig. 1236) .... *Leucorrhinia* **in part** (Page 248)

20'. Wings lacking basal dark spot (as in Fig. 1237) .... *Oligoclada* **in part** (Page 262)

basal dark spot

not broadly widened

1236 HW – *Leucorrhinia frigida*

not broadly widened

1237 HW – *Oligoclada xanthopleura*

21(16). Arculus in HW proximal to antenodal 2 (as in Fig. 1238) **22**

21'. Arculus in HW opposite to antenodal 2, or distal to 2 but closer to antenodal 2 than to 3 (as in Fig. 1239) **41**

1238 HW base *Oligoclada rhea*

proximal to antenodal 2
1 / 2 3

closer to antenodal 2 than to 3
1 2 / 3

1239 HW base *Orthemis attenuata*

1240 FW base *Nephepeltia* sp.

free

crossed

1241 FW base *Misagria parana*

22(21). Vulvar lamina not projected ventrally (as in Fig. 1242) **23**

22'. Vulvar lamina projected ventrally (Fig. 1243) **40**

female S8–10 lateral view

S8 S9 S10

vulvar lamina

1242 *Argyrothemis argentea*

S10
S9
S8
vulvar lamina

1243 *Erythrodiplax basalis*

female S8–10 ventral view

S8 S9

S10

vulvar lamina

1244 *Dythemis multipunctata*

S8 S9

S10

1245 *Erythrodiplax basalis*

23(22). Vulvar lamina less than 1/3 the length of S9 (Fig. 1244) **24**

23'. Vulvar lamina 1/3 of S9 length to as long as S9 (Fig. 1245) **39**

24(23). Crossveins in FW triangle present (as in Fig. 1241) **25**

24'. Crossveins in FW triangle absent (Fig. 1240) **33**

2 rows of cells

no Mspl

1246 FW – *Oligoclada abbreviata*

3 rows of cells

no large cells

Mspl

1247 FW – *Dythemis fugax*

25(24). Posterior lobe of prothorax in dorsal view widest at base and narrower than other prothoracic lobes (as in Fig. 1248), in lateral view bent caudally (as in Fig. 1250) **26**

25'.   Posterior lobe of prothorax in dorsal view constricted at base and wider than other prothoracic lobes (as in Fig. 1249a), in lateral view upright (as in Fig. 1249b) **32**

prothorax

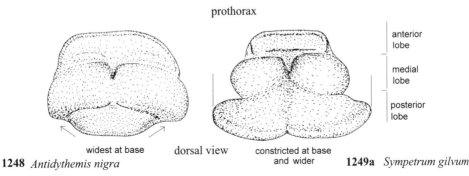

1248 *Antidythemis nigra*   dorsal view   **1249a** *Sympetrum gilvum*

widest at base

constricted at base and wider

anterior lobe

medial lobe

posterior lobe

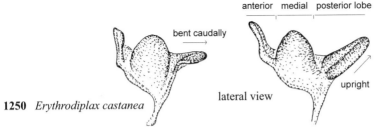

anterior   medial   posterior lobe

1249b *Sympetrum gilvum*

bent caudally

1250 *Erythrodiplax castanea*   lateral view

upright

26(25). Mspl in FW indistinct (as in Fig. 1246) ... **27**

26'.   Mspl in FW distinct (as in Fig. 1247) **28**

27(26). Cell rows at base of FW discoidal field numbering 2 (as in Fig. 1246) ....
.... ***Macrothemis* in part** (Page 253)

27'.   Cell rows at base of FW discoidal field numbering 3, 4 or more (as in Fig. 1247) ....
.... ***Elasmothemis* in part** (Page 238)

28(26). HW very broad at base (as in Figs. 1251-1252) **29**

28'.   HW not broadly widened at base (as in Figs. 1236-1237) **30**

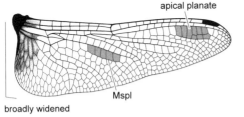

apical planate

broadly widened

Mspl

1251 HW – *Tauriphila australis*

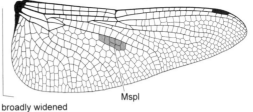

broadly widened

Mspl

1252 HW – *Paltothemis lineatipes*

29(28). Median planate consisting of 1 row of cells throughout (as in Fig. 1251) ....
.... ***Dythemis* in part / *Tauriphila* in part**

29'.   Median planate consisting of 2 or more rows of cells (as in Fig. 1252) ....
.... ***Paltothemis* in part** (Page 266)

Dythemis (Page 235): Apical planate of wings weakly developed *and* never accompanied by large cells (as in Fig. 1247)

Tauriphila (Page 275): Apical planate of wings well developed *and* usually with 3-4 large cells at its base (as in Fig. 1251) (except for some *T. argo*, which have only 1 large cell at its base)

30(28). Wings with basal dark spot not reaching nodus (as in Fig. 1253) .... *Dythemis* in part / *Tauriphila* in part

Dythemis (Page 235): Apical planate of wings weakly developed *and* never accompanied by large cells (Fig. 1253)

30'.　Wings lacking basal dark spot (as in Fig. 1254) **31**

Tauriphila (Page 275): Apical planate of wings well developed (Fig. 1254) *and* usually with 3-4 large cells at its base (except for some *T. argo*, which have only 1 large cell at its base)

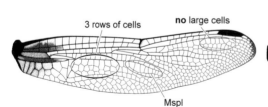

1253　FW – *Dythemis fugax*

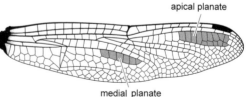

1254　FW – *Tauriphila asutralis*

31(30). FW discoidal field widening distally (Fig. 1258) .... *Dythemis* in part / *Elasmothemis* in part / *Brechmorhoga* in part

31'.　FW discoidal field parallel sided (as in Fig. 1259) .... *Dythemis* in part / *Elasmothemis* in part

female S8–10

1255a　dorsal view

Dythemis multipunctata

1255b　latero-ventral view

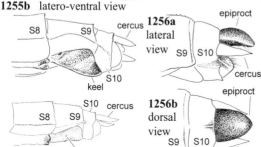

1257　*Brechmorhoga praedatrix* latero-ventral view　*Brechmorhoga nubecula*

1256a lateral view

1256b dorsal view

Brechmorhoga (Page 228): S9 sternum smoothly rounded (Fig. 1257), *if* keeled (*B. nubecula*), *then* epiproct as long as or longer than cerci, greatly enlarged, and polished (Figs. 1256a-b)

Dythemis/Elasmothemis: S9 sternum keeled (Fig. 1255b) and epiproct shorter than cerci, never polished or greatly enlarged (Fig. 1255a)

Dythemis (Page 235): Ventral carina in S4-5 well developed (Fig. 1261)

Elasmothemis (Page 238): Ventral carina in S4-5 absent or vestigial (Fig. 1262)

31".　FW discoidal field narrowing distally (Fig. 1260) ..... *Dythemis* in part / *Scapanea* in part

1258　FW – *Brechmorhoga archboldi*

1259　FW – *Elasmothemis cannacrioides*

1260　FW – *Scapanea frontalis*

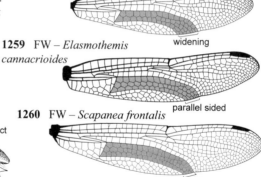

female S4–5 latero-ventral view　**1261**　*Dythemis sterilis*

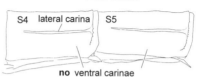

1262　*Elasmothemis* sp.

Dythemis (Page 235): Posterior angle of S3-4 ventral terga right (Fig. 1263); angle between claw and pretarsus acute (Fig. 1265)

Scapanea (Page 272): Posterior angle of S3-4 ventral terga obtuse (Fig. 1264); angle between claw and pretarsus right to obtuse (Fig. 1266)

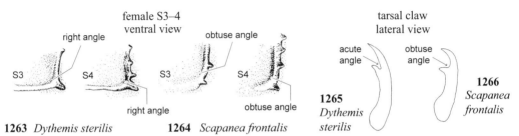

1263 *Dythemis sterilis*   **1264** *Scapanea frontalis*

1265 *Dythemis sterilis*

1266 *Scapanea frontalis*

32(25). Mspl in FW indistinct (as in Fig. 1253); CuP in HW distinctly separated from anal angle of triangle (as in Fig. 1267) .... *Oligoclada* **in part** (Page 262)

32'.   Mspl in FW distinct (Fig. 1254); CuP in HW arising at or near anal angle of triangle (as in Fig. 1268) .... *Sympetrum* **in part** (Page 274)

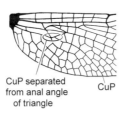

1267 HW base – *Oligoclada rhea*

1268 HW base – *Misagria parana*

33(24). CuP in HW arising at or near anal angle of triangle (as in Fig. 1268); posterior lobe of prothorax in dorsal view widest at base (as in Fig. 1269), in lateral view bent caudally (as in Fig. 1270b) **34**

33'.   CuP in HW distinctly separated from anal angle of triangle (as in Fig. 1267); posterior lobe of prothorax in dorsal view constricted at base, in lateral view upright (as in Fig. 1271) **38**

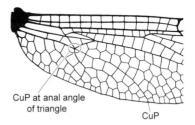

1269
Antidythemis nigra

prothorax
dorsal view

1270a
Erythrodiplax castanea

anterior lobe

medial lobe

posterior lobe

widest at base

constricted at base

anterior   medial   posterior lobe

bent caudally

upright

lateral view

1270b *Erythrodiplax castanea*

1271 *Sympetrum gilvum*

34(33). Vertex smoothly rounded (as in Figs. 1273a-b) **35**

34'. Vertex with tubercles (as in Figs. 1272a-b) **37**

tubercles   head

1272a
antero-ventral view

no tubercles

1273a
lateral view

Uracis ovipositrix

tubercles

1272b
dorsal view

no tubercles   *Nothodiplax dendrophila*

1273b
dorsal view

35(34). HW very broad at base (as in Fig. 1274) **36**

35'. HW not broadly widened at base (as in Fig. 1275) .... ***Gynothemis*** in part / ***Macrothemis*** in part

We have been unable to diagnose females of these two genera [see discussion on Page 246]

1274 HW base – *Miathyria simplex*

broadly widened

not
broadly widened

1275 HW base – *Gynothemis venipunctata*

36(35). Spurious vein 1 begins proximal to or at 1/2 of pterostigma (as in Fig. 1276) .... ***Macrothemis*** in part (Page 253)

36'. Spurious vein 1 begins at distal 1/2 of pterostigma (Fig. 1277) .... ***Miathyria*** (Page 255)

spurious
vein 1

1276 FW base – *Macrothemis tessellata*

spurious
vein 1

1277 FW base – *Miathyria simplex*

37(34). Mspl in FW indistinct (as in Fig. 1280) (except in *M. griseofrons*)* .... ***Gynothemis*** in part / ***Macrothemis*** in part (Page 253)

We have been unable to diagnose females of these two genera [see discussion on Page 246]

37'. Mspl in FW distinct (as in Fig. 1281) .... ***Brechmorhoga*** in part (Page 228)

Brechmorhoga: Short inner tarsal claw (Fig. 1278); relatively wide discoidal field in FW, with discoidal index approximately 1.4-1.8 [index consisting of ratio of distance between MA and MP at wing margin divided by distance at proximal portion] (Fig. 1280); longer body (40-62 mm)

***Macrothemis**: Long inner tarsal claw (as in Fig. 1279) (except for members of *M. tessellata* group of species); relatively narrow discoidal field in FW, with discoidal index <1.0-1.3 (as in Fig. 1281); shorter body (21-42 mm)

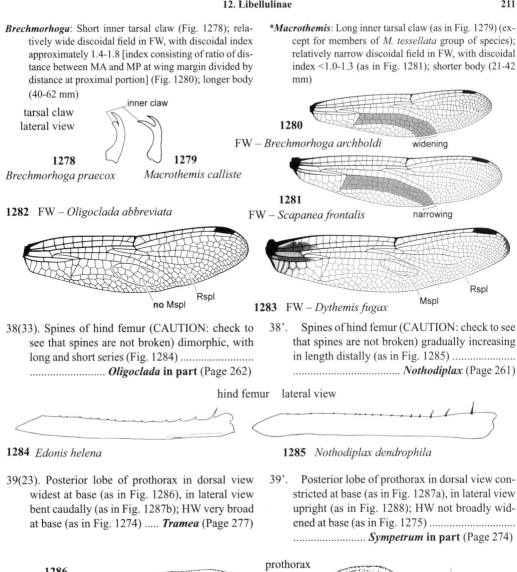

tarsal claw
lateral view

inner claw

1278
Brechmorhoga praecox

1279
Macrothemis calliste

1280
FW – *Brechmorhoga archboldi*   widening

1281
FW – *Scapanea frontalis*   narrowing

1282   FW – *Oligoclada abbreviata*

no Mspl   Rspl

1283   FW – *Dythemis fugax*

Rspl   Mspl

38(33).   Spines of hind femur (CAUTION: check to see that spines are not broken) dimorphic, with long and short series (Fig. 1284) ....
.... ***Oligoclada* in part** (Page 262)

38'.   Spines of hind femur (CAUTION: check to see that spines are not broken) gradually increasing in length distally (as in Fig. 1285) ....
.... ***Nothodiplax*** (Page 261)

hind femur   lateral view

1284   *Edonis helena*

1285   *Nothodiplax dendrophila*

39(23).   Posterior lobe of prothorax in dorsal view widest at base (as in Fig. 1286), in lateral view bent caudally (as in Fig. 1287b); HW very broad at base (as in Fig. 1274) ..... ***Tramea*** (Page 277)

39'.   Posterior lobe of prothorax in dorsal view constricted at base (as in Fig. 1287a), in lateral view upright (as in Fig. 1288); HW not broadly widened at base (as in Fig. 1275) ....
.... ***Sympetrum* in part** (Page 274)

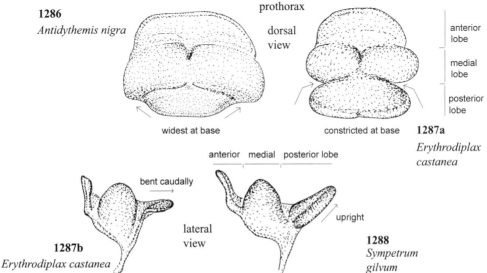

1286
Antidythemis nigra

prothorax
dorsal
view

anterior
lobe

medial
lobe

posterior
lobe

widest at base

constricted at base   **1287a**
*Erythrodiplax
castanea*

anterior   medial   posterior lobe

bent caudally

lateral
view

1287b
Erythrodiplax castanea

upright

1288
*Sympetrum
gilvum*

40(22). Vulvar lamina scoop shaped (Figs. 1289a-b); posterior lobe of prothorax in lateral view bent caudally (as in Fig. 1292b), in dorsal view narrower than other prothoracic lobes (as in Figs. 1291, 1292a) ....
.... *Crocothemis/ Erythrodiplax* **in part**

40'.   Vulvar lamina bilobed (Fig. 1290); posterior lobe of prothorax in lateral view upright (as in Fig. 1293b), in dorsal view wider than other prothoracic lobes (as in Fig. 1293a) ....
.... *Sympetrum* **in part** (Page 274)

female S8–10

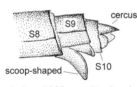

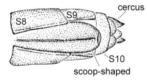

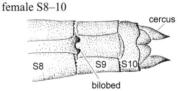

lateral view **1289a**   *Erythrodiplax basalis* **1289b**   ventral view   **1290** *Sympetrum villosum*

prothorax   dorsal view

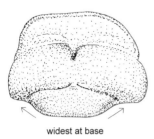

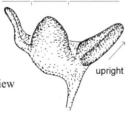

anterior lobe

medial lobe

posterior lobe

1291 *Antidythemis nigra*   **1292a** *Erythrodiplax castanea*   **1293a** *Sympetrum gilvum*

widest at base   constricted at base and narrower   wider

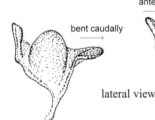

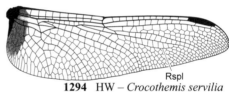

1294 HW – *Crocothemis servilia*

lateral view

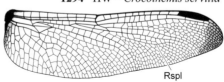

1295 HW – *Erythrodiplax umbrata*

1292b *Erythrodiplax castanea*   **1293b** *Sympetrum gilvum*

Crocothemis (Page 232): Occurring *only* in Florida in US and Cuba (as of 2005) *and* large size (total length more than 40 mm) *and* 1 row of cells in the radial planate (as in Fig. 1294)

Erythrodiplax (Page 241): *If* from Florida in US or Cuba *and* small size (total length 36 mm or less, except *E. umbrata* and *E. funerea*) *and* 2 rows of cells in radial planate (Fig. 1295)

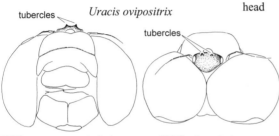

head

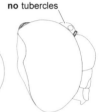

1297a lateral view   **1297b** dorsal view

1296a antero-ventral view   **1296b** dorsal view

Nothodiplax dendrophila

41(21). Vertex smoothly rounded (as in Figs. 1297a-b) **42**

41'.   Vertex with tubercles (as in Figs. 1296a-b) **43**

42(41). CuP in HW arising at or near anal angle of triangle (as in Fig. 1301); posterior lobe of prothorax in dorsal view widest at base (as in Fig. 1291); position of nodus in FW distal to 1/2 of wing length (as in Fig. 1302) .... *Gynothemis* in part / *Macrothemis* in part (Page 253)

We have been unable to diagnose females of these two genera [see discussion on Page 246]

42'.   CuP in HW distinctly separated from anal angle of triangle (as in Fig. 1300); posterior lobe of prothorax in dorsal view constricted at base (as in Figs. 1292a, 1293a); position of nodus in FW at 1/2 of wing length or proximal (as in Fig. 1303) .... *Anatya* (Page 224)

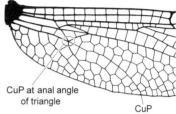

1301 HW base – *Misagria parana*

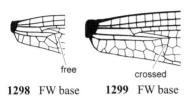

1298 FW base *Nephepeltia* sp.

free

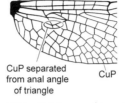

1299 FW base *Misagria parana*

crossed

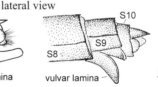

CuP separated from anal angle of triangle

CuP

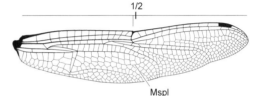

CuP at anal angle of triangle

CuP

1300 HW base – *Oligoclada rhea*

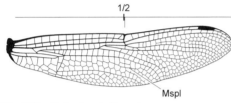

1/2

Mspl

1302 FW – *Paltothemis lineatipes*

1/2

Mspl

1303 FW – *Tramea abdominalis*

43(41). Vulvar lamina not projected ventrally (as in Fig. 1304) **44**

43'.   Vulvar lamina projected ventrally (Fig. 1305) .... *Erythrodiplax* in part (Page 241)

female S8–10   lateral view

1304

Argyrothemis argentea

S8   S9   S10

vulvar lamina

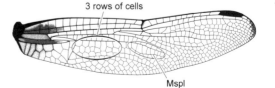

S10

S9

S8

vulvar lamina

1305

Erythrodiplax basalis

44(43). Crossveins in FW triangle present (as in Fig. 1299) **45**

44'.   Crossveins in FW triangle absent (Fig. 1298) **52**

45(44). Posterior lobe of prothorax in dorsal view widest at base (as in Fig. 1291) **46**

45'.   Posterior lobe of prothorax in dorsal view constricted at base (as in Figs. 1292a, 1293a) .... *Rhodopygia* (Page 272)

46(45). Mspl in FW indistinct (as in Fig. 1306) **47**

46'.   Mspl in FW distinct (as in Fig. 1307) **48**

2 rows of cells

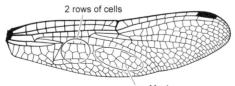

no Mspl

1306 FW – *Oligoclada abbreviata*

3 rows of cells

Mspl

1307 FW – *Dythemis fugax*

47(46). Cell rows at base of FW discoidal field numbering 2 (as in Fig. 1306) .... *Macrothemis* in part (Page 253)

47'.   Cell rows at base of FW discoidal field numbering 3, 4 or more (as in Fig. 1307) .... *Elasmothemis* in part (Page 238)

48(46). HW very broad at base (as in Fig. 1308) ....
.... **49**

48'. HW not broadly widened at base (as in Fig. 1309) **50**

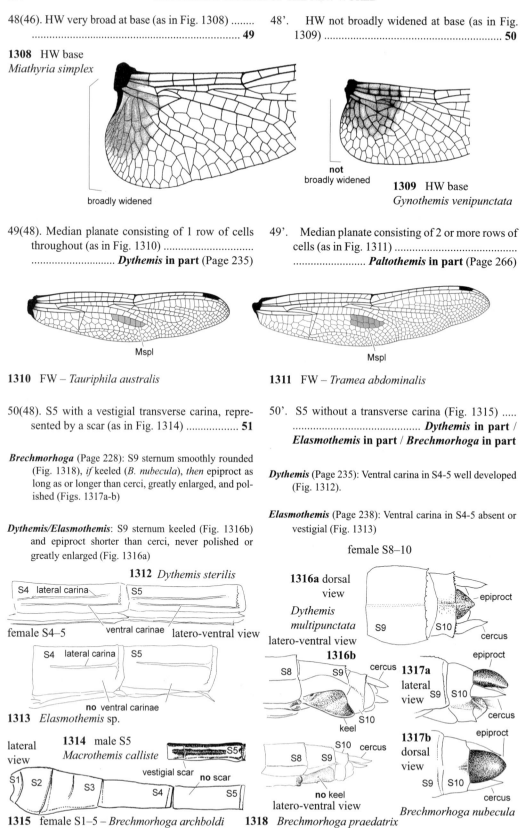

1308 HW base
Miathyria simplex

broadly widened

not broadly widened

1309 HW base
Gynothemis venipunctata

49(48). Median planate consisting of 1 row of cells throughout (as in Fig. 1310) ....
.... *Dythemis* **in part** (Page 235)

49'. Median planate consisting of 2 or more rows of cells (as in Fig. 1311) ....
.... *Paltothemis* **in part** (Page 266)

Mspl

1310 FW – *Tauriphila australis*

Mspl

1311 FW – *Tramea abdominalis*

50(48). S5 with a vestigial transverse carina, represented by a scar (as in Fig. 1314) **51**

50'. S5 without a transverse carina (Fig. 1315) .....
.... *Dythemis* **in part /** *Elasmothemis* **in part /** *Brechmorhoga* **in part**

Brechmorhoga (Page 228): S9 sternum smoothly rounded (Fig. 1318), *if* keeled (*B. nubecula*), *then* epiproct as long as or longer than cerci, greatly enlarged, and polished (Figs. 1317a-b)

Dythemis/Elasmothemis: S9 sternum keeled (Fig. 1316b) and epiproct shorter than cerci, never polished or greatly enlarged (Fig. 1316a)

Dythemis (Page 235): Ventral carina in S4-5 well developed (Fig. 1312).

Elasmothemis (Page 238): Ventral carina in S4-5 absent or vestigial (Fig. 1313)

female S8–10

1312 *Dythemis sterilis*

S4 lateral carina   S5

female S4–5   ventral carinae   latero-ventral view

S4 lateral carina   S5

no ventral carinae

1313 *Elasmothemis* sp.

lateral view

1314 male S5
Macrothemis calliste

S5

vestigial scar   **no** scar

S1 S2   S3   S4   S5

1315 female S1–5 – *Brechmorhoga archboldi*

1316a dorsal view

Dythemis multipunctata
latero-ventral view

epiproct

S9   S10

cercus

1316b

cercus

S8   S9

S10

keel

1317a lateral view

epiproct

S9   S10

cercus

1317b dorsal view

epiproct

S9   S10

cercus

S10   cercus

S8   S9

no keel
latero-ventral view

1318 *Brechmorhoga praedatrix*

Brechmorhoga nubecula

51(50). FW discoidal field widening distally (Fig. 1323) .... **Dythemis** in part / **Brechmorhoga** in part

Dythemis: S9 sternum keeled (Fig. 1316b) and epiproct shorter than cerci, never polished or greatly enlarged (Fig. 1316a)

Brechmorhoga (Page 228): S9 sternum smoothly rounded (Fig. 1318), *if* keeled (*B. nubecula*), *then* epiproct as long as or longer than cerci, greatly enlarged, and polished (Figs. 1317a-b)

51'.   FW discoidal field parallel sided (as in Fig. 1324) .... **Dythemis** in part (Page 235)

51".   FW discoidal field narrowing distally (Fig. 1325) ..... **Dythemis** in part / **Scapanea** in part

Dythemis (Page 235): Posterior angle of S3-4 ventral terga right (Fig. 1319); angle between claw and pretarsus acute (Fig. 1321)

Scapanea (Page 272): Posterior angle of S3-4 ventral terga obtuse (Fig. 1320); angle between claw and pretarsus right to obtuse (Fig. 1322)

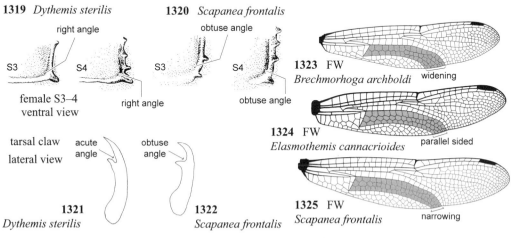

1319 *Dythemis sterilis*

right angle

S3   S4

female S3–4 ventral view   right angle

tarsal claw lateral view   acute angle

1321 *Dythemis sterilis*

1320 *Scapanea frontalis*

obtuse angle

S3   S4

obtuse angle

obtuse angle

1322 *Scapanea frontalis*

1323 FW *Brechmorhoga archboldi*   widening

1324 FW *Elasmothemis cannacrioides*   parallel sided

1325 FW *Scapanea frontalis*   narrowing

52(44). Mspl in FW indistinct (as in Fig. 1326) .... **Gynothemis** in part / **Macrothemis** in part (Page 253)

We have been unable to diagnose females of these two genera [see discussion on Page 246]

52'.   Mspl in FW distinct (as in Fig. 1327) .... **Brechmorhoga** in part (Page 228)

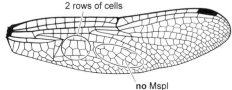

2 rows of cells

no Mspl

1326 FW – *Oligoclada abbreviata*

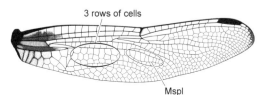

3 rows of cells

Mspl

1327 FW – *Dythemis fugax*

53(11). Aspl in HW more or less straight (as in Fig. 1328) .... **Micrathyria** in part (Page 256)

53'.   Aspl in HW bent at heel level of anal loop at an obtuse angle (as in Fig. 1329) **54**

53".   Aspl in HW bent at heel level of anal loop at a right (90 degree) to acute angle (as in Fig. 1330) .... **Antidythemis** (Page 225)

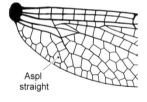

Aspl straight

1328 HW – *Micrathyria dido*

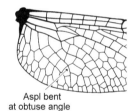

Aspl bent at obtuse angle

1329 HW – *Micrathyria aequalis*

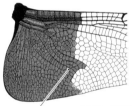

Aspl bent at acute angle

1330 HW *Antidythemis nigra*

54(53). Last antenodal in FW complete (as in Fig. 1331) **55**

54'.   Last antenodal in FW incomplete (as in Fig. 1332) **58**

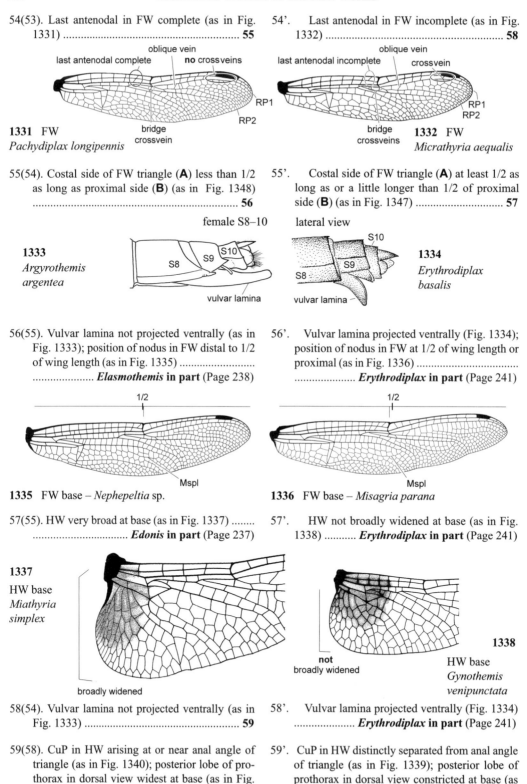

1331 FW
Pachydiplax longipennis

bridge crossvein

last antenodal complete | oblique vein | **no** crossveins | RP1 | RP2

last antenodal incomplete | oblique vein | crossvein | RP1 | RP2

bridge crossveins

1332 FW
Micrathyria aequalis

55(54). Costal side of FW triangle (**A**) less than 1/2 as long as proximal side (**B**) (as in Fig. 1348) **56**

55'.   Costal side of FW triangle (**A**) at least 1/2 as long as or a little longer than 1/2 of proximal side (**B**) (as in Fig. 1347) **57**

female S8–10

1333
Argyrothemis argentea

S8 | S9 | S10

vulvar lamina

lateral view

S10

S8 | S9

vulvar lamina

1334
Erythrodiplax basalis

56(55). Vulvar lamina not projected ventrally (as in Fig. 1333); position of nodus in FW distal to 1/2 of wing length (as in Fig. 1335) ....
.... ***Elasmothemis* in part** (Page 238)

56'.   Vulvar lamina projected ventrally (Fig. 1334); position of nodus in FW at 1/2 of wing length or proximal (as in Fig. 1336) ....
.... ***Erythrodiplax* in part** (Page 241)

1/2

Mspl

1335 FW base – *Nephepeltia* sp.

1/2

Mspl

1336 FW base – *Misagria parana*

57(55). HW very broad at base (as in Fig. 1337) ....
.... ***Edonis* in part** (Page 237)

57'.   HW not broadly widened at base (as in Fig. 1338) .... ***Erythrodiplax* in part** (Page 241)

1337
HW base
Miathyria simplex

broadly widened

not broadly widened

1338
HW base
Gynothemis venipunctata

58(54). Vulvar lamina not projected ventrally (as in Fig. 1333) **59**

58'.   Vulvar lamina projected ventrally (Fig. 1334) .... ***Erythrodiplax* in part** (Page 241)

59(58). CuP in HW arising at or near anal angle of triangle (as in Fig. 1340); posterior lobe of prothorax in dorsal view widest at base (as in Fig. 1341); position of nodus in FW distal to 1/2 of wing length (as in Fig. 1335) ....
.... ***Elasmothemis* in part** (Page 238)

59'.   CuP in HW distinctly separated from anal angle of triangle (as in Fig. 1339); posterior lobe of prothorax in dorsal view constricted at base (as in Fig. 1342); position of nodus in FW at 1/2 of wing length or proximal (as in Fig. 1336) ....
.... ***Micrathyria* in part** (Page 256)

1339 HW base – *Oligoclada rhea*

1340 HW base – *Misagria parana*

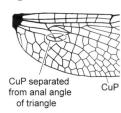

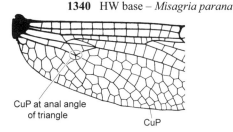

CuP separated
from anal angle
of triangle

CuP

CuP at anal angle
of triangle

CuP

prothorax
dorsal view

1341

*Antidythemis
nigra*

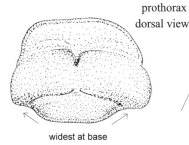

anterior
lobe

medial
lobe

posterior
lobe

1342

*Erythrodiplax
castanea*

widest at base

constricted at base

60(2). Last antenodal in FW complete (as in Fig. 1331) **61**

61(60). Bridge crossveins numbering 1 (as in Fig. 1331); S5 without a transverse carina (Fig. 1344) **62**

60'. Last antenodal in FW incomplete (as in Fig. 1332) **65**

61'. Bridge crossveins numbering 2 or more (as in Fig. 1332); S5 with a vestigial carina, represented by a scar (as in Fig. 1343) **64**

lateral view

1343 male S5
Macrothemis calliste

female S8–10   lateral view

S5

vestigial scar   no scar

S1 S2   S3   S4   S5

1344 female S1–5 – *Brechmorhoga archboldi*

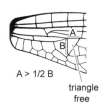

S8   S9   S10

vulvar lamina   sternum of S9

1345
Dasythemis esmeralda

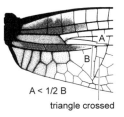

S10

S9

sternum of S9

1346
Brechmorhoga nubecula

62(61). Ventral side of S10 prolonged to about distal tip of cerci (Fig. 1345); posterior lobe of prothorax in dorsal view widest at base (as in Fig. 1341) .... ***Dasythemis*** (Page 233)

63(62). Crossveins under proximal 1/2 of pterostigma in FW absent (as in Fig. 1331); costal side of FW triangle (**A**) less than 1/2 as long as proximal side (**B**) (as in Fig. 1348); crossveins in FW triangle present (as in Fig. 1348) .... ***Pachydiplax*** **in part** (Page 265)

62'. Ventral side of S10 not prolonged distally (as in Fig. 1346); posterior lobe of prothorax in dorsal view constricted at base (as in Fig. 1342) **63**

63'. Crossveins under proximal 1/2 of pterostigma in FW present (as in Fig. 1332); costal side of FW triangle (**A**) at least 1/2 as long as or a little longer than 1/2 of proximal side (**B**) (as in Fig. 1347); crossveins in FW triangle absent (as in Fig. 1347) .... ***Elga*** **in part** (Page 239)

A

B

1347 FW base
Oligoclada abbreviata

A > 1/2 B

triangle free

A

B

A < 1/2 B

triangle crossed

1348 FW base
Dythemis fugax

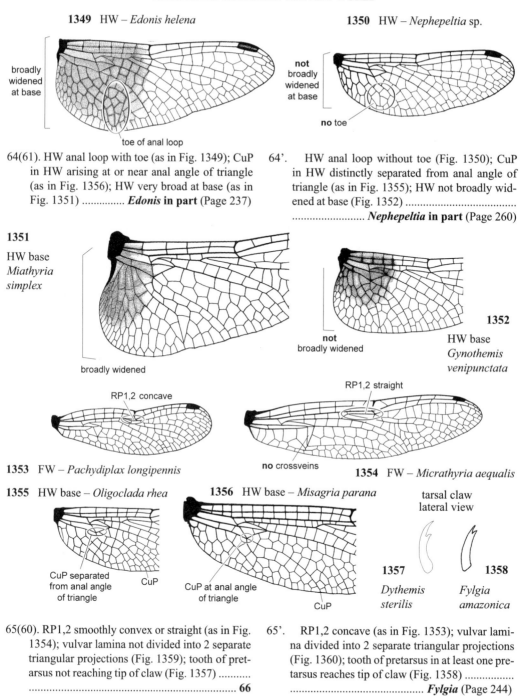

1349 HW – *Edonis helena*

broadly widened at base

toe of anal loop

1350 HW – *Nephepeltia* sp.

not broadly widened at base

no toe

64(61). HW anal loop with toe (as in Fig. 1349); CuP in HW arising at or near anal angle of triangle (as in Fig. 1356); HW very broad at base (as in Fig. 1351) .... *Edonis* **in part** (Page 237)

64'. HW anal loop without toe (Fig. 1350); CuP in HW distinctly separated from anal angle of triangle (as in Fig. 1355); HW not broadly widened at base (Fig. 1352) ....
.... *Nephepeltia* **in part** (Page 260)

1351

HW base *Miathyria simplex*

broadly widened

not broadly widened

1352

HW base *Gynothemis venipunctata*

RP1,2 concave

RP1,2 straight

no crossveins

1353 FW – *Pachydiplax longipennis*

1354 FW – *Micrathyria aequalis*

1355 HW base – *Oligoclada rhea*

CuP separated from anal angle of triangle

CuP

1356 HW base – *Misagria parana*

CuP at anal angle of triangle

CuP

tarsal claw lateral view

1357

Dythemis sterilis

1358

Fylgia amazonica

65(60). RP1,2 smoothly convex or straight (as in Fig. 1354); vulvar lamina not divided into 2 separate triangular projections (Fig. 1359); tooth of pretarsus not reaching tip of claw (Fig. 1357) ....
.... **66**

65'. RP1,2 concave (as in Fig. 1353); vulvar lamina divided into 2 separate triangular projections (Fig. 1360); tooth of pretarsus in at least one pretarsus reaches tip of claw (Fig. 1358) ....
.... *Fylgia* (Page 244)

female S8–10

1359

Tholymis citrina

vulvar lamina **not** completely divided

cercus

ventral view

vulvar lamina separate projections

S10

1360

Fylgia amazonica

S8 S9 S10

S8 S9

sternum of S9

66(65). Sternum of S9 projected distally to about posterior margin of S10 (Fig. 1361) **67**

66'. Sternum of S9 projected distally beyond tip of cerci (Fig. 1362) **69**

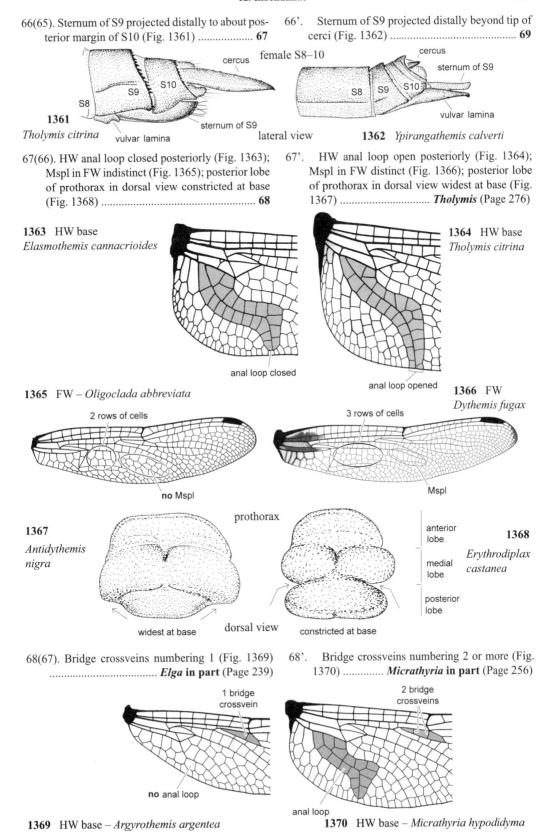

1361
Tholymis citrina

female S8–10

lateral view

1362 *Ypirangathemis calverti*

67(66). HW anal loop closed posteriorly (Fig. 1363); Mspl in FW indistinct (Fig. 1365); posterior lobe of prothorax in dorsal view constricted at base (Fig. 1368) **68**

67'. HW anal loop open posteriorly (Fig. 1364); Mspl in FW distinct (Fig. 1366); posterior lobe of prothorax in dorsal view widest at base (Fig. 1367) .... *Tholymis* (Page 276)

1363 HW base
Elasmothemis cannacrioides

1364 HW base
Tholymis citrina

1365 FW – *Oligoclada abbreviata*

1366 FW
Dythemis fugax

1367
Antidythemis nigra

prothorax

dorsal view

1368

Erythrodiplax castanea

68(67). Bridge crossveins numbering 1 (Fig. 1369) .... *Elga* in part (Page 239)

68'. Bridge crossveins numbering 2 or more (Fig. 1370) .... *Micrathyria* in part (Page 256)

1369 HW base – *Argyrothemis argentea*

1370 HW base – *Micrathyria hypodidyma*

69(66). Vertex smoothly rounded (as in Figs. 1372a-b) **70**

69'. Vertex with tubercles (as in Figs. 1371a-b) .... ***Uracis*** (Page 279)

Uracis ovipositrix

tubercles

1371a antero-ventral view

tubercles

1371b dorsal view

head

Nothodiplax dendrophila

no tubercles

no tubercles

1372a lateral view

1372b dorsal view

70(69). Costal side of FW triangle (**A**) less than 1/2 as long as proximal side (**B**) (as in Fig. 1374); crossveins in FW triangle present (as in Fig. 1374); vulvar lamina projected distally beyond tip of cerci (Fig. 1376) .... ***Ypirangathemis*** (Page 280)

70'. Costal side of FW triangle (**A**) at least 1/2 as long as or a little longer than 1/2 of proximal side (**B**) (as in Fig. 1373); crossveins in FW triangle absent (as in Fig. 1373); vulvar lamina less than 1/3 the length of S9 (as in Fig. 1375) .... ***Elga* in part** (Page 239)

A > 1/2 B

triangle free

1373 FW base
Oligoclada abbreviata

A < 1/2 B

triangle crossed

1374 FW base – *Dythemis fugax*

S8   S9

S10

vulvar lamina

1375 *Dythemis multipunctata* ventral view

female S8–10   S10

sternum of S9

S8   S9

vulvar lamina

1376 *Ypirangathemis calverti* lateral view

71(1). FW costa undulate (Figs. 1377-1378) **72**

71'. FW costa not undulate (Figs. 1381-1382) **73**

A < 1/2 B

C undulate

1377 FW base
Diastatops obscura

supratriangle crossed

A > 1/2 B

C undulate

supratriangle free

1378 FW base
Zenithoptera viola

72(71). Eyes on top of head separated (as in Fig. 1379); FW supratriangle with 1 or more crossveins (as in Fig. 1377); costal side of FW triangle (**A**) less than 1/2 as long as proximal side (**B**) (as in Fig. 1377) .... ***Diastatops*** (Page 234)

72'. Eyes on top of head contiguous (as in Fig. 1380); FW supratriangle without crossveins (as in Fig. 1378); costal side of FW triangle (**A**) at least 1/2 as long as or a little longer than 1/2 of proximal side (**B**) (as in Fig. 1378) .... ***Zenithoptera*** (Page 281)

head

1379

Diastatops pullata

1380

Nothodiplax dendrophila

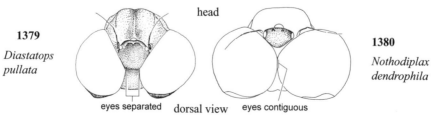

eyes separated   dorsal view   eyes contiguous

73(71). Bridge crossveins numbering 1 (as in Fig. 1381) **74**

73'. Bridge crossveins numbering 2 or more (as in Fig. 1382) **88**

last antenodal complete   1 bridge crossvein

last antenodal incomplete   2 bridge crossveins

1381 FW – *Oligoclada rhea*

1382 FW – *Perithemis lais*

74(73). Last antenodal in FW complete (as in Fig. 1381) **75**

74'. Last antenodal in FW incomplete (as in Fig. 1382) **81**

75(74). Aspl in HW more or less straight (as in Fig. 1383) **76**

75'. Aspl in HW bent at heel level of anal loop at an obtuse angle (as in Fig. 1384) **77**

1383 HW base
Planiplax erythropyga

Aspl straight

1384 HW base
Libellula herculea

Aspl obtusely bent

76(75). Radial planate consisting of 1 row of cells throughout (as in Fig. 1385); Mspl in FW indistinct (as in Fig. 1385) .... ***Oligoclada* in part** (Page 262)

76'. Radial planate consisting of 2 or more rows of cells (as in Fig. 1386); Mspl in FW distinct (as in Fig. 1386) .... ***Planiplax* in part** (Page 270)

Rspl
no Mspl

Rspl
Mspl

1385 FW – *Oligoclada abbreviata*

1386 FW – *Dythemis fugax*

77(75). HW very broad at base (as in Fig. 1387) .... ***Macrodiplax*** (Page 252)

77'. HW not broadly widened at base (as in Fig. 1388) **78**

1387 HW base
Miathyria simplex

not
broadly widened   **1388** HW base
Gynothemis venipunctata

broadly widened

78(77). Costal side of FW triangle (**A**) less than 1/2 as long as proximal side (**B**) (as in Fig. 1374) **79**

78'. Costal side of FW triangle (**A**) at least 1/2 as long as or a little longer than 1/2 of proximal side (**B**) (as in Fig. 1373) **80**

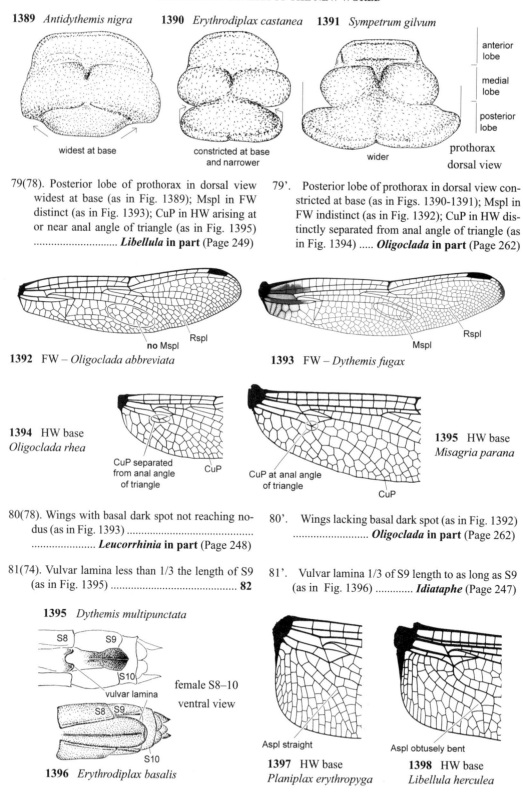

1389 *Antidythemis nigra*

1390 *Erythrodiplax castanea*

1391 *Sympetrum gilvum*

anterior lobe

medial lobe

posterior lobe

widest at base

constricted at base and narrower

wider

prothorax dorsal view

79(78). Posterior lobe of prothorax in dorsal view widest at base (as in Fig. 1389); Mspl in FW distinct (as in Fig. 1393); CuP in HW arising at or near anal angle of triangle (as in Fig. 1395) .... *Libellula* **in part** (Page 249)

79'. Posterior lobe of prothorax in dorsal view constricted at base (as in Figs. 1390-1391); Mspl in FW indistinct (as in Fig. 1392); CuP in HW distinctly separated from anal angle of triangle (as in Fig. 1394) ..... *Oligoclada* **in part** (Page 262)

Rspl

no Mspl

1392 FW – *Oligoclada abbreviata*

Rspl

Mspl

1393 FW – *Dythemis fugax*

1394 HW base
Oligoclada rhea

CuP separated from anal angle of triangle

CuP

CuP at anal angle of triangle

CuP

1395 HW base
Misagria parana

80(78). Wings with basal dark spot not reaching nodus (as in Fig. 1393) .... *Leucorrhinia* **in part** (Page 248)

80'. Wings lacking basal dark spot (as in Fig. 1392) .... *Oligoclada* **in part** (Page 262)

81(74). Vulvar lamina less than 1/3 the length of S9 (as in Fig. 1395) **82**

81'. Vulvar lamina 1/3 of S9 length to as long as S9 (as in Fig. 1396) .... *Idiataphe* (Page 247)

1395 *Dythemis multipunctata*

S8　　S9

S10

vulvar lamina

female S8–10 ventral view

S8　S9

S10

1396 *Erythrodiplax basalis*

Aspl straight

1397 HW base
Planiplax erythropyga

Aspl obtusely bent

1398 HW base
Libellula herculea

82(81). Aspl in HW more or less straight (as in Fig. 1397) **83**

82'. Aspl in HW bent at heel level of anal loop at an obtuse angle (as in Fig. 1398) **86**

83(82). Vertex smoothly rounded (as in Figs. 1400a-b) **84**

83'.　Vertex with tubercles (as in Figs. 1399a-b) **85**

Uracis ovipositrix　　　head　　　　*Nothodiplax dendrophila*

tubercles　　　　tubercles　　　　no tubercles　　　no tubercles

1399a antero-ventral view　**1399b** dorsal view　　**1400a** dorsal view　**1400b** lateral view

84(83). Mspl in FW indistinct; radial planate consisting of 1 row of cells throughout (as in Fig. 1392) .... *Oligoclada* **in part** (Page 262)

84'.　Mspl in FW distinct; radial planate consisting of 2 or more rows of cells (as in Fig. 1393) .... *Planiplax* **in part** (Page 270)

85(83). Mspl in FW indistinct (as in Fig. 1392); CuP in HW distinctly separated from anal angle of triangle (as in Fig. 1394) .... *Oligoclada* **in part** (Page 262)

85'.　Mspl in FW distinct (as in Fig. 1393); CuP in HW arising at or near anal angle of triangle (as in Fig. 1395) .... *Celithemis* (Page 231)

86(82). Posterior lobe of prothorax in dorsal view widest at base and narrower than other prothoracic lobes (as in Fig. 1389), in lateral view bent caudally (as in Fig. 1401) .... *Libellula* **in part** (Page 249)

86'.　Posterior lobe of prothorax in dorsal view constricted at base and wider than other prothoracic lobes (as in Fig. 1391); in lateral view upright (as in Fig. 1402) **87**

anterior　medial　posterior lobe

bent caudally

upright

1401

Erythrodiplax castanea

prothorax lateral view

1402

Sympetrum gilvum

87(86). Mspl in FW indistinct (as in Fig. 1392); CuP in HW distinctly separated from anal angle of triangle (as in Fig. 1394) .... *Oligoclada* **in part** (Page 262)

87'.　Mspl in FW distinct (as in Fig. 1393); CuP in HW arising at or near anal angle of triangle (as in Fig. 1395) .... *Brachymesia* (Page 227)

88(73). Aspl in HW more or less straight (as in Fig. 1397); posterior lobe of prothorax in dorsal view constricted at base (as in Fig. 1390), in lateral view upright (as in Fig. 1402) **89**

88'.　Aspl in HW bent at heel level of anal loop at an obtuse angle (as in Fig. 1398); posterior lobe of prothorax in dorsal view widest at base (as in Fig. 1389), in lateral view bent caudally (as in Fig. 1401) .... *Libellula* **in part** (Page 249)

89(88). Vertex smoothly rounded (as in Figs. 1400a-b) .... *Planiplax* **in part** (Page 270)

89'.　Vertex with tubercles (as in Fig. 1399a-b) .... *Perithemis* (Page 268)

Anatya Kirby, 1889: 263, 293.

[♂ pp. 193, couplet 72; ♀ pp. 213, couplet 42]

Type species: *Anatya anomala* Kirby, 1889 [by original designation]

2 species:

guttata (Erichson, 1848) [*Libellula*]*
  syn *anomala* Kirby, 1889
  syn *difficilis* (Selys, 1879) [*Agrionoptera*]
  syn *normalis* Calvert, 1899
  syn *theresiae* Selys, 1900
januaria Ris, 1911 – L [Santos, 1973c]

References: Ris, 1911a; De Marmels, 1992d.
Distribution: Central Mexico south to S Brazil, Paraguay and Bolivia.

Map 79. Distribution of *Anatya* spp.

Medium libellulines (about 30–40 mm); postfrons yellow or black to metallic blue, and eyes bright blue in mature males; pterothorax pale blue, mesepisternum with black reticulations to largely black, sides with variable extension of white and black spots; abdomen largely black with laterodorsal light blue to yellow paired spots on S1-7; base of cercus black, remainder yellow. Wings (Fig. 1403) hyaline; HW relatively narrow; arculus in HW opposite to or beyond antenodal 2; costal side of FW triangle bent at distal portion; Mspl indistinct; Rspl distinct and radial planate with 1 row of cells throughout; cell rows at midlength of HW anal field numbering 1. Posterior lobe of prothorax constricted at base; anterior lamina entire (Fig 1407b); posterior hamule bifid (Fig.1407a-b); tip of cercus longer than basal portion and upturned (Fig. 1406); vulvar lamina less than 1/3 the length of S9 and not projected ventrally. **Unique characters**: Distal segment of vesica spermalis with an apical sclerotized rim, c-shaped in lateral view (Fig. 1405).

Status of classification: Number of species not determined. Several names have been applied to populations of *A. guttata* with differing lengths of the cercus; short appendage forms receiving the name *A. normalis*, and long appendage forms receiving the name *A. guttata*. The condition appears to be variable; populations are known from Venezuela with intermediate morphology (De Marmels, 1992d), from S Peru with both short and long cerci (Paulson *pers. comm.*), and a specimen from Paraguay shows characteristics of *A. normalis* rather than *A. guttata* (Calvert, 1899), as well as one from central Brazil (Paulson *pers. comm.*).

Potential for new species: Likely.

Habitat: Margins of partially shaded or exposed ponds and canals. Adults perch in obelisk position in the sun on tips of grass or twigs with wings fully set. *Anatya guttata* taken from canopy-fogging samples in Manu Reserve, Peru (Louton, Garrison, and Flint, 1996).

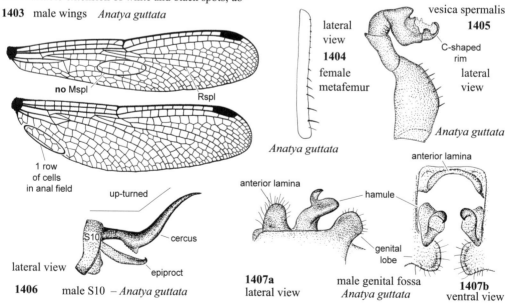

1403 male wings *Anatya guttata*

no Mspl / Rspl

1 row of cells in anal field

1404 lateral view / female metafemur / *Anatya guttata*

vesica spermalis **1405** / C-shaped rim / lateral view / *Anatya guttata*

up-turned / cercus / S10 / epiproct / lateral view / **1406** male S10 – *Anatya guttata*

1407a lateral view / anterior lamina / hamule / genital lobe / male genital fossa *Anatya guttata* / anterior lamina / **1407b** ventral view

Antidythemis Kirby, 1889: 258, 267.

[♂ pp. 195, couplet 79; ♀ pp. 215, couplet 53]

Type species: *Antidythemis trameiformis* Kirby, 1889 [by original designation]

2 species:

nigra Buchholz, 1952*
trameiformis Kirby, 1889

References: Ris, 1913a; Buchholz, 1952.
Distribution: Amazonian region of South America from Venezuela south to Bolivia.

Map 80. Distribution of *Antidythemis* spp.

Large libellulines (about 48–50 mm); brown to black postfrons and vertex, pterothorax dark reddish brown sometimes with metallic blue-green luster, and dark brown to black abdomen. Wings (Fig. 1408) hyaline with large brown anal spot on HW extending to distal level of toe; HW very broad at base; FW with more than 1 cubito-anal crossvein; FW subtriangle not defined; pterostigma long, surmounting 4-5 cells in FW; arculus in HW opposite to or beyond antenodal 2; Mspl and Rspl distinct; their planates with more than 2 rows of cells. Posterior lobe of prothorax widest at base (Fig. 1409b), its posterior margin bent caudad (Fig. 1409a); S4 with a transverse carina; anterior lamina entire (Fig. 1412a); posterior hamule bifid, with inner branch larger than vestigial outer branch (Figs. 1412a-b); vulvar lamina 1/3 of S9 to as long as S9. Superficially similar to *Tramea* but easily distinguished from that genus by the long pterostigma (short in *Tramea*). **Unique characters**: Aspl in HW bent at heel level of anal loop at a right (90°) to acute angle (Fig. 1403); distal segment of vesica spermalis with a pair of long pointed spines and a medial semihyaline blade (Figs. 1412b-d).

Status of classification: Unclear whether one or two species are involved; De Marmels (1989; 1993) suggested that color differences ascribed to both species may be age dependent, and stated (De Marmels, 1993) that the genitalia of *A. nigra* from Venezuela were intermediate between illustrations of both species given by Buchholz (1952).

Potential for new species: Unlikely.

Habitat: De Marmels (1993) observed males perching 2 meters above the ground and flying over a dirt road with rain puddles in partially deforested areas. De Marmels (1989) reports oviposition by a female in a rain puddle on side of a road. Specimens scarce in collections.

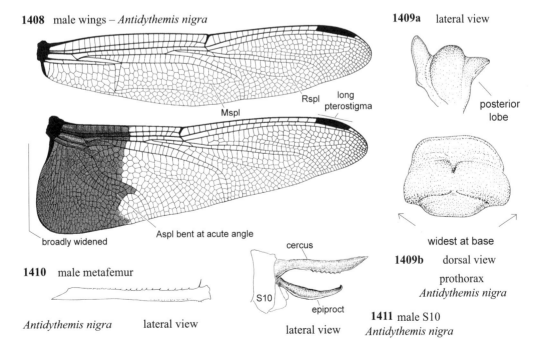

1408 male wings – *Antidythemis nigra*

Rspl long pterostigma

Mspl

broadly widened

Aspl bent at acute angle

1410 male metafemur

Antidythemis nigra   lateral view

cercus

S10

epiproct

lateral view

1411 male S10
Antidythemis nigra

1409a lateral view

posterior lobe

widest at base

1409b dorsal view

prothorax
Antidythemis nigra

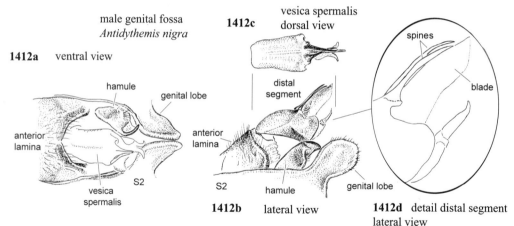

male genital fossa
Antidythemis nigra

1412a   ventral view

hamule

genital lobe

anterior
lamina

anterior
lamina

S2

S2

vesica
spermalis

hamule

vesica spermalis
1412c   dorsal view

distal
segment

spines

blade

genital lobe

1412b   lateral view

1412d   detail distal segment
lateral view

Argyrothemis Ris, 1909a: 26.

[♂ pp. 183, couplet 38; ♀ pp. 204, couplet 10]

Type species: *Argyrothemis argentea* Ris, 1909 [by monotypy]

[NOTE: The citation commonly used for this species is Ris, 1911a, but according to Article 12.1 of the ICZN (1999) the correct year is 1909, because the genus and species names were mentioned and diagnosed in the key included in the first volume of the Libellulinen monograph (Ris, 1909a).]

1 species:

argentea Ris, 1909* – **L** [Fleck, 2003a]

Map 81. Distribution of *Argyrothemis* sp.

References: Ris, 1909a; 1911a; Geijskes, 1971.
Distribution: Amazonian region of South America (Guyanas, Venezuela, Peru, and Brazil).

Medium libellulines (about 26–39 mm); brown with dark metallic blue postfrons and vertex, mesepisternum in mature males covered with blue-gray pruinosity, abdomen largely brown with laterodorsal pale paired spots on S2-5 and S7; cercus black. Wings (Fig. 1413) entirely hyaline or with brown tips, or with a broad band between nodus and pterostigma in some populations (Surinam); HW relatively narrow; costal side of FW triangle bent at about midlength; 1 row of cells at base of FW discoidal field; Mspl and Rspl indistinct; HW anal loop indistinct. Posterior lobe of prothorax constricted at base; anterior lamina entire (Fig. 1415a); posterior hamule not bifid (Figs. 1415a-b); vulvar lamina (Figs. 1414a-b) projected distally beyond tip of cerci (shared with *Uracis* and *Ypirangathemis*). **Unique characters**: Posterior hamule tip long, straight, cylindrical, pointed and caudally directed from its base (Figs.1415a-b).
Status of classification: Well known species.
Potential for new species: Unlikely.
Habitat: Males found in deep forest perching on leaves or sticks along small shaded rivulets (De Marmels, 1989). Adults collected in palm swamp in lowland Peru (Louton, Garrison, and Flint, 1996). Larvae camouflage themselves with detritus (Fleck, 2003a).

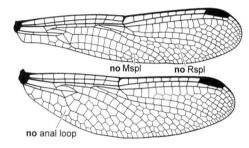

no Mspl   **no** Rspl

no anal loop

1413   male wings – *Argyrothemis argentea*

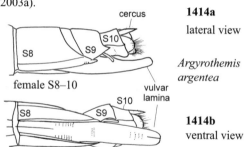

cercus

S10

S8   S9

female S8–10

vulvar
lamina

S8   S9   S10

1414a

lateral view

*Argyrothemis
argentea*

1414b

ventral view

1415a   ventral view   male genital fossa   **1415b**   lateral view
                                  Argyrothemis argentea

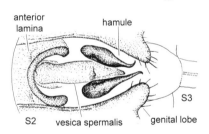

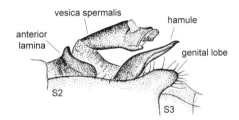

Brachymesia Kirby, 1889: 262, 280.

[♂ pp. 200, couplet 101; ♀ pp. 223, couplet 87]

Type species: *Brachymesia australis* Kirby, 1889 [by original designation]

  syn *Cannacria* Kirby, 1889: 262, 300.

    Type species: *Cannacria batesii* Kirby, 1889 [by original designation]

3 species:

furcata (Hagen, 1861) [*Erythemis*]* – **L** [Geijskes, 1934; Needham, Westfall, and May, 2000]
  syn *australis* Kirby, 1889
  syn *smithii* (Kirby, 1894) [*Cannacria*]
gravida (Calvert, 1890) [*Lepthemis*]* – **L** [Byers, 1936; Needham, Westfall, and May, 2000]
herbida (Gundlach, 1889) [*Libellula*]* – **L** [Needham, Westfall, and May, 2000]
  syn *batesii* (Kirby, 1889) [*Cannacria*]
  syn *fumipennis* (Currie, 1901) [*Cannacria*]

References: Ris, 1912; Needham, Westfall, and May, 2000.
Distribution: S United States south through N Argentina and Chile.

Medium to large libellulines (about 38–54 mm); all red (*B. furcata*) to brown or black with concolorous pterothorax and greatly swollen S1-3. Wings (Fig. 1417) hyaline or with diffuse brown markings along costal field or black spots between nodus and tip; sectors of arculus in FW not stalked; Mspl and Rspl distinct. Posterior lobe of prothorax constricted at base, upright and wider than other prothoracic lobes; anterior lamina strongly bifid (Fig.1416a) and as high

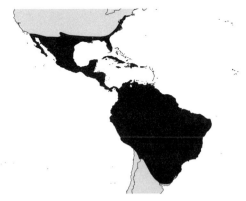

Map 82. Distribution of *Brachymesia* spp.

or higher than posterior hamule (Fig. 1416b) (shared with *Pantala*); posterior hamule not bifid (Figs. 1416a-b); vulvar lamina less than 1/3 the length of S9 and not projected ventrally. **Unique characters**: None known.
Status of classification: Good. Two species (*B. gravida* and *B .herbida*) were placed in the genus *Cannacria* Kirby, but this genus was synonymyzed by Ris (1912).
Potential for new species: Unlikely.
Habitat: Adults active in sunshine at ponds, swamps, pools and brackish estuaries where they perch on emergent reeds and sticks. Adults often wary and difficult to capture. Paulson (1999) provides excellent detailed information on habits of adults, and clarifies past descriptions of larvae described within this genus.

male genital fossa

Brachymesia
furcata

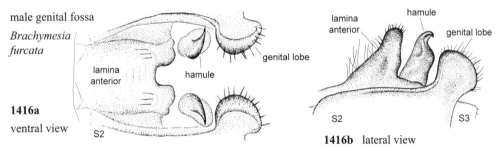

1416a
ventral view

1416b   lateral view

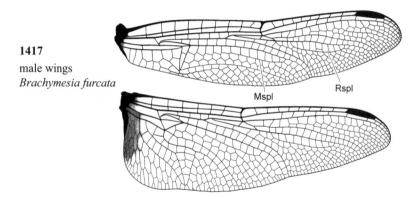

1417
male wings
Brachymesia furcata

Brechmorhoga Kirby, 1894: 264.
[♂ pp. 190, 195, couplets 64, 78; ♀ pp. 208, 210-211, 214-215, couplets 31, 37, 50, 51, 52]

Type species: *Brechmorhoga grenadensis* Kirby, 1894 [by original designation]
    syn *Nothemis* Navás, 1915: 146.
    Type species: *Nothemis apollinaris* Navás, 1915 [by original designation]
16 species:

archboldi (Donnelly,1970) [*Scapanea*]*
diplosema Ris, 1913
flavoannulata Lacroix, 1920
flavopunctata (Martin, 1897) [*Dythemis*]
innupta Rácenis, 1954
latialata González-Soriano, 1999*
mendax (Hagen, 1861) [*Dythemis*]* – **L** [Needham and Fisher, 1936]
neblinae De Marmels, 1989*
nubecula (Rambur, 1842) [*Libellula*]* – **L** [Santos, 1969b]
    syn *catharina* (Karsch, 1890) [*Macrothemis*]
pertinax eurysema Ris, 1913*
pertinax pertinax (Hagen, 1861) [*Dythemis*]* – **L** [Novelo-Gutiérrez, 1995b]
    syn *sallaei* (Selys, 1868) [*Libellula* (*Dythemis*)]
pertinax peruviana Ris, 1913
praecox grenadensis Kirby, 1894*
praecox postlobata Calvert, 1898*
praecox praecox (Hagen, 1861) [*Dythemis*]* – **L** [Novelo-Gutiérrez, 1995a]
    syn *columba* (Hagen 1861) (*nomen nudum*)
praedatrix Calvert, 1909* – **L** [Fleck, 2004]
rapax crocosema Ris, 1913*
rapax rapax Calvert, 1898* – **L** [De Marmels, 1982a]
    syn *apollinaris* (Navás, 1915) [*Nothemis*]
tepeaca Calvert, 1908*,
travassosi Santos, 1946* – **L** [Santos and Costa, 1999]
vivax Calvert, 1906* – **L** [De Marmels, 1982a]

References: Calvert, 1898; Ris, 1913a.

Map 83. Distribution of *Brechmorhoga* spp.

Distribution: SW United States south through N Argentina.

Medium to large libellulines (40–62 mm); elongate, slender, black body with irregular patterns of striping on thorax and white to yellow spots on several abdominal segments. Wings (Figs. 1421-1424) hyaline; FW discoidal sector widening distally; Mspl and Rspl distinct. Posterior lobe of prothorax widest at base, its posterior margin bent caudad; male hind femur (Fig. 1420b) with short, stout spines directed proximally (shared with *Macrothemis* and *Scapanea*). Anterior lamina entire; posterior hamule not bifid (Figs. 1432a-b); in most species male S7-9 widened and flattened, with widest point less than twice as wide as base of S7 (shared with *Nannothemis*, *Nephepeltia* and some species of *Macrothemis* and *Micrathyria*); vulvar lamina less than 1/3 the length of S9 and not projected ventrally (Figs. 1427-1431a). **Unique characters**: None known.

Status of classification: Consensus among odonatologists is that the genus is related to *Macrothemis* based on similarity of stature, shape, and habits (Garrison and von Ellenrieder, 2006). *Brechmorhoga* and

Macrothemis can be further distinguished by larval characters (Ramírez and Novelo-Gutiérrez, 1999). *Brechmorhoga, Macrothemis, Gynothemis* and *Scapanea* are very likely related, and probably belong to the same natural group. A phylogenetic analysis is needed in order to determine if the genera as currently defined are monophyletic or not. *Brechmorhoga* is in need of revision as status of a few species is unknown. Ris (1913a) assigned subspecies names to various allopatric populations of a few species. Based on the examination of types, Philippe Machet (*in litt.*) considers *B. flavoannulata* Lacroix, 1920, a junior synonym of *B. flavopunctata* (Martin, 1897), and De Marmels (1999) suggests that *B. innupta* Rácenis (known only from the female holotype) is a probable synonym of *B. flavopunctata*. We include both names as valid pending further study of this group.

Potential for new species: Likely. One new species from Brazil known to the authors, and several others probably to be discovered due to cryptic habits of adults.

Habitat: Exclusively stream inhabiting species where adults course up and down sections of fast, rocky laden creeks and rivulets. Adults perch in pendent position in forest clearings or in brush near preferred habitats; occasionally found sitting on river banks. Larvae found in streams and rivers with no or little canopy cover, commonly in areas of rocky substrate, shallow water and rapid flow (Novelo-Gutiérrez, 1995a, b). Larvae of *B. nubecula* reared from 2-4 m streams in primary and secondary tropical lowland forests in Peru (Louton, Garrison, and Flint, 1996). The larva of *B. praedatrix* is apparently associated with the aquatic plant *Mourera fluviatilis* Aublet (Podostemaceae), which grows only in fast moving currents (Fleck, 2004).

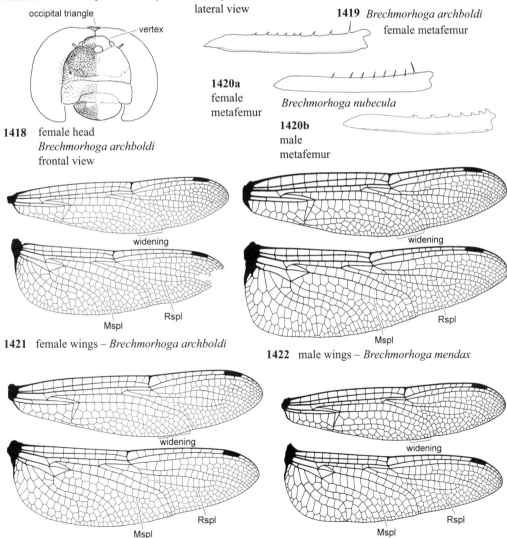

1418 female head *Brechmorhoga archboldi* frontal view

1419 *Brechmorhoga archboldi* female metafemur

1420a female metafemur

Brechmorhoga nubecula

1420b male metafemur

1421 female wings – *Brechmorhoga archboldi*

1422 male wings – *Brechmorhoga mendax*

1423 female wings – *Brechmorhoga praecox grenadensis*

1424 male wings – *Brechmorhoga nubecula*

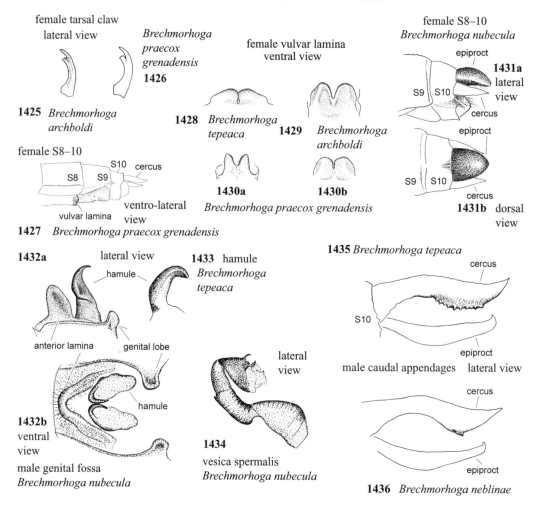

female tarsal claw
lateral view

1425 *Brechmorhoga archboldi*

Brechmorhoga praecox grenadensis
1426

female S8–10

vulvar lamina

ventro-lateral view

1427 *Brechmorhoga praecox grenadensis*

female vulvar lamina
ventral view

1428 *Brechmorhoga tepeaca*

1429 *Brechmorhoga archboldi*

1430a   **1430b**
Brechmorhoga praecox grenadensis

female S8–10
Brechmorhoga nubecula

epiproct

1431a lateral view

cercus

epiproct

cercus

1431b dorsal view

1432a lateral view

hamule

anterior lamina   genital lobe

hamule

1432b ventral view

male genital fossa
Brechmorhoga nubecula

1433 hamule
Brechmorhoga tepeaca

lateral view

1434

vesica spermalis
Brechmorhoga nubecula

1435 *Brechmorhoga tepeaca*

cercus

S10

epiproct

male caudal appendages   lateral view

cercus

epiproct

1436 *Brechmorhoga neblinae*

Cannaphila Kirby, 1889: 261, 305.
[♂ pp. 177, 179, couplets 14, 23; ♀ p. 203, couplets 6, 8]
Type species: *Cannaphila insularis* Kirby, 1889 [by original designation]
3 species:

insularis Kirby, 1889* – **L** [Klots, 1932]
    syn *angustipennis* (Rambur (*nec* Stephens, 1835), 1842) [*Libellula*]
    syn *funerea* (Carpenter, 1897) [*Misagria*]
mortoni Donnelly, 1992*
vibex (Hagen, 1861) [*Libellula*]* – **L** [Limongi, 1991]
    syn *merida* (Selys, 1868) [*Libellula*]

References: Ris, 1910; Donnelly, 1992a.
Distribution: SW United States south through N Argentina.

Map 84. Distribution of *Cannaphila* spp.

Medium (36–46 mm) libellulines; postfrons and vertex brown to metallic blue in males; thorax with stripes becoming obscured by pruinosity with age in males; abdomen pale brown to reddish brown with yellow streaks in females, black in males. Wings hyaline or sometimes infuscate; HW relatively narrow; last antenodal in FW complete; CuP in HW distinctly separated from anal angle of triangle. Posterior lobe of prothorax widest at base, its posterior margin bent caudad; anterior lamina entire or slightly bifid (Fig. 1438b); posterior hamule bifid (Figs. 1438a-b), with inner branch smaller than outer branch; female S8 with lateral flanges (shared with *Libellula*, *Misagria* and *Orthemis*); vulvar lamina less than 1/3 the length of S9 and not projected ventrally. **Unique characters**: None known.

Status of classification: Fair. The status of the two subspecies of *C. insularis* is controversial. Garrison (1986a) and Needham, Westfall, and May (2000) diagnosed both subspecies but Donnelly (1992a) preferred not to recognize them.

Potential for new species: Unlikely.

Habitat: Adults perch on twigs and snags in open and forested areas where they can be abundant. *Cannaphila vibex* collected in Peru at elevations from 250-1450 m in heavily forested areas (Louton, Garrison, and Flint, 1996).

1437 male wings – *Cannaphila vibex*

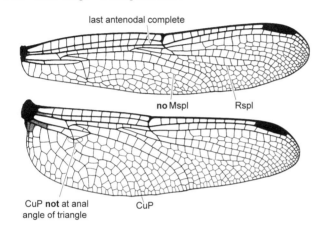

1438a lateral view

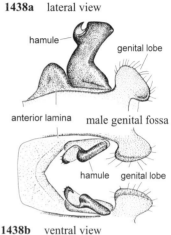

1438b ventral view

Cannaphila vibex

Celithemis Hagen, 1861: 147.

[♂ pp. 199, couplet 98; ♀ pp. 223, couplet 85]

Type species: *Libellula eponina* Drury, 1773 [Kirby, 1889 by subsequent designation]

8 species:

amanda (Hagen, 1861) [*Diplax*]* – **L** [Needham, Westfall, and May, 2000]
     syn *pulchella* (Burmeister (*nec* Drury, 1773), 1839) [*Libellula*]
bertha Williamson, 1922* – **L** [Needham, Westfall, and May, 2000]
     syn *bertha leonora* Westfall, 1952
elisa (Hagen, 1861) [*Diplax*]* – **L** [Needham, 1901]
eponina (Drury, 1773) [*Libellula*]* – **L** [Needham, 1901]
     syn *camilla* (Rambur, 1842) [*Libellula*]
     syn *lucilla* (Rambur, 1842) [*Libellula*]
fasciata Kirby, 1889* – **L** [Broughton, 1928; Leonard, 1934]
     syn *monomelaena* Williamson, 1910
martha Williamson, 1922*
ornata (Rambur, 1842) [*Libellula*]* – **L** [Broughton, 1928]
verna Pritchard, 1935* – **L** [Needham, Westfall, and May, 2000]

References: Williamson, 1922; Needham, Westfall, and May, 2000.

Distribution: SE Canada, E United States south to Cuba, the Bahamas, and Nuevo Leon State in Mexico.

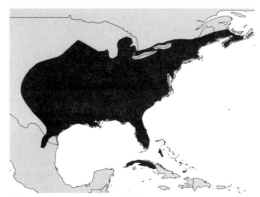

Map 85. Distribution of *Celithemis* spp.

Small to medium libellulines (26–43 mm), with relatively broad hind wings often marked with patterns of orange and brown; postfrons and vertex yellow to red or black in mature males, pterothorax pale to red with black stripes or all dark, abdomen blue or black with yellow or red spots in some species. Sectors of arculus in FW not stalked; Aspl in HW more or less straight (Fig. 1439). Posterior lobe of prothorax constricted at base, upright and wider than other prothoracic lobes; anterior lamina entire (Fig. 1440b); posterior hamule bifid, with inner branch larger than vestigial outer branch (Figs. 1440a-b); vulvar lamina shorter than 1/3 of S9 length. **Unique characters**: None known.

Status of classification: Good. A well-known genus, superficially similar to the Old World tropical genus *Rhyothemis*. Some species that exhibit clinal variation have received names; at least one species (*C. bertha*) polymorphic as to wingtip markings.

Potential for new species: Unlikely.

Habitat: Adults perch on tips of sedges, cattails or other emergent vegetation at edges of ponds, swamps, or pools. Extensive descriptions of habitats for some species given by Paulson (1999).

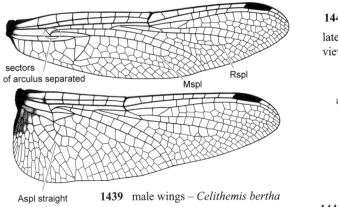

sectors of arculus separated

Rspl

Mspl

Aspl straight **1439** male wings – *Celithemis bertha*

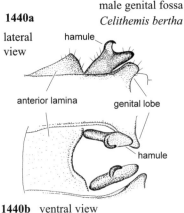

male genital fossa

1440a *Celithemis bertha*

lateral view

hamule

anterior lamina

genital lobe

hamule

1440b ventral view

Crocothemis Brauer, 1868: 367, 736.

[♂ pp. 186, couplet 50; ♀ pp. 212, couplet 40]

Type species: *Libellula erythraea* Brullé, 1832 [Kirby, 1889 by subsequent designation]

[NOTE: Fraser (1936) in designating *Libellula servilia* Drury, 1773 as type species ("No genotype appears to have been cited for *Crocothemis*, so that I have chosen *servilia* as the first described species.") apparently overlooked Kirby's (1889) designation of *Libellula erythraea* Brullé, 1832 as the type species.]

    syn *Beblecia* Kirby, 1900: 71

        Type species: *Beblecia adolescens* Kirby, 1900 [by monotypy]

10 species; 1 species introduced into the New World:

servilia servilia (Drury, 1773) [*Libellula*]* – **L** [Needham, Westfall, and May, 2000]

    syn *ferruginea* (Fabricius, 1781) [*Libellula*]

    syn *reticulata* Kirby, 1886

    syn *soror* (Rambur, 1842) [*Libellula*]

References: Lohmann, 1981.

Distribution: An Old World species introduced to Florida (Paulson, 1978) and expanding its range, now found in Cuba (Flint, 1996; Trapero Quintana and Naranjo López, 2001). Due to its ability to colonize new habitats in the Old World tropics, *C. servilia* will most likely continue to expand its range in the New World.

Map 86. Distribution of *Crocothemis* spp.

Medium libellulines (41–43 mm); largely reddish concolorous pterothorax and abdomen. Wings (Fig. 1441) hyaline with reddish basal spots; costal side of FW triangle bent at distal portion; CuP in HW distinctly separated from anal angle of triangle. Posterior lobe of prothorax widest at base, its posterior margin bent caudad; anterior lamina entire (Fig. 1443b); posterior hamule bifid with inner branch smaller than outer branch (Figs. 1443a-b); distal segment of vesica spermalis long and cylindrical (Figs. 1442a-b); vulvar lamina scoop shaped and projecting ventrally (shared with *Erythemis*, *Nannothemis*, *Pseudoleon*

and some *Erythrodiplax*), and longer than 1/3 of S9.
Unique characters: None known.
Status of classification: Good. We have been unable
to find any consistent character with which to separate *Crocothemis servilia* from *Erythrodiplax*. Of the
sympatric species of *Erythrodiplax*, only *E. umbrata*
and *E. funerea* share its larger size (total length more
than 40 mm); the remaining sympatric species (*E.
berenice, E. bromeliicola, E. fervida, E. justiniana*)
are smaller (total length 36 mm or less). *Erythodiplax*

funerea and *E. umbrata* have 2 rows of cells in the
radial planate, whereas *C. servilia* has only 1.
Potential for new species: Extremely unlikely.
Habitat: Common at ponds, slowly moving streams
and irrigation canals. Adults perch in sunshine at almost any available habitat. This species has been the
subject of numerous studies on different aspects of
its biology in the Old World (*i.e.* Aguesse, 1959; Higashi, 1969; Dalchetti and Utzeri, 1974; Siva-Jothy,
1984; Convey, 1992).

1441 male wings – *Crocothemis servilia*

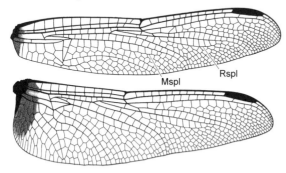

Mspl   Rspl

1443a lateral view

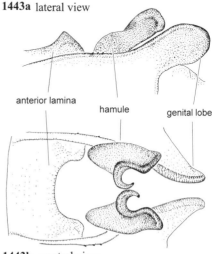

anterior lamina   hamule   genital lobe

1442a lateral view   **1442b** dorsal view

vesica spermalis – *Crocothemis servilia*

1443b ventral view
Crocothemis servilia   male genital fossa

Dasythemis Karsch, 1889: 251.
[♂ pp. 177, 179, couplets 16, 24; ♀ pp. 217, couplet 42]
Type species: *Dasythemis liriopa* Karsch, 1889 [by
monotypy]
   syn *Malamarptis* Karsch, 1890: 380
      Type species: *Malamarptis mincki* Karsch,
      1890 [by monotypy]
4 species:

esmeralda Ris, 1910*
essequiba Ris, 1919*
mincki clara Ris, 1908*
mincki mincki (Karsch, 1890) [*Malamarptis*]* – **L**
[Carvalho, Werneck-de-Carvalho, and Calil, 2002]
venosa (Burmeister, 1839) [*Libellula*]* – **L**
[Carvalho, Werneck-de-Carvalho, and Calil, 2002]
   syn *liriopa* Karsch, 1889
   syn *macrostigma* Förster, 1907

References: Ris, 1910.
Distribution: Exclusively South American genus occurring from Guyanas south through Argentina.

Map 87. Distribution of *Dasythemis* spp.

Medium libellulines (31–39 mm); postfrons and
vertex brown to metallic blue in males, pterothorax
yellow to black, sometimes with pale lateral stripes,
abdomen brown with yellow streaks to black, both
pterothorax and abdomen becoming obscured by pru-

inosity in males. Wings hyaline (Figs. 1444-1445); last antenodal in FW complete; costal side of FW triangle at least 1/2 as long as proximal side; Mspl indistinct; Rspl distinct; radial planate with 1 row of cells throughout. Posterior lobe of prothorax widest at base, its posterior margin bent caudad; anterior lamina entire (Fig. 1446b); posterior hamule bifid, with inner branch larger than outer branch (Figs. 1446a-b). **Unique characters**: Ventral side of female S10 prolonged to about distal tip of cerci; female sternum S9 projected distally to about posterior margin of S10, rectangular and smoothly con-

vex, with distal margin fringed with long hairs (Figs. 1447a-b).

Status of classification: Fair. Last species described by Ris (1910). Species are interspecifically variable as to HW length and development of anal loop (Figs. 1444-1445).

Potential for new species: Moderate. One known to the authors from Venezuela.

Habitat: Adults perch on twigs and vegetation at ponds and marshes or in road clearings and partially shaded fields. *Dasythemis esmeralda* collected at swamps and seeps in primary lowland forests in Peru (Louton, Garrison, and Flint, 1996).

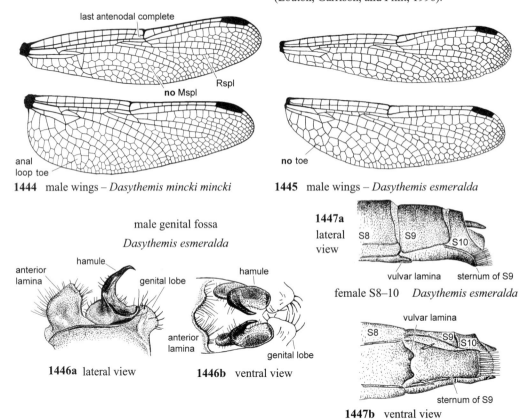

1444 male wings – *Dasythemis mincki mincki*

1445 male wings – *Dasythemis esmeralda*

male genital fossa
Dasythemis esmeralda

1446a lateral view   **1446b** ventral view

1447a lateral view

female S8–10   *Dasythemis esmeralda*

1447b ventral view

Diastatops Rambur, 1842: 135.
[♂ pp. 197, couplet 86; ♀ pp. 220, couplet 72]
Type species: *Libellula pullata* Burmeister, 1839 [Kirby, 1889 by subsequent designation]
8 species:

dimidiata (Linnaeus, 1758) [*Libellula*]*
    syn *fenestrata* Hagen, 1855 (*nomen nudum*)
    syn *marginata* (De Geer, 1773) [*Libellula*]
emilia Montgomery, 1940*
estherae Montgomery, 1940*
intensa Montgomery, 1940* – **L** [Costa, Souza-Franco, and Takeda, 1999b]

maxima Montgomery, 1940
nigra Montgomery, 1940*
obscura (Fabricius, 1775) [*Libellula*]* – **L** [Santos, Costa, and Pujol-Luz, 1993]
    syn *tincta* Rambur, 1842
pullata (Burmeister, 1839) [*Libellula*]* – **L** [Fleck, 2003b]
    syn *fuliginea* Rambur, 1842

References: Montgomery, 1940.
Distribution: Exclusively South American occurring from Colombia and Venezuela south through the lower Paraná basin in Argentina.

Map 88. Distribution of *Diastatops* spp.

Small to medium (25–32 mm), relatively broad-winged libellulines, possessing partial patterns of brown banding but more often with completely darkened wings with various irregular patches of yellow to red veins at base of HW, often accompanied by dense, reticulate venation; body brown. FW costa undulate (shared with *Zenithoptera*); sectors of arculus

in FW not stalked; FW subtriangle not defined; FW supratriangle with 1 or more crossveins; HW broad to very broad at base; Aspl in HW more or less straight (shared with *Celithemis*, *Planiplax*, *Perithemis*) (Fig. 1448). Posterior lobe of prothorax constricted at base; transverse carina on S4 (shared with *Zenithoptera*); anterior lamina entire (Fig. 1450b); posterior hamule not bifid (Figs. 1450a-b); vulvar lamina shorter than 1/3 of S9. **Unique characters**: Eyes separated dorsally (Fig. 1449).

Status of classification: Well defined species. Undulate costal condition shared by the exclusively neotropical genus *Zenithoptera* and by the Old World tropical genus *Palpopleura*.

Potential for new species: Possible. However none has been added since Montgomery's revision (1940).
Habitat: Ponds and marshes where adults perch on grass and reeds; De Marmels (1989) recorded *D. nigra* from a cold stream in deep forest. Adults often sit with wings asymmetrically set. Reproductive behavior of *D. intensa* was described by Wildermuth (1994).

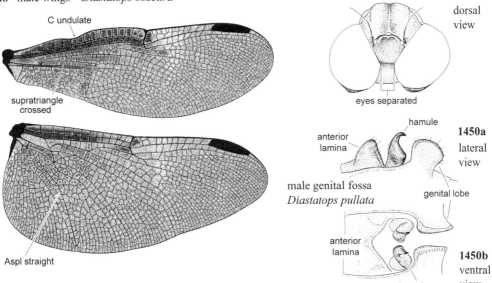

1448 male wings – *Diastatops obscura*

C undulate

supratriangle crossed

Aspl straight

1449 head   *Diastatops pullata*

dorsal view

eyes separated

hamule

anterior lamina

1450a lateral view

genital lobe

male genital fossa
Diastatops pullata

anterior lamina

1450b ventral view

hamule

Dythemis Hagen, 1861: 162.
[♂ pp. 176, 179, 186, 189, 191, 194, couplets 11, 20, 48, 57, 67, 76; ♀ pp. 205, 207-208, 214-215, couplets 19, 29, 30, 31, 49, 50, 51]
Type species: *Libellula rufinervis* Burmeister, 1839 [Kirby, 1889 by subsequent designation]
7 species:

fugax Hagen, 1861* – **L** [Geijskes, 1946]
maya Calvert, 1906* – **L** [Novelo and González, 2004]
multipunctata multipunctata Kirby, 1894* – **L** [De Marmels, 1982a; Westfall, 1988]
    syn *nigra* Martin, 1897

    syn *tabida (nomen nudum)* Hagen, 1861
multipunctata reducta De Marmels, 1989*
nigrescens Calvert, 1899* – **L** [Young and Bayer, 1979]
rufinervis (Burmeister, 1839) [*Libellula*]* – **L** [Klots, 1932]
    syn *conjuncta* (Rambur, 1842) [*Libellula*]
    syn *vinosa* (Scudder, 1866) [*Libellula*]
sterilis Hagen, 1861* – **L** [Geijskes, 1946; Novelo and González, 2004]
    syn *broadwayi* Kirby, 1894
velox Hagen, 1861* – **L** [Needham, 1904]

References: Ris, 1913a; Ris, 1919; Needham, West-fall, and May, 2000.

Distribution: S United States south through N Argentina and Chile.

Map 89. Distribution of *Dythemis* spp.

Medium (33-50 mm), slender libellulines; postfrons and vertex pale yellow, red, brown to metallic blue; color of thorax variable often with striping; abdomen red to brown or black with series of pale spots; en-tire thorax and abdomen become covered with pruinosity in at least one species. Wings hyaline or with reddish or brown basal spot in HW; last antenodal in FW complete or incomplete; Mspl distinct, median planate with 1 row of cells throughout; Rspl distinct, radial planate with 2 rows of cells (Fig. 1451). Posterior lobe widest at base, its posterior margin bent caudad; anterior lamina entire (Fig. 1452b); posterior hamule bifid with inner branch larger than vestigial outer branch (Figs. 1452a-b); S9 sternum keeled (Figs. 1453a-b); ventral carina in female S4-5 well developed (Fig. 1456); posterior angle of S3-4 female ventral terga right (Fig. 1455); vulvar lamina less than 1/3 the length of S9 and not projected ventrally (Figs. 1453a-b). **Unique characters**: None known.

Status of classification: Well known species.

Potential for new species: Probably unlikely as last species was described in 1906.

Habitat: Streams and pools in streams, in open or shaded areas. One species, *D. multipunctata*, seems to prefer shaded parts of streams in forest; reared from 1-3 m streams in lowland tropical forest (Louton, Garrison, and Flint, 1996).

1451 male wings – *Dythemis fugax*

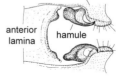

male genital fossa

Dythemis multipunctata

1452a lateral view

1452b ventral view

female S8–10 *Dythemis multipunctata*

1453a ventral view

1453b latero-ventral view

1453c dorsal view

Dythemis sterilis

lateral view

1454 tarsal claw

Dythemis sterilis

1455 female ventral terga S3–4

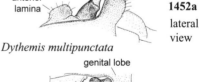

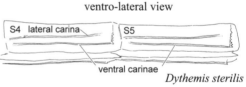

ventro-lateral view

Dythemis sterilis

1456 female ventral terga S4–5

Edonis Needham, 1905b: 114.

[♂ pp. 180, couplet 27; ♀ pp. 203, 216, 218, couplets 7, 57, 64]

Type species: *Edonis helena* Needham, 1905 [by original designation]

1 species:

helena Needham, 1905*

References: Needham, 1905b; Ris, 1911a.
Distribution: S Brazil to NE Argentina.

Map 90. Distribution of *Edonis* spp.

1457 male wings – *Edonis helena*

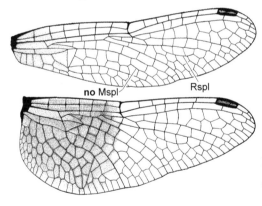

no Mspl   Rspl

Small libellulines (29.3–32 mm); body brown. Wings (Fig. 1457) hyaline with a broad orange basal spot at HW base; last antenodal in FW complete; 2 or more bridge crossveins; Mspl indistinct; Rspl distinct and radial planate with 1 row of cells throughout. Posterior lobe of prothorax with a slight constriction at base, its dorsal margin erect and fringed with long hairs (Fig. 1458); male hind femora with spines short and directed distally, only slightly increasing in length distally (Fig. 1459). Anterior lamina strongly bifid and with anterior surface covered with strong setae (Fig. 1461a, c); posterior hamule bifid with inner branch about as large as outer branch (Fig. 1461a, c). Similar to *Nephepeltia* from which it can be separated by HW triangle base slightly proximal to arculus (opposite to arculus in *Nephepeltia*) and HW base broadly widened with more than 4 cell rows between margin and anal loop (not broadly widened in *Nephepeltia* with a maximum of 3 rows of cells in anal field). **Unique characters**: Distal segment of vesica spermalis with a pair of flat, triangular, anteriorly directed processes (Figs. 1461b-c).

Status of classification: Species poorly known; female genitalia still unknown.
Potential for new species: Unlikely.
Habitat: Unknown. Larva unknown.

1458 posterior lobe of prothorax
Edonis helena
dorsal view

1459 metafemur
Edonis helena
lateral view

male S10   *Edonis helena*

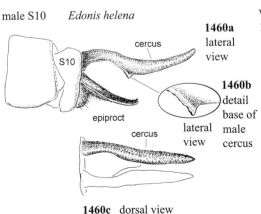

1460a
lateral
view

cercus

S10

1460b
detail
base of
male
cercus

epiproct

cercus

lateral
view

1460c dorsal view

ventral view   **1461b**
1461a

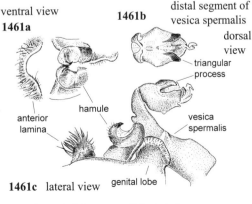

distal segment of
vesica spermalis

dorsal
view

triangular
process

anterior
lamina

hamule

vesica
spermalis

1461c lateral view   genital lobe

male genital fossa   *Edonis helena*

Elasmothemis Westfall, 1988: 422.
[♂ pp. 175, 178, 180-181, 188, 193, 196, couplets 9, 19, 28, 30, 54, 73, 82; ♀ pp. 205, 207-208, 213-214, 216, couplets 19, 27, 31, 47, 50, 56, 59]

Type species: *Dythemis cannacrioides* Calvert, 1906 [Westfall, 1988 by original designation]
6 species:

alcebiadesi (Santos, 1945) [*Dythemis*]
cannacrioides (Calvert, 1906) [*Dythemis*]* – **L** [Westfall, 1988]
constricta (Calvert, 1898) [*Dythemis*]* – **L** [Pujol-Luz, 1990]
   syn *willinki* (Fraser, 1947) [*Macrothemis*]
kiautai (De Marmels, 1989) [*Dythemis*]
schubarti (Santos, 1945) [*Dythemis*]*
williamsoni (Ris, 1919) [*Dythemis*]* – **L** [Westfall, 1988]

References: Westfall, 1988.
Distribution: Mexico south to N Argentina.

Map 91. Distribution of *Elasmothemis* spp.

Medium to large libellulines (43–55 mm); orange brown to brown body, face yellow to brown, post-frons and vertex brown to metallic blue, some with a metallic shine in pterothorax (*E. constricta*), or becoming pruinose with age (*E. williamsoni*). Wings (Fig. 1462) hyaline; last antenodal in FW complete or incomplete; Mspl distinct or indistinct, when distinct, median planate with 1 or 2 rows of cells; Rspl distinct, radial planate with 1 or 2 rows of cells. Posterior lobe of prothorax widest at base, its posterior margin bent caudad; posterior hamule not bifid (Figs. 1465a-b); vulvar lamina less than 1/3 of S9 and not projected ventrally. **Unique characters**: Anterior lamina (Figs. 1465a-b) entire and as high as 4/5 of posterior hamule or higher (in *Brachymesia* and *Pantala* it is also high but strongly bifid, and in *Micrathyria catenata* is higher than posterior hamule); distal segment of vesica spermalis (Figs. 1465a-b) with a medio-ventral pouch or keel usually accompanied by 2 cornua (cornua absent in *E. williamsoni*).
Status of classification: Good.
Potential for new species: Likely. *E. cannacrioides* may include more than one species, with the eastern Mexican populations being much larger than the more southerly ones, and the Amazonian specimens different than the northern Venezuelan ones. Females of *Elasmothemis* can be separated from both *Brechmorhoga* and *Dythemis* by the development of the ventral carina in S4-5 (Fig. 1464), which is absent or vestigial in *Elasmothemis* and well developed in *Brechmorhoga* and *Dythemis*. *Elasmothemis* further differs from *Dythemis* by the lateral spines on S6-9 of the larvae (spines present only on S8-9 in *Dythemis*).
Habitat: Streams and rivers; adults perch on twigs with wings set. Females of *E. cannacrioides* deposit egg strands on floating roots of a liana (González-Soriano, 1987). *Elasmothemis cannacrioides* reared from sandy 5 m stream under partial canopy in lowland Peru (Louton, Garrison, and Flint, 1996)

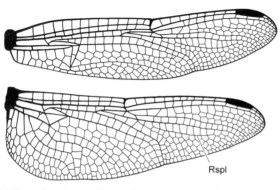

1462 male wings – *Elasmothemis cannacrioides*

1463 hind femur   lateral view

Elasmothemis cannacrioides

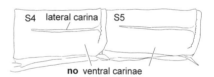

1464 female ventral terga S4–5
    Elasmothemis sp.
    ventro-lateral view

1465a

lateral view

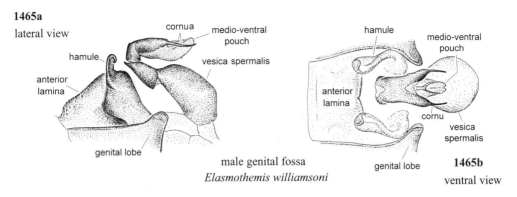

male genital fossa
Elasmothemis williamsoni

1465b
ventral view

Elga Ris, 1909a: 27.
[♂ pp. 174, 184, couplets 4, 42; ♀ pp. 217, 219-220, couplets 63, 68, 70]
Type species: *Elga leptostyla* Ris, 1909 [by mono-
typy]
[NOTE: The citation commonly used for this species is Ris,
1911a, but according to Article 12.1 of the ICZN (1999) the
correct year is 1909, because the genus and species names
were mentioned and diagnosed in the key included in the
first volume of the Libellulinen monograph (Ris, 1909a).]
2 species:

leptostyla Ris, 1909* – **L** [De Marmels, 1990a;
Fleck, 2003b]
   syn *santosi* Machado, 1954
newtonsantosi Machado, 1992

References: Ris, 1909a; 1911a; Machado, 1954;
1992.
Distribution: Panama south to Peru and Brazil.

Small libellulines (26–28 mm); largely black with
yellow markings on sides of thorax and S1-7, frons
brown to metallic blue. Wings (Fig. 1466) hyaline,
with slight yellow tinge at the base on males; costal
side of FW triangle bent at midlength or distal end;
CuP in HW distinctly separated from anal angle of
triangle; Mspl indistinct; Rspl indistinct or weakly
developed. Posterior lobe of prothorax constricted at
base; anterior lamina entire (Fig. 1468a); posterior
hamule not bifid (Figs. 1468a-b); female sternum of

Map 92. Distribution of *Elga* spp.

S9 projected distally to about posterior margin of S10,
triangular and smoothly convex (shared with *Fylgia*,
Nephepeltia and some *Micrathyria* and *Zenithoptera*). **Unique characters**: Cerci narrowly cylindrical
(Figs. 1469a-b); distal segment of vesica spermalis
with a pair of laterodistal cornua ending in a small
recurved hook (Figs.1467a-b).
Status of classification: Ris (1911a) considered it related to *Nephepeltia* according to venational characters.
Potential for new species: Likely. One known to the
authors from Venezuela.
Habitat: Lotic environments whithin forests (Machado, 1954).

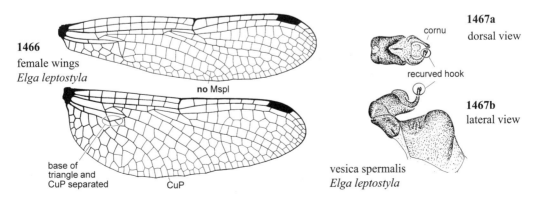

1466
female wings
Elga leptostyla

no Mspl

base of
triangle and
CuP separated   CuP

1467a
cornu
dorsal view

recurved hook

1467b
lateral view

vesica spermalis
Elga leptostyla

male genital fossa

Elga leptostyla

1468a
ventral
view

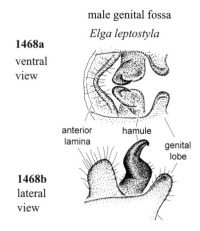

anterior
lamina   hamule
          genital
          lobe

1468b
lateral
view

male S10

1469a
lateral view

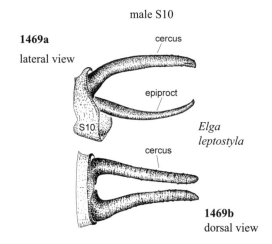

cercus

epiproct

S10

*Elga
leptostyla*

cercus

1469b
dorsal view

Erythemis Hagen, 1861: 168.
[♂ pp. 184, couplet 41; ♀ pp. 204, couplet 14]
Type species: *Libellula peruviana* Rambur, 1842
[Kirby, 1889 by subsequent designation]
[NOTE: Although Kirby (1889: 305) gives *Libellula peruviana* Rambur, 1842 as type species, that name was not among the original three names (*furcata* Hagen, 1861; *bicolor* Erichson, 1848; *longipes* Hagen, 1861) included under *Erythemis*. However Hagen (1861: 169), under *bicolor* did state "Is it different from *Libellula peruviana* Rambur?"— perhaps suggesting synonymy between these two names.]
   syn *Mesothemis* Hagen, 1861: 170
      Type species: *Libellula simplicicollis* Say,
      1840 [Kirby, 1889 by subsequent designation]
   syn *Lepthemis* Hagen, 1861: 160
      Type species: *Libellula vesiculosa* Fabricius,
      1775 [Kirby, 1889 by subsequent designation]
10 species:

attala (Selys *in* Sagra, 1857) [*Libellula*]* – **L**
[Rodrigues Capítulo, 1983b]
   syn *annulata* (*nec* Beauvois, 1805) (Rambur,
   1842) [*Libellula*]
   syn *annulosa* (Selys *in* Sagra, 1857) [*Libellula*]
   syn *mithra* (Selys *in* Sagra, 1857) [*Libellula*]
carmelita Williamson, 1923*
collocata (Hagen, 1861) [*Mesothemis*]* – **L**
[Pritchard and Smith, 1956]
credula (Hagen, 1861) [*Diplax*]* – **L** [Calvert, 1928;
Klots, 1932; Santos, 1969c]
haematogastra (Burmeister, 1839) [*Libellula*]*
mithroides (Brauer, 1900) [*Mesothemis*]* – **L** [Costa
and Pujol-Luz, 1993]
peruviana (Rambur, 1842) [*Libellula*]* – **L** [Calvert,
1928; Klots, 1932]
   syn *?rubriventris* (Blanchard, 1845) [*Libellula*]
   syn *bicolor* (Erichson, 1848) [*Libellula*]
plebeja (Burmeister, 1839) [*Libellula*]* – [Calvert,
1928; Klots, 1932; Needham and Westfall, 1955]
   syn *verbenata* (Hagen, 1861) [*Lepthemis*]

simplicicollis (Say, 1840) [*Libellula*]* – **L**
[Needham, 1901; Bick, 1941]
   syn *?imbuta* (Say, 1840) [*Libellula*]
   syn *caerulans* (Rambur, 1842) [*Libellula*]
   syn *gundlachii* (Scudder, 1866) [*Mesothemis*]
   syn *maculiventris* (Rambur, 1842) [*Libellula*]
vesiculosa (Fabricius, 1775) [*Libellula*]* – **L** [Klots,
1932; Needham and Westfall, 1955]
   syn *acuta* (Say, 1840) [*Libellula*]

References: Kennedy, 1923a; Williamson, 1923b.
Distribution: S Canada to Uruguay and central Argentina.

Map 93. Distribution of *Erythemis* spp.

Medium to large (35–62 mm) libellulines; variously colored (red, blue, brown, black) with swollen to greatly swollen basal abdominal segments; remainder segments broad to narrow. Antefrons and vertex not metallic, pterothorax largely concolorous and, in some species, becoming entirely pruinose. Wings

(Fig. 1470) hyaline or with a basal black spot; last antenodal incomplete; costal side of FW triangle straight or bent at distal portion; CuP in HW arising at or near anal angle of triangle or distinctly separated from anal angle of triangle; Mspl and Rspl distinct, median and radial planates with 1 row of cells throughout. Posterior lobe of prothorax constricted at base, upright and wider than other prothoracic lobes; anterior lamina entire (Fig. 1473a); posterior hamule bifid with inner branch smaller than outer branch (Figs. 1473a-b); vulvar lamina scoop shaped and projecting ventrally (shared with *Crocothemis*, *Nannothemis*, *Pseudoleon* and some *Erythrodiplax*), shorter or longer than 1/3 of S9. **Unique characters**: Hind femur of males and females with 3-4 long and strong spines on its distal 1/2, numerous short and distally directed spines on its basal 1/2 (Figs. 1471-1472).

Status of classification: Fair. Resolution of species fairly easy due to differences in abdominal morphology and overall coloration.

Potential for new species: Likely. One new species allied to *E. mithroides* from Middle America and a possible new species known only from females from the neotropical region are known to the authors.

Habitat: Adults active in sunshine where they squat on flat substrates including margins of ponds, leaves, or perch on reeds and twigs next to their habitats, which include ponds, pools, marshes and slow streams. Often wary and difficult to approach. Behavioral and ecological studies carried out by Currie (1963), McVey (1985), DeMarco, Latini, and Ribeiro (2002) and Rodrigues Capítulo (2000).

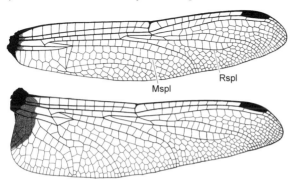

1470 male wings – *Erythemis plebeja*

lateral view

1471 male hind femur
Erythemis attala

1472 female hind femur
Erythemis attala

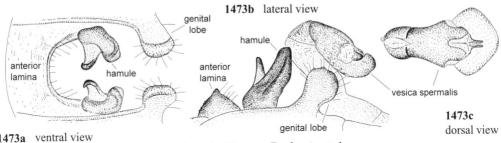

1473b lateral view

genital lobe

hamule

anterior lamina

anterior lamina

hamule

vesica spermalis

1473c
dorsal view

1473a ventral view

genital lobe

male genital fossa *Erythemis attala*

Erythrodiplax Brauer, 1868: 368, 722.
[♂ pp. 176-177, 179, 181, 186, 192, 197, couplets 13-15, 22-24, 29, 31, 50, 68-69, 83; ♀ pp. 205, 212-213, 216, couplets 17, 40, 43, 56-58]
Type species: ***Libellula plebeja*** Rambur, 1842 [Kirby, 1889 by subsequent designation]
    syn *Nadiplax* Navás, 1916a: 72
        Type species: *Nadiplax diversa* Navás, 1916
        [by original designation]
56 species:

abjecta (Rambur, 1842) [*Libellula*]*
    syn *ponderosa* Karsch, 1891
acantha Borror, 1942
amazonica amazonica Sjöstedt, 1918* – **L** [De

Marmels, 1992d]
    syn *lenti* Ris, 1919
amazonica melanica Borror, 1942*
anatoidea Borror, 1942*
andagoya Borror, 1942*
angustipennis Borror, 1942*
anomala (Brauer, 1865) [*Libellula* (*Diplax*)]* – **L**
    [Carvalho, Ferreira, and Nessiminian, 1991]
atroterminata Ris, 1911*
attenuata (Kirby, 1889) [*Trithemis* (?)]*
    syn *attenuata* forma *hyalina* Sjöstedt, 1918
basalis avittata Borror, 1942*
basalis basalis (Kirby, 1897) [*Micrathyria*]* – **L**

[Costa, Vieira, and Lourenço, 2001]
   syn ?*erratica* (Erichson, 1848) [*Libellula*]
basifusca (Calvert, 1895) [*Trithemis*]*
berenice berenice (Drury, 1773) [*Libellula*]* – L
[Calvert, 1904]
   syn *histrio* (Burmeister, 1839) [*Libellula*]
berenice naeva (Hagen, 1861) [*Dythemis*]* – L
[Klots, 1932; Needham and Westfall, 1955]
branconensis Sjöstedt, 1929
bromeliicola Westfall *in* Needham, Westfall, and
May, 2000
castanea (Burmeister, 1839) [*Libellula*]*
cauca Borror, 1942*
chromoptera Borror, 1942*
cleopatra Ris, 1911*
clitella Borror, 1942*
connata (Burmeister, 1839) [*Libellula*]*
   syn *chloropleura* (Brauer, 1865) [*Libellula*
   (*Diplax*)]
   syn *communis* (Rambur, 1842) [*Libellula*]
   syn *fraterna* (Hagen, 1873) [*Diplax*]
   syn *leontina* (Brauer, 1865) [*Libellula*]
corallina (Brauer, 1865) [*Libellula* (*Diplax*)]*
   syn *medium* (Navás, 1916) [*Sympetrum*]
   syn *nutrina* Förster, 1914
   syn *plebeia* (Rambur (*nec* Burmeister, 1839),
   1842) [*Libellula*]
diversa (Navás, 1916) [*Nadiplax*]
famula (Erichson, 1848) [*Libellula*]*
   syn *ochracea aequatorialis* Ris, 1911
fervida (Erichson, 1848) [*Libellula*]*
   syn *justina* (Selys *in* Sagra, 1857) [*Libellula*]
fulva Borror, 1957*
funerea (Hagen, 1861) [*Libellula*]* – L [Needham
and Westfall, 1955; Pritchard and Smith, 1956]
   syn *affinis* (Kirby, 1889) [*Neurothemis*]
   syn *tyleri* (Kirby, 1899) [*Trithemis*]
fusca (Rambur, 1842) [*Libellula*]* – L [Santos,
1967a]
   syn *incompta* (Rambur, 1842) [*Libellula*]
gomesi Santos, 1946
hyalina Förster, 1907*
ines Ris, 1911*
juliana Ris, 1911* – L [Carvalho, Ferreira, and
Nessiminian, 1991]
justiniana (Selys *in* Sagra, 1857) [*Libellula*]* – L
[Klots, 1932; Needham and Westfall, 1955]
   syn ?*portoricana* (Kolbe, 1888) [*Diplax*]
   syn *ambusta* (Hagen, 1868) [*Diplax*]
kimminsi Borror, 1942*
latimaculata Ris, 1911* – L [Costa, Vieira, and
Lourenço, 2001]
lativittata Borror, 1942*
laurentia Borror, 1942*
leticia Machado, 1996
longitudinalis (Ris, 1919) [*Anatya*]*

luteofrons Santos, 1956
lygaea Ris, 1911* – L [Costa, Vieira, and Lourenço,
2001]
maculosa (Hagen, 1861) [*Nannophya*]*
   syn *friedericella* (Förster, 1905) [*Diplacodes*]
media Borror, 1942*
melanorubra Borror, 1942* – L [Limongi, 1991]
minuscula (Rambur, 1842) [*Libellula*]*
nigricans (Rambur, 1842) [*Libellula*]* – L [von
Ellenrieder and Muzón, 2000]
   syn *vilis* (Rambur, 1842) [*Libellula*]
nivea Borror, 1942
ochracea (Burmeister, 1839) [*Libellula*]* – L
[Carvalho, Ferreira, and Nessiminian, 1991]
   syn *distinguenda* (Rambur, 1842) [*Libellula*]
pallida (Needham, 1904) [*Micrathyria*] – L [Costa,
Vieira, and Lourenço, 2001]
paraguayensis (Förster, 1905) [*Diplacodes*]* – L
[Muzón and Garré, 2005]
parvimaculata Borror, 1942*
solimaea Ris, 1911*
tenuis Borror, 1942*
transversa Borror, 1957* – L [De Marmels, 1992b]
umbrata (Linnaeus, 1758) [*Libellula*]* – L [Calvert,
1928; Costa, Vieira, and Lourenço, 2001]
   syn *fallax* (Burmeister, 1839) [*Libellula*]
   syn *flavicans* (Rambur, 1842) [*Libellula*]
   syn *fuscofasciata* (Blanchard, 1845) [*Libellula*]
   syn *montezuma* (Calvert, 1899) [*Trithemis*]
   syn *ruralis* (Burmeister, 1839) [*Libellula*]
   syn *subfasciata* (Burmeister, 1839) [*Libellula*]
   syn *tripartita* (Burmeister, 1839) [*Libellula*]
   syn *unifasciata* (De Geer, 1773) [*Libellula*]
unimaculata (De Geer, 1773) [*Libellula*]*
   syn *erichsoni* (Kirby, 1894) [*Trithemis*]
   syn *hemimelaena* (Karsch, 1890) [*Micrathyria*]
   syn *pulla* (Burmeister, 1839) [*Libellula*]
venusta (Kirby, 1897) [*Micrathyria*]*
   syn ?*erratica* (Erichson, 1848) [*Libellula*]

References: Borror, 1942; 1957; Paulson, 2003.
Distribution: S Canada to S Chile and Argentina.

Map 94. Distribution of *Erythrodiplax* spp.

Very small to medium (20–50 mm) libellulines; brown, red or black to metallic blue postfrons and vertex; body yellow or brown to red and black often covered with bluish pruinescence in mature males. Wings (Figs. 1478-1481) hyaline to variously colored with black, red, or orange; last antenodal of FW incomplete (except in some specimens of *E. tenuis*); usually 1 bridge crossvein (rarely 2-3) and 1 cubito-anal crossvein (rarely 2-4); supratriangles usually free; Mspl usually indistinct, but in some species distinct with up to 1 (*E. umbrata*) or 2 (*E. funerea*) rows of cells. Posterior lobe of prothorax bent caudad (Figs. 1474c, 1476b), slightly to greatly narrowed basally (Fig. 1476a), with sides parallel (Figs. 1474a, 1475), or diverging (Fig. 1474b); abdomen triangular in crossection. Anterior lamina entire (Fig. 1485b); posterior hamule bifid, with inner branch smaller than outer branch (Figs. 1485a-b); vulvar lamina (Figs. 1477a-b) scoop-shaped and ventrally directed (shared with *Crocothemis*, *Erythemis*, *Nannothemis* and *Pseudoleon*), shorter or longer than 1/3 of S9. Distal segment of vesica spermalis (Figs. 1482-1484) long and cylindrical, with complex and largely hidden distal lobes (shared with *Nannothemis*, *Pseudoleon*, *Uracis* and *Ypirangathemis*), gradually widening to the rounded or bluntly angulated tip (shared with *Uracis* and *Ypirangathemis*), except for some specimens of *E. berenice*, where it is cylindrical throughout as in *Nannothemis*. **Unique characters**: None known.

Status of classification: The genus as currently conceived could be paraphyletic, as it possesses several character states of the tibial spination and of various venational characters found in other genera. Neither Ris (1911) nor Borror (1942) diagnosed the genus. According to Ris (1911) and Borror (1942) it is ap-parently related to *Erythemis* and *Micrathyria* based on wing venation and to *Rhodopygia* and *Uracis* based on structure of the vesica spermalis. We found the morphology of the vesica spermalis very close to that of *Uracis* and *Ypirangathemis*, not *Rhodopygia*. Males are very similar to *Uracis* males [see couplet 68 in page 192]. Angulation of transverse carina on S2 easily separates them. Borror (1947) also compared *Ypirangathemis* with *Erythrodiplax* and updated (1957) his previous key (1942) with the addition of two new species. We have been unable to find any consistent character with which to separate females of *Erythrodiplax* from the introduced *Crocothemis servilia* (see comment under *Crocothemis*, page 232). Borror (1942) considered several taxa as subspecies and although the status of some of these has changed to full species (Paulson, 2003), others such as *E. amazonica melanica* Borror, and *E. basalis avittata* Borror, may also require elevation to species after further studies.

Potential for new species: Very likely. At least three new species from Venezuela, Brazil and Argentina known to the authors.

Habitat: Temporary ponds, lakes, stream pools, phytotelmata, marshes and one species (*E. berenice*) has been known to breed in undiluted seawater in mangrove swamps and saline lakes. Adults perch with wings set on tips of grass or low stems around pools or emergent vegetation; usually found foraging in fields. Some species (*E. amazonica*) are shy forest species which alight on sunlit leaves and fly successively to higher leaves when disturbed (De Marmels, 1989). Paulson (1998a) studied some behavioral aspects of *E. umbrata*.

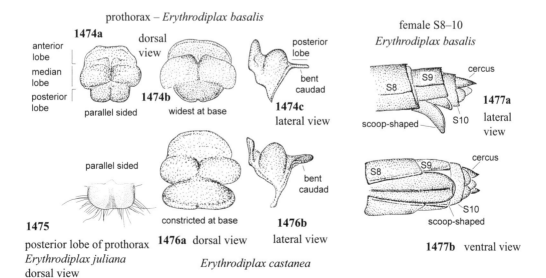

prothorax – *Erythrodiplax basalis*

1474a dorsal view

anterior lobe
median lobe
posterior lobe

parallel sided

1474b widest at base

posterior lobe

bent caudad

1474c lateral view

parallel sided

1475 posterior lobe of prothorax *Erythrodiplax juliana* dorsal view

constricted at base

1476a dorsal view

bent caudad

1476b lateral view

Erythrodiplax castanea

female S8–10
Erythrodiplax basalis

cercus
S9
S8
1477a lateral view
scoop-shaped
S10

cercus
S8
S9
S10
scoop-shaped
1477b ventral view

1478 male wings – *Erythrodiplax angustipennis*

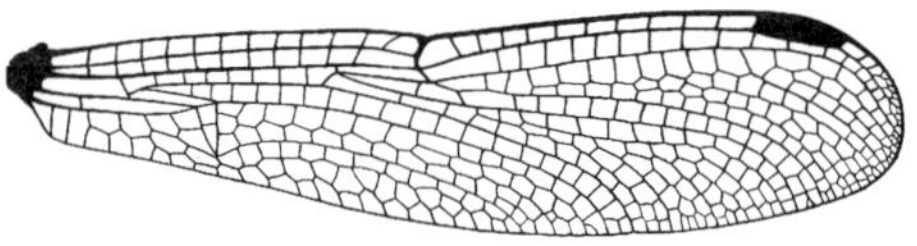

1479 male wings – *Erythrodiplax fusca*

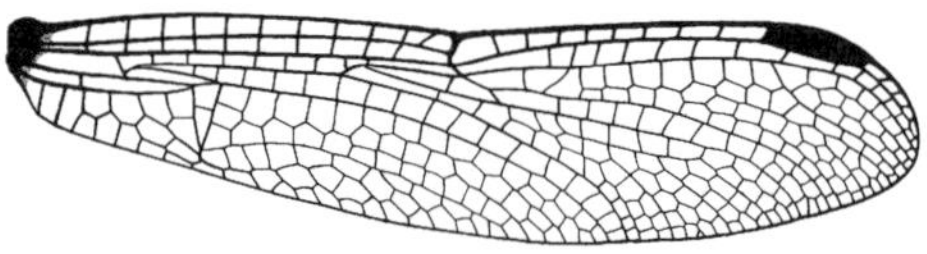

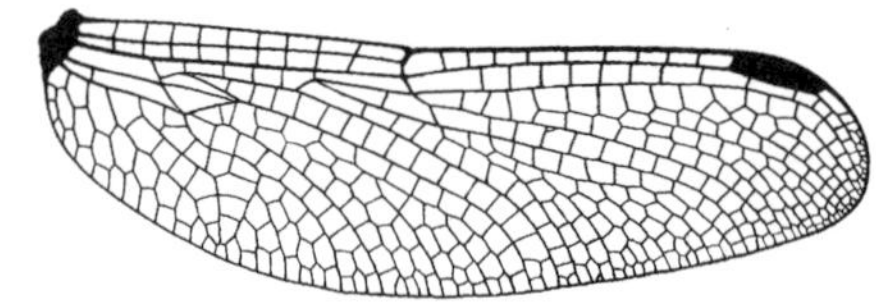

no Mspl   Rspl

Mspl   Rspl

1480 male wings
Erythrodiplax paraguayensis

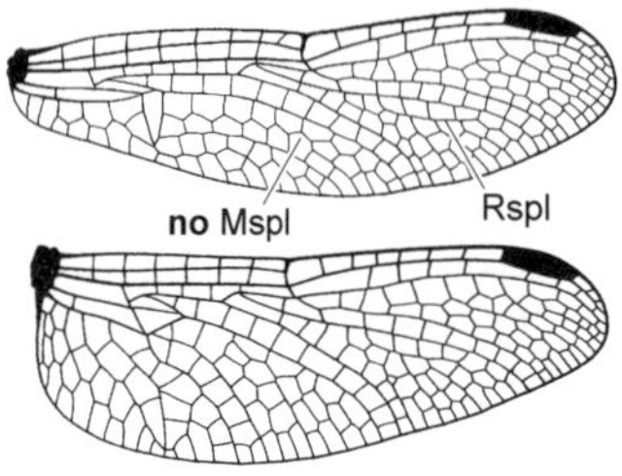

1481 female wings – *Erythrodiplax umbrata*

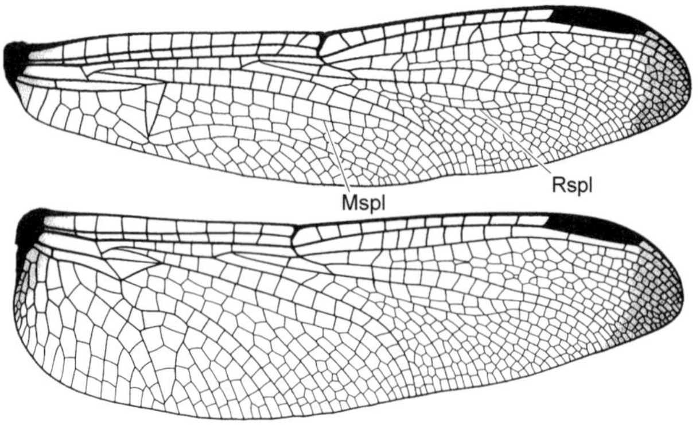

male genital fossa

1483 *Erythrodiplax basifusca*
lateral view

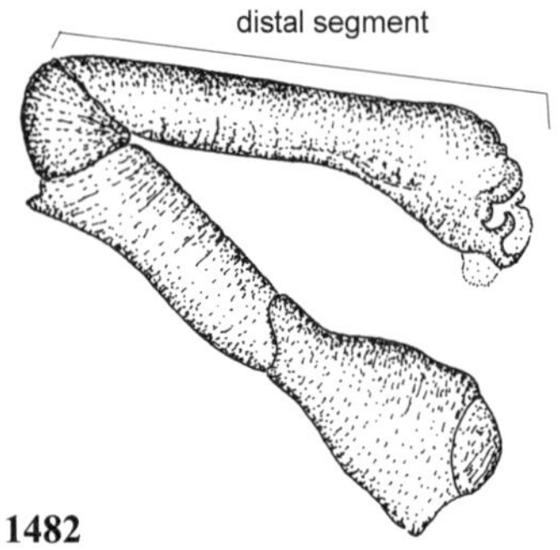

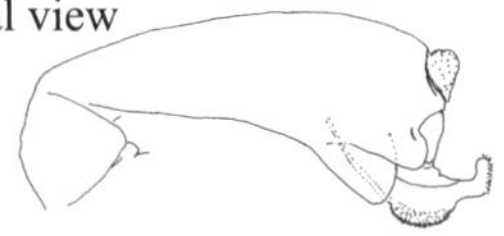

distal segment

anterior
lamina   hamule   genital
lobe

1484a
lateral
view

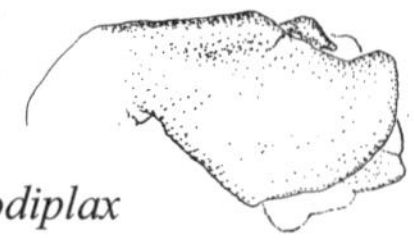

1485a
lateral view

*Erythrodiplax
juliana*

*Erythrodiplax
castanea*

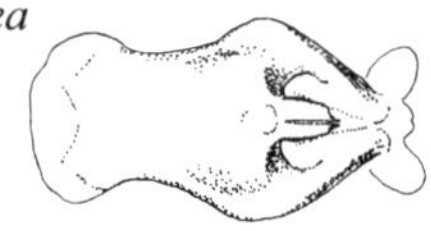

hamule   genital
lobe

anterior
lamina

1482
vesica spermalis – *Erythrodiplax juliana*
lateral view

1484b dorsal view
vesica spermalis distal segment

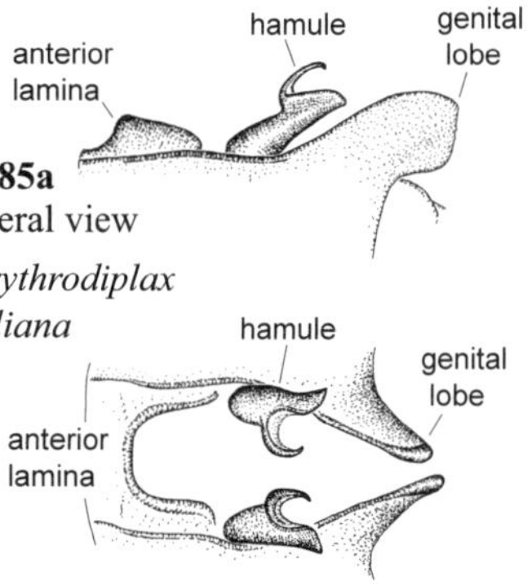

1485b ventral view

Fylgia Kirby, 1889: 259, 312.
[♂ pp. 197, couplet 84; ♀ pp. 218, couplet 65]
Type species: *Fylgia amazonica* Kirby, 1889 [by
original designation]
1 species:

amazonica amazonica Kirby, 1889*
 syn *semiaurea* (Karsch, 1889) [*Nannothemis*]
amazonica lychnitina De Marmels, 1989* – **L** [De
Marmels, 1992d]

References: Kirby, 1889; De Marmels, 1989.
Distribution: Amazonian region from Venezuela to
Peru and Brazil.

Very small (19–22 mm) libellulines. Eyes white
dorsally in life; males black with dorsum of S3-7 to
tip red; females with pterothorax green with brown
stripes and abdomen orange-brown with dark markings. Wings (Fig. 1486) hyaline to smoky; last an-

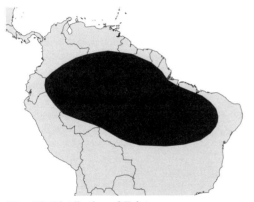

Map 95. Distribution of *Fylgia* sp.

tenodal in FW incomplete; costal side of FW triangle bent at about midlength; 1 row of cells at FW discoidal field; Mspl and Rspl distinct; medial and radial planates with 1 row of cells throughout; Aspl in HW more or less straight; anal loop well developed without a toe. Spines of male hind femur very thin and gradually increasing in length distally; tooth of

pretarsus as long as tip and almost apical (Fig. 1487); abdomen short, stout; anterior lamina entire (Fig. 1489b); posterior hamule not bifid (Figs. 1489a-b); female sternum S9 (Fig. 1488) projected distally to about posterior margin of S10, semicircular and smoothly convex (shared with *Elga*, *Nephepeltia* and some *Micrathyria* and *Zenithoptera*). **Unique characters**: RP1,2 slightly concave (Fig. 1486); vulvar lamina divided into 2 separate triangular projections (Fig. 1488).

Status of classification: This genus seems to have no clearly related taxa in the New World although Ris (1909) placed it in his group 5 which included the New World genera *Anatya*, *Argyrothemis*, *Nannothemis*, *Micrathyria*, *Nephepeltia* and others.

Potential for new species: Unlikely. De Marmels (1989) described a new subspecies based on extension of red on abdomen (7 to 10 on *F. a. amazonica*, 3 to 10 in *F. amazonica lychnitina*).

Habitat: Small, clear, stagnant pools with leaf litter in deep forest. Males land on broad leaves overhanging water's edge in dappled sunlight.

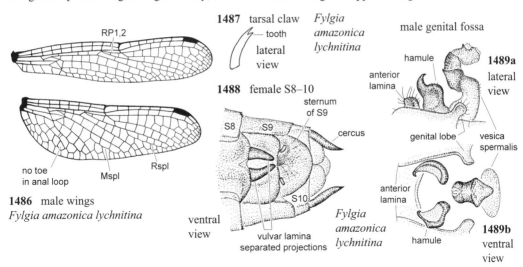

1487 tarsal claw — tooth — lateral view

1488 female S8–10

1486 male wings
Fylgia amazonica lychnitina

Fylgia amazonica lychnitina

male genital fossa

1489a lateral view

1489b ventral view

Fylgia amazonica lychnitina

Gynothemis Calvert *in* Ris, 1909a: 34.
[♂ pp. 190, 195, couplets 62, 77; ♀ pp. 210, 213, 215, couplets 35, 37, 42, 52]
Type species: *Gynothemis venipunctata* Calvert *in* Ris, 1909 [by original designation]
4 species:

aurea Navás, 1933
pumila Karsch, 1890* – **L** [Fleck, 2004]
    syn var. *axillata* Navás, 1924
uniseta Geijskes, 1972* – **L** [Geijskes, 1972]
venipunctata Calvert *in* Ris, 1909*
[NOTE: The citation commonly used for this species is Calvert, 1909, but genus and species names were first mentioned and diagnosed by Ris (based on the manuscript names from Calvert) in his key included in the first volume of the Libellulinen monograph (Ris, 1909a), before Calvert's description (1909) was published that same year.]

Map 96. Distribution of *Gynothemis* spp.

References: Ris, 1909a; 1913a; Geijskes, 1972; Garrison and von Ellenrieder, 2006.
Distribution: Venezuela south to Mato Grosso in Brazil.

Moderately small (22–27 mm), delicately built libellulines. Antefrons and vertex brown to metallic blue in mature males; pterothorax relatively small, dark brown with yellow stripes (Fig. 1492) or with yellow lateral sides (Fig. 1493); abdomen cylindrical, brown with sterna pale and pale tergal stripes. Hind femur in males with spines gradually increasing in length distally (Fig. 1498, *G. venipunctata*), or with extremely short to absent, distally directed spines at basal 1/2 and 1-6 slender, long spines increasing distally in length at distal 1/2 of femur (Fig. 1496, *G. pumila*, and Fig. 1497, *G. uniseta*); tooth of pretarsus shorter than claw (Fig. 1495), or apical and as long as claw (Fig. 1494, *G. pumila*); posterior lobe of prothorax widest at base, its posterior margin bent caudad. Wings (Figs. 1490-1491) hyaline or with golden yellow spot at wing base; FW triangle and supratriangle free; Mspl indistinct. Anterior lamina entire (Fig. 1499b); posterior hamule not bifid (Figs. 1499a-b); vulvar lamina (Figs. 1502-1503) less than 1/3 the length of S9 and not projected ventrally. **Unique characters**: None known.
Status of classification: Similar to some *Macrothemis* in vesica spermalis morphology (see Figs. 1525-1526a-b) and development of pretarsal tooth (in *G. pumila*). Distinction from male *Macrothemis* by morphology of femoral spines (short, stout and directed proximally in *Macrothemis*; distally directed in *Gynothemis*); distinction from female *Macrothemis* not possible based on examined characters. Placement of females to genus can only be done by species identification. According to Geijskes (1972), the larva of *G. uniseta* is more similar to the larva of the Antillean endemic genus *Scapanea* than it is to *Macrothemis*. Fleck (2004) provided a detailed comparison of the larva of *G. pumila* with those of *Macrothemis* and *Brechmorhoga*, and showed that it differs from all of them; however it agrees well with the description of *G. uniseta* Geijskes (1972).
The status of *G. aurea* is uncertain; known only from the original description based on a single female. It could well be placed in *Macrothemis*. We have left it in *Gynothemis* but have excluded it from the data set for this genus pending further study.
Potential for new species: Likely.
Habitat: Adults have been taken perching on twigs on small creeks within the forest (Geijskes, 1972). Garrison (1983) reported *G. venipunctata* flying like a *Macrothemis* about 1.5 m above the ground and around high tree-tops. *Gynothemis pumila* float over clearings in little mini-swarms at head height and above, and breed in sandy rain-forest streams, and perhaps in muddy pools (Paulson, *pers. comm.*).

1490 male wings – *Gynothemis venipunctata*

1491 male wings – *Gynothemis pumila*

Rspl
no Mspl

pterothorax   lateral view

male tarsal claw
lateral view

male hind femur   lateral view

Gynothemis venipunctata

1496 *Gynothemis pumila*

1495

1497 *Gynothemis uniseta*

1492
Gynothemis uniseta

1493
Gynothemis venipunctata

1494
Gynothemis pumila

1498 *Gynothemis venipunctata*

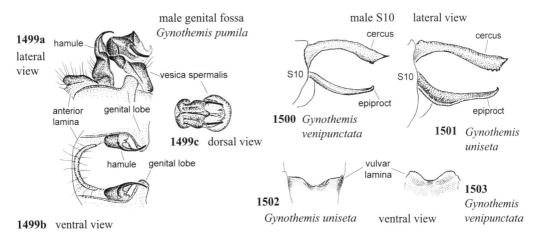

1499a lateral view — male genital fossa *Gynothemis pumila*

hamule · vesica spermalis · anterior lamina · genital lobe · hamule · genital lobe

1499c dorsal view

1499b ventral view

male S10   lateral view

cercus · S10 · epiproct

1500 *Gynothemis venipunctata*

cercus · S10 · epiproct

1501 *Gynothemis uniseta*

vulvar lamina

1502 *Gynothemis uniseta* — ventral view

1503 *Gynothemis venipunctata*

Idiataphe Cowley, 1934b: 243.

[♂ pp. 201, couplet 103; ♀ pp. 222, couplet 81]

Type species: *Erythemis longipes* Hagen, 1861 [Cowley, 1934b by original designation]

　syn *Ephidatia* Kirby, 1889: 262, 283

　　Type species: *Erythemis longipes* Hagen, 1861 [Kirby, 1889 by subsequent designation]

[NOTE: *Idiataphe* is a replacement name for *Ephidatia* Kirby, 1889, a junior primary homonym of *Ephidatia* Lecoq, 1862, an emendation of *Ephydatia* Lamouroux, 1816, in Porifera.]

4 species:

amazonica (Kirby, 1889) [*Ephidatia*]*
batesi (Ris, 1913) [*Ephidatia*]*
cubensis (Scudder, 1866) [*Macromia*]* – **L**
[Needham and Fisher, 1936; García-Diaz, 1938; Paulson, 1999]

　syn *specularis (nomen nudum)* (Hagen, 1867) [*Erythemis*]

longipes (Hagen, 1861) [*Erythemis*]*

References: Rácenis, 1969.
Distribution: Florida in United States and Antilles south through NE Argentina and SE Brazil.

Map 97. Distribution of *Idiataphe* spp.

Medium libellulines (34–42 mm); orange-brown to metallic blue or violet postfrons and vertex; pterothorax brown with blue or green metallic luster in some species; abdomen black usually with orange or yellow stripes on lateral margins. Wings hyaline but more often with a small orange basal spot; sectors of arculus in FW not stalked; last antenodal in FW incomplete; Mspl and Rspl distinct; medial and radial planates with 1 row of cells throughout; wing tips slightly pointed (Fig. 1505). Posterior lobe of prothorax widest at base, its posterior margin bent caudad; hind legs very long and thin, and in both sexes with spines absent or microscopic except for 1 slender distal spine (Fig. 1506); male cerci (except in *I. batesi*) with a ventral mediolongitudinal flange (Fig. 1504); posterior hamule bifid with inner branch smaller than outer branch (Figs. 1507a-b); vulvar lamina about 1/3 the length of S9 and not projected ventrally. **Unique characters**: Anterior lamina bifid with a tuft of stiff setae directed anteriorly on each side (Fig. 1507a-b).

Status of classification: Species resolution is problematic for *I. cubensis* (De Marmels, *pers. comm.*).
Potential for new species: Unlikely.
Habitat: Adults fly over marshes and exposed ponds where they perch on tips of grass or reeds at water's edge. Adults dart out over water often hovering for a few seconds before returning to perch. Oviposition is by female alone or in tandem; females remain away from water except for reproduction (Paulson, 1999; Dunkle, 2000); females remain away from water except for reproduction (Paulson, 1999).

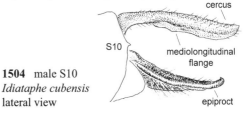

cercus · S10 · mediolongitudinal flange · epiproct

1504 male S10 *Idiataphe cubensis* lateral view

1505 male wings – *Idiataphe cubensis*

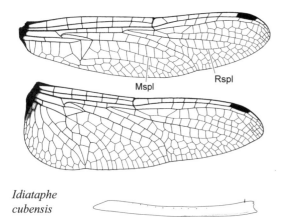

Mspl
Rspl

male genital fossa
Idiataphe cubensis

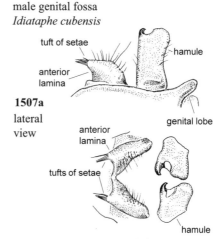

tuft of setae
hamule
anterior lamina

1507a
lateral view

anterior lamina
genital lobe

tufts of setae

hamule

1507b ventral view

Idiataphe cubensis

1506 male hind femur   lateral view

Leucorrhinia Brittinger, 1850: 333.
[♂ pp. 178, 199, couplets 17, 94; ♀ pp. 206, 222, couplets 20, 80]
Type species: *Libellula albifrons* Burmeister, 1839 [Kirby, 1889 by subsequent designation]
   syn *Coenotia* Buchecker, 1876: 10
      Type species: *Libellula caudalis* Charpentier, 1840 [Kirby, 1889 by subsequent designation]
14 species; 7 New World species:

borealis Hagen, 1890* – **L** [Walker, 1916; Walker and Corbet, 1975]
frigida Hagen, 1890* – **L** [Walker, 1913; Walker and Corbet, 1975]
glacialis Hagen, 1890* – **L** [Walker and Corbet, 1975; Kenner, 2001]
hudsonica (Selys, 1850) [*Libellula*]* – **L** [Walker, 1914; Walker and Corbet, 1975; Kenner, Cannings, and Cannings, 2000]
      syn *hageni* Calvert, 1890
intacta (Hagen, 1861) [*Diplax*]* – **L** [Needham, 1901; Walker and Corbet, 1975]
patricia Walker, 1940* – **L** [Kenner, Cannings, and Cannings, 2000]
proxima Calvert, 1890* – **L** [Walker, 1916; Walker and Corbet, 1975]

References: Ris, 1912; Walker and Corbet, 1975; Needham, Westfall, and May, 2000.
Distribution: Holarctic.

Small to medium libellulines (24–46 mm); face and anterior half or more of postfrons white, posterior half and vertex dark; pterothorax hairy; body largely black, with pale areas of bright yellow to red. Wings hyaline with small dark basal spot; pterostigma short and thick; last antenodal complete; sectors of arculus in FW stalked or not stalked; costal side of FW triangle bent distally or straight; FW discoidal field

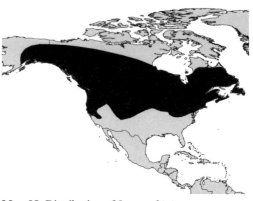

Map 98. Distribution of *Leucorrhinia* spp.

widening distally; Mspl indistinct or distinct; Rspl distinct (Fig. 1508). Short eye-seam; posterior lobe of prothorax constricted at base, upright and wider than other prothoracic lobes. Anterior lamina (Fig. 1509b) entire to strongly bifid; posterior hamule bifid with inner branch larger than outer branch (Figs. 1509a-b); vulvar lamina less to more than 1/3 of S9 and not projected ventrally. **Unique characters**: None known.
Status of classification: Species well diagnosed. Westman, Johansson, and Nilsson (2000) provided a phylogenetic assessment of the members of this genus.
Potential for new species: Unlikely.
Habitat: Cold marshy bogs at northern latitudes. Adults of some species (*L. intacta*) at ponds or along slow moving canals. Excellent biological notes in Walker and Corbet (1975); reproductive behavior of *L. hudsonica* by Hilton (1985) and territoriality of *L. intacta* by Wolf and Waltz (1984). Although the larvae of all the New World species are known, the characters used to diagnose them are in some cases variable and extant keys should be used with caution (DuBois, 2003).

1508   male wings – *Leucorrhinia frigida*

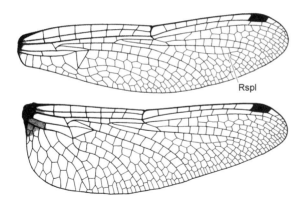

Rspl

male genital fossa   *Leucorrhinia frigida*

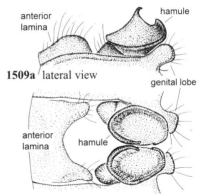

anterior
lamina   hamule

1509a lateral view   genital lobe

anterior
lamina   hamule

1509b   ventral view

Libellula Linnaeus, 1758: 543.
[♂ pp. 198, 200-201, couplets 93, 102, 104; ♀ pp. 222-223, couplets 79, 86, 88]

Type species: *Libellula depressa* Linnaeus, 1758 [Westwood, 1840 by subsequent designation]

[NOTE: Controversy exists as to whether *Libellula quadrimaculata* Linnaeus, 1758 or *L. depressa* Linnaeus, 1758 is to be considered the type species of *Libellula*. The earliest mention of a "type" for this genus is "*Libellula 4-maculata*, Fab." by Latreille (1810) under his "Table des genres avec l'indicationde l'espèe qui leur sert de type". There has been controversy as to whether Latreille's listings actually constitute type designations; ICZN Opinion 11 supports this interpretation, but supplementary opinion No. 136 further states that Opinion 11 validating Latreille's type designations is valid *only* in those cases in which the species selected by Latreille as 'type' was one named by him — a condition not met in this case. Consequently, Westwood's (1840) selection of *Libellula depressa* as the type of *Libellula* would be the earliest valid type designation, and, as did Kirby (1889), we follow this usage here.]

  syn *Platetrum* Newman, 1833: 511
    Type species: ?
[NOTE: Originally included species were "[*Libellula*] *Depressum*, Linn." and "[*L.*] *Conspurcatum*, Linn. [*sic*]"; = *L. fulva* Müller 1764. No type species appears to have been selected for this genus.]

  syn *Leptetrum* Newman, 1833: 511
    Type species: *Libellula quadrimaculata* Linnaeus, 1758 [Kirby, 1889 by subsequent designation]

  syn *Plathemis* Hagen, 1861: 149
    Type species: *Libellula lydia* Drury, 1773 [Kirby, 1889 by subsequent designation]

  syn *Pigiphila* Buchecker, 1876: 11
    Type species: *Libellula depressa* Linnaeus, 1758 [by original designation]

  syn *Belonia* Kirby, 1889: 260, 288
    Type species: *Belonia foliata* Kirby, 1839 [Kirby, 1889 by original designation]

  syn *Holotania* Kirby, 1889: 261, 288
    Type species: *Libellula axilena* Westwood,

1837 [Kirby, 1889 by original designation]

  syn *Ladona* Needham, 1897a: 146
    Type species: *Libellula exusta* Say, 1840 [by original designation]

  syn *Eolibellula* Kennedy, 1922b: 111
    Type species: *Libellula semifasciata* Burmeister, 1839 [Kennedy, 1922b by original designation]

  syn *Syntetrum* Kennedy, 1922b: 111
    Type species: *Libellula angelina* Selys, 1883 [Kennedy, 1922b by original designation]

  syn *Eurothemis* Kennedy, 1922b: 111
    Type species: *Libellula fulva* Müller, 1764 [Kennedy, 1922b by original designation]

  syn *Neotetrum* Kennedy, 1922b: 111
    Type species: *Libellula forensis* Hagen, 1861 [Kennedy, 1922b by original designation]

  syn *Eotania* Carle & Kjer, 2002b: 3
    Type species: *Mesothemis composita* Hagen, 1861 [Carle and Kjer, 2002 by original designation]

30 species; 27 New World species:

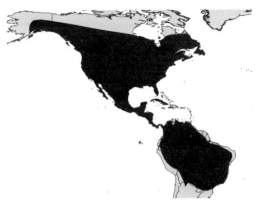

Map 99. Distribution of *Libellula* spp.

auripennis Burmeister, 1839* – **L** [Needham, 1901]
   syn *costalis* Rambur, 1842
axilena Westwood, 1837* – **L** [Needham, 1903]
   syn *leda* Say, 1840
comanche Calvert, 1907* – **L** [Needham and
Westfall, 1955; Pritchard and Smith, 1956]
   syn *flavida* Hagen (*nec* Rambur 1842), 1861
composita (Hagen, 1873) [*Mesothemis*]* – **L**
[Musser, 1962]
croceipennis Selys, 1868* – **L** [Needham and
Westfall, 1955]
   syn *saturata aliasignata* Muttkowski, 1910
   syn *uniformis* (Kirby, 1889) [*Belonia*]
cyanea Fabricius, 1775* – **L** [Needham, 1901]
   syn *bistigma* Uhler, 1857
   syn *quadrupla* Say, 1840
deplanata Rambur, 1842* – **L** [Needham, 1897a]
exusta Say, 1840*
flavida Rambur, 1842* – **L** [Needham, 1903 as *L.
plumbea*]
   syn *plumbea* Uhler, 1857
foliata (Kirby, 1889) [*Belonia*]*
forensis Hagen, 1861* – **L** [Musser, 1962; Walker
and Corbet, 1975]
gaigei Gloyd, 1938*
herculea Karsch, 1889* – **L** [De Marmels, 1982a;
1992d]
   syn *longipennis* (Kirby, 1889) [*Belonia*]
incesta Hagen, 1861* – **L** [Byers, 1927a; Walker
and Corbet, 1975]
jesseana Williamson, 1922*
julia Uhler, 1857* – **L** [Needham, 1901; Walker and
Corbet, 1975; Pilon and Desforges, 1989]
luctuosa Burmeister, 1839* – **L** [Needham, 1901;
Needham and Westfall, 1955; Pritchard and Smith,
1956]
   syn *basalis* Say (*nec* Stephens, 1835), 1840
   syn *odiosa* Hagen, 1861
lydia Drury, 1773* – **L** [Needham, 1901; Levine,
1957; Walker and Corbet, 1975]
   syn *serva* Fabricius, 1793
   syn *trimaculata* De Geer, 1773
mariae Garrison, 1992*
needhami Westfall, 1943* – **L** [Needham and
Westfall, 1955]
nodisticta Hagen, 1861*
pulchella Drury, 1773* – **L** [Needham, 1901;
Musser, 1962; Walker and Corbet, 1975]
   syn *confusa* Uhler, 1857
   syn *versicolor* Fabricius, 1775
quadrimaculata Linnaeus, 1758* – **L** [Needham,
1901; Musser, 1962; Walker and Corbet, 1975]
   syn *quadripunctata* Fabricius, 1781
   syn *ternaria* Say [in part], 1840
saturata Uhler, 1857* – **L** [Needham, 1904;
Needham and Westfall, 1955; Musser, 1962]

semifasciata Burmeister, 1839* – **L** [Bick, 1951a]
   syn *bifasciata (nomen oblitum?)* Fabricius, 1775
   syn *hersilia* Blanchard, 1849
   syn *maculata* Rambur, 1842
   syn *ternaria* Say [in part], 1840
subornata (Hagen, 1861) [*Plathemis*]*
– **L** [Needham and Westfall, 1955; Levine, 1957;
Needham, Westfall, and May, 2000]
vibrans Fabricius, 1793* – **L** [Needham and
Westfall, 1955]

References: Ris, 1910; Bennefield, 1965; Needham,
Westfall, and May, 2000.
Distribution: Holarctic genus with species occurring
in Europe and Asia. In the New World, distributed
from Alaska and N Canada south to N Argentina;
most diverse in North America with only one species
(*L. herculea*) widely distributed in South America.

Medium to very large (34–63 mm) libellulines; face
white, yellow, red, brown to black; thorax and abdomen pale yellow to orange, red, or dark brown
and often covered with pruinosity in mature males.
Wings (Fig. 1510) hyaline but more often with various patterns of yellow, orange, and brown; costal side
of FW triangle less than 1/2 as long as proximal side;
FW triangle crossed; FW subtriangle well defined
or not defined; sectors of arculus in FW not stalked;
FW discoidal field widening distally; pterostigma
long. Posterior lobe of prothorax widest at base, its
posterior margin bent caudad; male hind femur with
spines gradually increasing in length distally. Sternum S1 in males flat (Figz. 1513a-b) or with 2 lateral projections (subgenus *Plathemis*, Fig. 1512a-b);
posterior hamule bifid with outer branch larger than
inner branch (subgenera *Plathemis* and *Ladona*, Figs.
1512a-b) or subequal to inner branch (Fig. 1511, *L.
quadrimaculata*, Old World species *L. depressa*), or
smaller than inner branch to absent (Figs. 1513a-b,
subgenera *Belonia*, *Holotania*, *Neotetrum*, etc.). Vulvar lamina less than 1/3 of the length of S9, bilobed
and projected ventrally or not (subgenera *Plathemis*,
Ladona, Old World species *L. fulva*); S8 in female
with lateral flanges present or absent (subgenus *Ladona* and *L. quadrimaculata*). **Unique characters**:
None known. The four synapomorphies given by
Carle and Kjer (2002) are not unique to this genus
nor are they stable within *Libellula sensu lato* as
follows: 1. FW with 4-5 unmatched postnodals: we
found 2-3 unmatched postnodals in some specimens
and even between wings of the same specimen in *L.
auripennis*, *L. composita*, *L. comanche*, *L. cyanea*, *L.
depressa*, *L. flavida*, *L. fulva*, *L. gaigei*, *L. incesta,
L. jesseana, L. lydia, L. quadrimaculata, L. semifasciata, L. subornata*, and other libelluline genera
(e.g. *Dythemis, Orthemis, Paltothemis, Planiplax,*

Sympetrum, Zenithoptera) can have 4-5 unmatched postnodals. 2. Wings with 2-6 bridge crossveins: *Antidythemis, Diastatops, Edonis, Fylgia, Micrathyria, Nephepeltia, Perithemis, Zenithoptera,* and some individuals of *Elasmothemis, Erythrodiplax, Leucorrhina,* and *Misagria* have 2 or more bridge crossveins. 3. Basal brown area of FW extended to antenodal 1 (reversed at least three times): at least 12 species of *Libellula* (*L. auripennis, L. comanche, L. croceipennis, L. herculea, L. gaigei,* males of *L. incesta, L. jesseana, L. needhami, L. semifasciata, L. foliata, L. quadrimaculata, L. saturata*) lack this state, and there is a basal brown spot in the FW of *Macrodiplax, Pseudoleon,* some species of *Diastatops, Dythemis, Erythrodiplax, Leucorrhinia, Macrothemis, Paltothemis, Sympetrum,* and *Tramea.* 4. Abdomen wide (S5 wider than long, reversed at least four times): at least 12 species of *Libellula* (*L. auripennis, L. axilena, L. comanche, L. cyanea, L. flavida, L. gaigei, L. incesta, L. jesseana, L. needhami, L. foliata, L. quadrimaculata, L. vibrans*) lack this state and some *Brachymesia, Crocothemis, Erythrodiplax, Fylgia, Perithemis,* and *Planiplax* have a wide S5.

Status of classification: Species well diagnosed. Several subgenera have been proposed of which two (*Ladona* Needham and *Plathemis* Hagen) have been treated as genera by some authors (Needham and Westfall, 1955; May, 1992a; Needham, Westfall, and May, 2000; Artiss, 2001; Carle and Kjer, 2002). Based on mitochondrial CO1 and 16S rRNA sequence data, Artiss, Schultz, Polhemus, and Simon (2001) suggested that *Plathemis* and *Orthemis* would constitute the sister group of the remaining species of *Libellula.* However, according to studies by Kambhampati and Charlton (1999) based on 16S rRNA sequence data, and Carle and Kjer (2002) based on wing venation and vesica spermalis morphology, *Libellula* constitutes a monophyletic unit, and its subgenera are all monophyletic. Carle and Kjer (2002) placed *Libellula depressa* in the genus *Platetrum* as sister group of

Plathemis, and suggested that *Plathemis – Platetrum* constitute the sister group of *Ladona – Eurothemis.* Although the subgroups are clearly different from each other and easily distinguished by morphological and behavioral traits in some cases (Kennedy, 1922a; b; Needham, Westfall, and May, 2000), separating them as genera would be necessary *only* if it could be clearly demonstrated that some species or groups are more closely related to *Orthemis* or to some other genus than to the remaining *Libellula* species. Since *Libellula* seems to be monophyletic, it is a question of preference to choose between the two options. It would greatly upset nomenclature to allocate the type species of *Libellula* (*Libellula depressa* included in the subgenus *Platetrum*) to a different genus because the name *Libellula* would be restricted to that genus and the remaining species of *Libellula* (including 23-25 species according to Carle and Kjer, 2002) would need to be assigned a new name in accordance to the rules of nomenclature. Therefore, to avoid instability we prefer to treat *Ladona, Plathemis* and the remaining groups as subgenera of *Libellula* rather than as separate genera. According to Carle and Kjer (2002) *Orthemis* constitutes the sister group of *Libellula.*

Potential for new species: Likely in the neotropical region. One undescribed species from Brazil known to the authors.

Habitat: Adults common at almost any body of water in North America, where they perch along or on tips of reeds, branches, and snags. Members of the subgenera *Ladona* and *Plathemis* often land on flat surfaces. Males hover while guarding females as they oviposit by washing eggs into the water. *Libellula composita* oviposits in tandem. The genus is one of the best known in terms of behavior (Campanella and Wolf, 1974; Campanella, 1975; Williams, 1977; Hilton, 1983b; König and Albano, 1987; Moore, 1987; Alcock, 1989; DeBano, 1993; 1996; Artiss, 2001). Larvae sprawl on muddy bottoms of ponds and marshes where they ambush prey.

1510 male wings – *Libellula herculea*

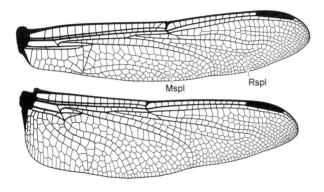

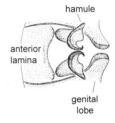

1511 male genital fossa *Libellula quadrimaculata* ventral view

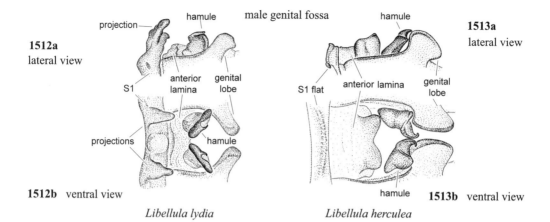

1512a lateral view

male genital fossa

1513a lateral view

1512b ventral view

Libellula lydia

1513b ventral view

Libellula herculea

Macrodiplax Brauer, 1868: 366, 737.

[♂ pp. 198, couplet 87; ♀ pp. 221, couplet 77]

Type species: *Diplax cora* Kaup *in* Brauer, 1867 [Kirby, 1889 by subsequent designation]

[NOTE: In describing *Macrodiplax*, Brauer cites "*cora* Kaup. Brau. Ceram, Philippinen." and "*paucinervis* Hg. coll. Java (? = Cora)." as included species. We have been unable to find any reference to *paucinervis* Hg [=Hagen?] in the literature; it is not mentioned in any of the catalogs of the region (*e.g.* Lieftinck, 1954; Pinhey, 1962) nor in Bridges (1994). We consider the name a *nomen nudum*.]

2 species; 1 New World species:

balteata (Hagen, 1861) [*Tetragoneuria*]* – **L** [Bick, 1955]

References: Ris, 1913b.

Distribution: Pantropical; in the New World from S United States, Bahamas, Antilles, Mexico, south through Trinidad and Venezuela.

[NOTE: the record from Buenos Aires, Argentina, by Rodrigues Capítulo and Muzón (1989) is probably based on a mislabeled specimen, since this species has never been taken south of Venezuela.]

and vertex yellow becoming dark brown in mature males; pterothorax yellow to brown, abdomen largely brown. Posterior lobe of prothorax widest at base, its posterior margin bent caudad; wings (Fig. 1514) with sparse venation; distal side of triangle convexly bent, subtriangle large and almost always with 1 curved transverse crossvein; paranal cells in FW large; sectors of arculus in FW not stalked. Abdomen cylindrical, slightly swollen basally; anterior lamina longer than wide (1.5-2 times as long as wide) (Fig. 1515b, shared with *Miathyria marcella* and *Tauriphila*); posterior hamule not bifid (Figs. 1515a-b); vulvar lamina less than 1/3 of S9, projected ventrally and bilobed.

Unique characters: Large paranal cells in combination with crossed subtriangle (Fig. 1514).

Status of classification: Good,. Only one easily recognized species in the New World.

Potential for new species: Unlikely.

Habitat: Adults hover over water and perch on tips of reeds, sometimes in obelisk position, bare branches and snags at edge of ponds and estuaries. They are often found at alkali ponds or along brackish coastal water areas.

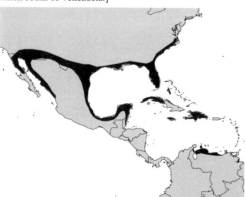

Map 100. Distribution of *Macrodiplax* sp.

Medium (34–44 mm) libellulines; yellow to brown with widened HW with a brown anal spot; postfrons

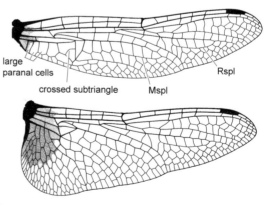

1514 female wings – *Macrodiplax balteata*

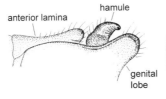
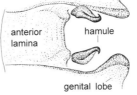

1515a
lateral view

1515b
ventral view

male genital fossa   *Macrodiplax balteata*

Macrothemis Hagen, 1868: 281.
[♂ pp. 189-190, 193-195, couplets 56, 61-62, 64, 72, 75, 77, 78; ♀ pp. 207, 210-211, 213, 215, couplets 27, 35-37, 42, 47, 52]
Type species: *Libellula celeno* Selys *in* Sagra, 1857 [Kirby, 1889 by subsequent designation]
    syn *Cendra* Navás, 1916a: 74
        Type species: *Cendra cearana* Navás, 1916 [by original designation]
    syn *Ophippus* Navás, 1916a: 76
        Type species: *Ophippus garbei* Navás, 1916 [by original designation]
40 species:

absimilis Costa, 1991*
aurimaculata Donnelly, 1984* – **L** [Ramírez and Novelo-Gutiérrez, 1999]
belliata Belle, 1987
brevidens Belle, 1983*
calliste (Ris, 1913) [*Gynothemis*]*
capitata Calvert, 1909
celeno (Selys *in* Sagra, 1857) [*Libellula*]* – **L** [Klots, 1932; Needham and Westfall, 1955; Ramírez and Novelo-Gutiérrez, 1999]
cynthia Ris, 1913
declivata Calvert, 1909*
delia Ris, 1913*
extensa Ris, 1913*
fallax May, 1998*
flavescens (Kirby, 1897) [*Miathyria*]*
griseofrons Calvert, 1909*
    syn *cearana* (Navás, 1916) [*Cendra*]
guarauno Rácenis, 1957
hahneli Ris, 1913*
hemichlora (Burmeister, 1839) [*Libellula*]*
    syn *cydippe* Calvert, 1898
heteronycha (Calvert, 1909) [*Brechmorhoga*]*
    syn *garbei* Navás, 1916 [*Ophippus*]
hosanai Santos, 1967
idalia Ris, 1919*
imitans imitans Karsch, 1890*
imitans leucozona Ris, 1913*
inacuta Calvert, 1898* – **L** [Novelo-Gutiérrez and Ramírez, 1998]
lauriana Ris, 1913
ludia Belle, 1987
lutea Calvert, 1909*
marmorata Hagen, 1868*

syn *tenuis* Karsch, 1890
mortoni Ris, 1913*
musiva Calvert, 1898* – **L** [Santos, 1970d]
    syn *uniseries* Calvert, 1909
newtoni Costa, 1990
nobilis Rácenis, 1957
pleurosticta (Burmeister, 1839) [*Libellula*]
polyneura Ris, 1913*
proterva Belle, 1987
pseudimitans Calvert, 1898* – **L** [Limongi, 1991]
rochai Navás, 1918
rupicola Rácenis, 1957*
tenuis Hagen, 1868*
tessellata tessellata (Burmeister, 1839) [*Libellula*]*
tessellata inequiunguis Calvert, 1895* – **L** [Ramírez and Novelo-Gutiérrez, 1999]
    syn *vulgipes* Calvert, 1898
ultima González-Soriano, 1992*
valida Navás, 1916

References: Ris, 1913a; 1919; May, 1998.
Distribution: S Texas and Arizona south through Uruguay and central Argentina
[NOTE: the record from 'Chile' by Martin (1921) is probably based on a mislabeled or misidentified specimen, since this genus has never been taken so far south on the Pacific coast.]

Map 101. Distribution of *Macrothemis* spp.

Small to medium (29–52 mm), delicately built libellulines, with relatively small thorax and spindle-shaped abdomen that, in some species, is broadened and flattened on S7-9. Postfrons and vertex brown to metallic blue in mature males; pterothorax and abdo-men dark brown to black often with yellowish green markings. Wings (Figs. 1516-1519) hyaline to heav-ily infumated in brown in some females; FW triangle free (except in *M. griseofrons*); Mspl in FW not de-veloped (except in *M. griseofrons*, Fig. 1518), often lacking to poorly developed in HW; FW discoidal field narrowing, parallel sided, or widening slightly toward wing margin; HW not widened at base (except in *M. flavescens*). Posterior lobe of prothorax widest at base, its posterior margin bent caudad; metafemora in males (Figs. 1520-1521) with spines short, stout, and directed proximally (shared with *Brechmorhoga* and *Scapanea*); tooth of pretarsus as long as tip and near middle of claw (Fig. 1522), except in *Macrothe-mis aurimaculata*, *M. brevidens*, *M.newtoni*, *M. tes-sellata*, and *M. valida* which have a tooth shorter than claw, and in *M. heteronycha* and *M. absimilis*, which have dimorphic pretarsi with outer tooth longer than claw and inner tooth shorter than claw in meso- and metathoracic legs (Figs. 1523a-b). Anterior lamina (Fig. 1531b) entire to strongly bifid; posterior hamule not bifid (Figs. 1529-1531a-b); vulvar lamina (Fig. 1524) less than 1/3 of S9 and not projected ventrally. **Unique characters**: None known.

Status of classification: This genus is probably related to *Brechmorhoga, Gynothemis* and *Scapanea*, shar-ing with them form and habits (Garrison and von El-lenrieder, 2006). Generic differences between *Mac-rothemis* and *Brechmorhoga* are discussed under lat-ter genus. As discussed under the account for *Gyno-*

themis, we have been unable to diagnose the females of *Macrothemis* and *Gynothemis*, and it would not be surprising if future studies prove *Macrothemis* to be paraphyletic, with some species more closely related to *Gynothemis* (i.e. *tessellata* group of species shares some aspects of larval morphology with the latter, see Fleck, 2004). This large genus is badly in need of revision; several species are poorly known and inadequately described. Another complicating factor is that females, which are more difficult to identify, are more often collected than males and some species were originally described from this sex. The latest species was described by May (1998) who provided a key of males to all known species.

Potential for new species: Very likely. New species are known from several collections and several oth-ers will probably be discovered pending a compre-hensive study of the genus. Collections often contain series of unidentifiable females.

Habitat: Stream-inhabiting species where adults hov-er or course up and down sections of moderate to fast, rocky laden creeks and rivulets. Adults often alight on sandy banks, rocks or boulders in streams or forest clearings or in brush near preferred habitats; adults (including tenerals) also found (sometimes in swarms with one or more species) over open pastures or glades in forest. Larvae found both in lotic and lentic environments, where they burrow into the substrate. In streams and rivers they prefer areas of closed cano-py, slow or still water and muddy substrate (Ramírez and Novelo-Gutiérrez, 1999).

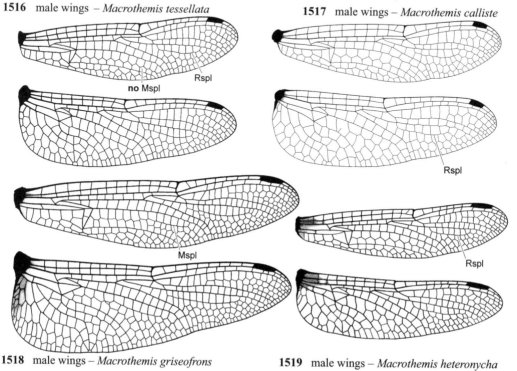

1516 male wings – *Macrothemis tessellata*

Rspl

no Mspl

1517 male wings – *Macrothemis calliste*

Rspl

Mspl

Rspl

Rspl

1518 male wings – *Macrothemis griseofrons*

1519 male wings – *Macrothemis heteronycha*

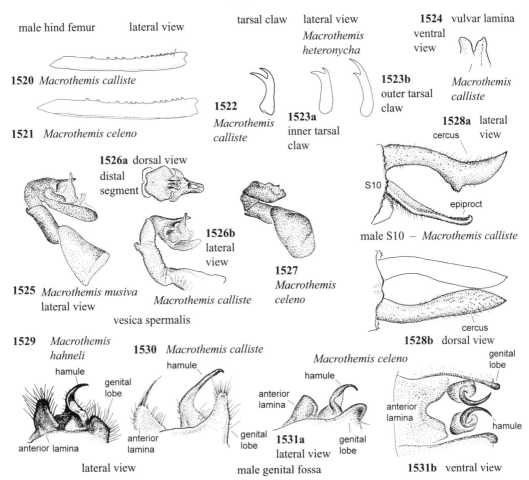

male hind femur   lateral view

1520 *Macrothemis calliste*

1521 *Macrothemis celeno*

tarsal claw   lateral view
Macrothemis heteronycha

1522 *Macrothemis calliste*

1523a inner tarsal claw

1523b outer tarsal claw

1524 vulvar lamina ventral view

Macrothemis calliste

1528a lateral view
cercus

S10

epiproct

male S10 – *Macrothemis calliste*

1526a dorsal view
distal segment

1526b lateral view

1525 *Macrothemis musiva* lateral view

Macrothemis calliste
vesica spermalis

1527 *Macrothemis celeno*

cercus

1528b dorsal view

1529 *Macrothemis hahneli*
hamule
genital lobe
anterior lamina
lateral view

1530 *Macrothemis calliste*
hamule
anterior lamina
genital lobe

Macrothemis celeno
hamule
anterior lamina
1531a lateral view
genital lobe
male genital fossa

genital lobe
anterior lamina
hamule
1531b ventral view

Miathyria Kirby, 1889: 258, 269.

[♂ pp. 182, couplets 35, 61; ♀ pp. 210, couplet 36]

Type species: *Libellula simplex* Rambur, 1842 [Kirby, 1889 by original designation]

    syn *Nothifixis* Navás, 1916b: 15

        Type species: *Nothifixis laxa* Navás, 1916 [by original designation]

2 species:

marcella (Selys *in* Sagra, 1857) [*Libellula*]* – **L** [Klots, 1932; Bick, 1953; Westfall, 1953]

simplex (Rambur, 1842) [*Libellula*]* – **L** [Limongi, 1991]

    syn *laxa* (Navás, 1916) [*Nothifixis*]

    syn *pusilla* Kirby, 1889

Map 102. Distribution of *Miathyria* spp.

References: Ris, 1913b.

Distribution: SE United States to central Argentina and Uruguay.

Small to medium (28–40 mm) libellulines; postfrons and vertex brown becoming red or metallic purple in mature males; thorax reddish brown to brown with light dusting of purple pruinosity on mesepisternum in *M. marcella*; abdomen orange to red with mid-dorsal black line on S5-10. Posterior lobe of prothorax widest at base, its posterior margin bent caudad; wings (Fig. 1532) hyaline with largely red to orange veins and with a brown anal spot in broad HW; venation sparce; sectors of arculus in FW stalked; last antenodal in FW incomplete; Mspl when present only in HW. Anterior lamina entire, longer than wide (1.5-2 times as long as wide) (Fig. 1533b, shared with *Macrodiplax balteata* and *Tauriphila*); posterior hamule not bifid (Figs. 1533a-b); vulvar lamina shorter than 1/3 of S9 and not projected ventrally. **Unique characters**: None known.

Status of classification: The two species are easily distinguishable.

Potential for new species: Unlikely.

Habitat: Lentic environments with floating vegetation such as water lettuce (*Pistia*) or water hyacinths (*Eichornia*), among which roots the larvae live. Adults usually fly in swarms about 3 m over fields or roads, in company with individuals of *Tauriphila* and *Tramea*. They are seldom observed perching but do so obliquely on sides of stems (Dunkle, 2000).

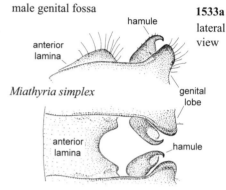

male genital fossa

1533a lateral view

hamule

anterior lamina

Miathyria simplex

genital lobe

anterior lamina

hamule

1533b ventral view

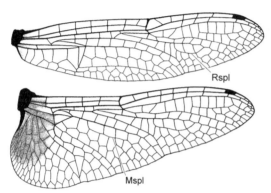

1532 male wings – *Miathyria simplex*

Micrathyria Kirby, 1889: 264, 303.
[♂ pp. 190, 195-197, couplets 79, 82-83; ♀ pp. 215-216, 219, couplets 53, 59, 68]

Type species: *Libellula didyma* Selys *in* Sagra, 1857 [Kirby, 1889 by original designation]

46 species:

aequalis (Hagen, 1861) [*Dythemis*]* – **L** [Needham and Westfall, 1955]
    syn *septima* Selys, 1900
almeidai Santos, 1945*
artemis Ris, 1911* – **L** [Santos, 1972]
athenais Calvert, 1909
atra (Martin, 1897) [*Dythemis*]* – **L** [Santos, 1978]
borgmeieri Santos, 1947 – **L** [Assis and Costa, 1994]
caerulistyla Donnelly, 1992
cambridgei Kirby, 1897*
catenata Calvert, 1909*
coropinae Geijskes, 1963*
debilis (Hagen, 1861) [*Dythemis*]*
dictynna Ris, 1919*
dido Ris, 1911*
didyma (Selys *in* Sagra, 1857) [*Libellula*]* – **L** [Needham, 1943a; Needham and Westfall, 1955]
    syn *dicrota* (Hagen, 1861) [*Dythemis*]
    syn *poeyi* (Scudder, 1866) [*Mesothemis*]
    syn *pruinosa* Kirby, 1894
dissocians Calvert, 1906* – **L** [Klots, 1932; Needham, 1943a]

divergens Westfall, 1992*
dunklei Westfall, 1992
duplicata Navás, 1922
dythemoides Calvert, 1909*
eximia Kirby, 1897*
hageni Kirby, 1890* – **L** [Needham, 1943a; Needham and Westfall, 1955]
hesperis Ris, 1911* – **L** [Assis and Costa, 1994]
hippolyte Ris, 1911*
hypodidyma Calvert, 1906* – **L** [Santos, 1968b]
    syn *protoe* Förster, 1907
iheringi Santos, 1946*
kleerekoperi Clavert, 1946
laevigata Calvert, 1909*
longifasciata Calvert, 1909* – **L** [Souza and Costa, 2002]
mengeri mengeri Ris, 1919* – **L** [Assis and Costa, 1994]
mengeri watsoni Dunkle, 1995*
occipita Westfall, 1992*
ocellata dentiens Calvert, 1909* – **L** [Assis and Costa, 1994]
    syn *carlota* Needham, 1942
ocellata ocellata Martin, 1897*
ocellata quicha Calvert, 1909*
paruensis Geijskes, 1963
pirassunungae Santos, 1953* – **L** [Assis and Costa, 1994]
pseudeximia Westfall, 1992*

pseudhypodidyma Costa, Lourenço and Viera, 2002
ringueleti Rodrigues, 1988*
romani Sjöstedt, 1918*
schumanni Calvert, 1906*
spinifera Calvert, 1909*
spuria (Selys, 1900) [*Anatya*]* – **L** [Souza and Costa, 2002]
    syn *macrocercis* Calvert, 1909
stawiarskii Santos, 1953 – **L** [Assis and Costa, 1994]*
surinamensis Geijskes, 1963*
sympriona Tennessen, 2000*
tibialis Kirby, 1897* – **L** [Souza and Costa, 2002]
ungulata Förster, 1907*
    syn *gerula (nomen nudum)* Hagen, 1861
    [*Dythemis*]
venezuelae De Marmels, 1989*

Map 103. Distribution of *Micrathyria* spp.

References: Ris 1911a; 1919; Westfall, 1992; Costa, Lourenço, and Viera, 2002 (Brazil).
Distribution: SE United States to central Argentina and Uruguay.

Small to medium (24–41 mm) libellulines; face pale, eyes turquoise in life (shared with *Nephepeltia*) and metallic postfrons and vertex in mature males; body black with pale yellow, green or light blue markings on thorax and abdomen. Wings (Figs. 1534-1536) hyaline (with a tinge of brown at base in *M. artemis* and *M. stawiarskii*) or tinged with pale brown in some females; last antenodal incomplete; Mspl indistinct; CuP in HW distinctly separated from anal angle of triangle. Posterior lobe of prothorax constricted at base; abdomen slender or with S7-9 widened and flattened (with widest point less than twice as wide as base of S7); sternum of S1 smooth (Figs. 1539a-b) or with lateral projections bearing denticles (*e.g. M. hypodidyma*, Figs. 1540a-b); anterior lamina (Figs. 1539a-1540a) usually lower than 4/5 of posterior hamule (higher than posterior hamule in *M. catenata*), entire or strongly bifid; posterior hamule (Figs. 1539a-b-1540a-b) simple or bifid, with inner branch smaller

or larger than outer branch; vulvar lamina not longer than 1/3 of S9 and not projected ventrally; venter of S9 usually not projected posteriorly, but in some species (*e.g. M. pseudeximia*) reaching posterior margin of S10, triangular and smoothly convex (shared with *Elga, Fylgia, Nephepeltia* and some *Zenithoptera*). Similar to *Nephepeltia*, from which it differs by the last antenodal incomplete (complete in *Nephepeltia*). **Unique characters**: None known.
Status of classification: This large genus is badly in need of revision. Several species are poorly known and inadequately described although many species groups are definable by combination of venational and genitalic characters (Ris, 1911a; 1919).
Potential for new species: Likely. The last species was described in 2002, and 4-5 undescribed species from South America as well as 2 from Mexico and Central America are known to the authors.
Habitat: Lakes, canals, sloughs and ponds, including temporary and saline ones. Males perch on tips of reeds and grass on shore vegetation with wings set. Females commonly found perching away from water on tips of twigs of bushes and trees. Activity patterns have been studied by May (1977; 1980), and oviposition by Paulson (1969).

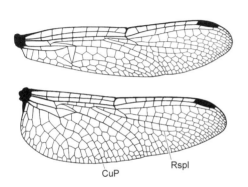

1534 male wings – *Micrathyria aequalis*

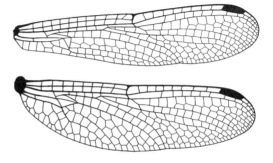

1535 male wings – *Micrathyria dido*

1536 male wings – *Micrathyria hypodidyma*

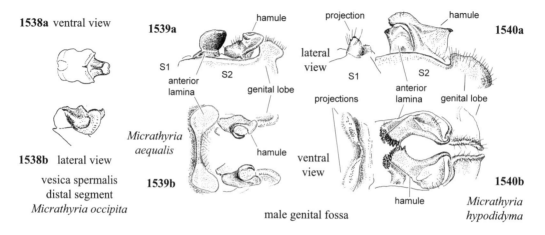

last antenodal incomplete

1537a dorsal view

cercus

1537b lateral view

cercus

S10   epiproct

male S10 – *Micrathyria divergens*

1538a ventral view

1538b lateral view

vesica spermalis distal segment *Micrathyria occipita*

1539a

S1   S2

anterior lamina   genital lobe

hamule

Micrathyria aequalis

1539b

hamule

male genital fossa

projection

lateral view

S1

projections

ventral view

hamule

hamule

anterior lamina   genital lobe

S2

1540a

1540b

Micrathyria hypodidyma

Misagria Kirby, 1889: 259, 296.

[♂ pp. 174, 180, 182, couplets 6, 26, 32; ♀ pp. 202, couplet 4]

Type species: *Misagria parana* Kirby, 1889 [by original designation]

4 species:

bimacula Kimmins, 1943*
calverti Geijskes, 1951*
divergens De Marmels, 1981
parana Kirby, 1889*

References: Geijskes, 1951; De Marmels, 1981b.
Distribution: Amazonian, from Venezuela south to Peru and Brazil.

Medium (34–45 mm) libellulines; postfrons and vertex brown to dark metallic blue in mature males, body dark with pale yellow markings. Wings hyaline with a tinge of gold at base, and females often with brown tips; arculus in HW from slightly proximal to distal to antenodal 3 (Fig. 1541, shared with some species

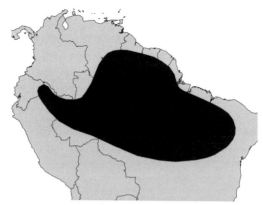

Map 104. Distribution of *Misagria* spp.

of *Uracis*); last antenodal complete; Mspl indistinct; anal loop short, without sole; cell rows at midlength of HW anal field numbering 1 or 2. Posterior lobe of prothorax widest at base, its posterior margin bent caudad; anterior lamina (Fig. 1543b) entire, to slightly or strongly bifid; posterior hamule prominent and bifid with inner branch smaller than outer branch (Figs. 1543a-b); vulvar lamina shorter than 1/3 of S9 and not projected ventrally; female S8 with lateral flanges (shared with *Cannaphila*, *Orthemis* and some *Libellula* species). **Unique characters**: Distal segment of vesica spermalis with a pair of ventral

flagella at least as long as twice the length of distal segment (Fig. 1542).
Status of classification: Fair.
Potential for new species: Likely.
Habitat: Adults perch on tips of branches and snags with wings set in partially shaded woods. Larvae unknown.

1541 male wings – *Misagria parana*

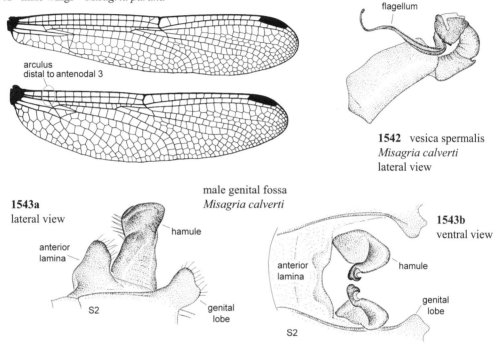

1542 vesica spermalis
Misagria calverti
lateral view

male genital fossa
Misagria calverti

1543a
lateral view

1543b
ventral view

Nannothemis Brauer, 1868: 369, 726.
[♂ pp. 197, couplet 84; ♀ pp. 204, couplet 10]
Type species: *Nannophya bella* Uhler, 1857 [Kirby, 1889 by subsequent designation]
   syn *Aino* Kirby, 1890: 113
      Type species: *Aino puella* Kirby, 1890 [Kirby, 1890 by original designation]
1 species:

bella (Uhler, 1857) [*Nannophya*]* – **L** [Weith and Needham, 1901]
   syn *puella* (Kirby, 1890) [*Aino*]
   syn *pygmaea* (Kirby, (*nec* Rambur, 1842)1889) [*Nannophya*]

References: Needham, Westfall, and May, 2000.
Distribution: SE Canada and E United States.

Very small libellulines (18–20 mm); head white dorsally; body black becoming pruinescent blue in males; black with yellow stripes in females. Wings (Fig. 1544) hyaline in males, with a broad

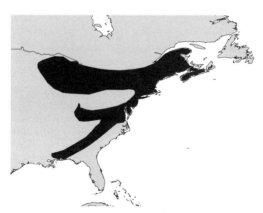

Map 105. Distribution of *Nannothemis* sp.

golden spot at base in females, sometimes with dark brown streak covering cubito-anal space; anal loop indistinct; last antenodal in FW incomplete; FW triangle bent at midlength; Mspl indistinct. Posterior lobe of prothorax widest at base, its posterior margin bent caudad; anterior lamina (Fig. 1546b) entire; posterior hamule bifid, with inner branch smaller than outer branch (Figs. 1546a-b); S7-9 of male widened and flattened, widest point less than twice as wide as base of S7 (shared with *Brechmorhoga*, *Nephepeltia*, and some species of *Macrothemis* and *Micrathyria*); distal segment of vesica spermalis (Fig. 1545) long and cylindrical (shared with *Erythrodiplax*, *Pseudoleon*, *Uracis* and *Ypirangathemis*) not widening distally (shared only with some specimens of *Erythrodiplax berenice*), with complex but largely hidden distal lobes;

vulvar lamina scoop shaped and projected ventrally (shared with *Crocothemis*, *Erythemis*, *Pseudoleon* and *Erythrodiplax*), and longer than S9. **Unique characters**: RP1,2 strongly concave (Fig. 1544). Status of classification: Unique species easily diagnosed.
Potential for new species: Extremely unlikely.
Habitat: Adults forage at marshy areas and stagnant pools where they perch on low vegetation; oviposition is in tandem. Larvae live on marshy areas covered by a few centimeters of water, clinging to grass or debris.

1544 male wings – *Nannothemis bella*

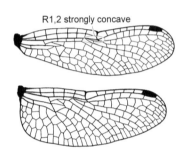

R1,2 strongly concave

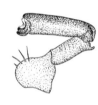

1545 vesica spermalis *Nannothemis bella* lateral view

1546a lateral view

anterior lamina   hamule

Nannothemis bella

genital lobe

anterior lamina

1546b ventral view

hamule

male genital fossa

Nephepeltia Kirby, 1889: 259, 310.
[♂ pp. 182, couplet 32; ♀ pp. 218, couplet 64]
Type species: *Libellula phryne* Perty, 1834 [Kirby, 1889 by original designation]
   syn *Neothemis* Karsch, 1889: 256
      Type species: *Neothemis flavifrons* Karsch, 1889 [by monotypy]
6 species:

aequisetis Calvert, 1909*
berlai Santos, 1950*
chalconota Ris, 1919*
flavifrons (Karsch, 1889) [*Neothemis*]*
   syn *inermis (nomen nudum)* (Hagen, 1867) [*Nannothemis*]
   syn *prodita (nomen nudum)* (Hagen, 1861) [*Nannothemis*]
leonardina Rácenis, 1953*
phryne phryne (Perty, 1834) [*Libellula*]* – **L** [De Marmels, 1990a]
   syn *apicalis (nomen nudum)* (Hagen, 1861) [*Dythemis*]
   syn *semiaurea (nomen nudum)* (Hagen, 1861) [*Nannophya*]
phryne tupiensis Santos, 1950*

References: Santos, 1950; Rácenis, 1953.
Distribution: S Mexico to NE Argentina.

Small libellulines (19–28 mm); white face and me-

Map 106. Distribution of *Nephepeltia* spp.

tallic postfrons and vertex; eyes turquoise in mature males; thorax dark dorsally, laterally black to dark metallic blue with pale stripes; abdomen black with paired yellow spots on S1-7. Wings (Fig. 1547) hyaline or amber at wing base, costal side of FW triangle broken near middle or at distal end; last antenodal in FW complete; 2 or more bridge crossveins; Mspl indistinct. Posterior lobe of prothorax constricted at base; male S7-9 widened and flattened, widest point less than twice as wide as base of S7 (shared with *Brechmorhoga*, *Nannothemis*, and some species of *Macrothemis* and *Micrathyria*); anterior lamina (Fig. 1549b) entire; posterior hamule bifid, with inner branch as large as or larger than outer branch (Figs. 1549a-b). Female sternum of S9 projected distally to about posterior margin of S10, triangular and smoothly convex (shared with *Elga*, *Fylgia* and some *Micrathyria* and *Zenithoptera*). **Unique characters**: Male venter of thorax with tubercle or spine (Fig. 1548), except in *N. leonardina* and *N. aequisetis*.

Status of classification: *Nephepeltia* is considered a relative of *Micrathyria*. Similar to *Edonis* from which it can be separated by the HW triangle base opposite to arculus (slightly proximal to arculus in *Edonis*) and the HW base not broadly widened at base with a maximum of 3 rows of cells in anal field (broadly

widened in *Edonis* with more than 4 cell rows between margin and anal loop), and to *Micrathyria*, from which it differs by the last antenodal complete (incomplete in *Micrathyria*). Latest key to species is Ris (1911a); a comparative table and picture key are given by Santos (1950). Species identification fairly reliable.

Potential for new species: Likely. At least one new species from Peru and Brazil known to the authors.

Habitat: Adults perch with wings set on tips of stems along ponds, canals, marshes in sunshine; where males often assume the obelisk position.

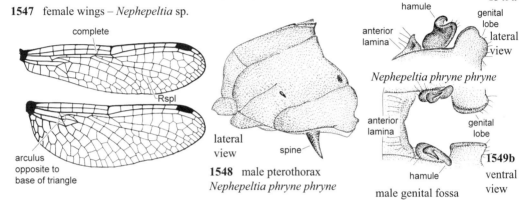

1547 female wings – *Nephepeltia* sp.

complete

Rspl

arculus opposite to base of triangle

lateral view — spine

1548 male pterothorax
Nephepeltia phryne phryne

hamule — **1549a**
genital lobe
anterior lamina — lateral view

Nephepeltia phryne phryne

anterior lamina — genital lobe

1549b
hamule — ventral view
male genital fossa

Nothodiplax Belle, 1984a: 6.

[♂ pp. 186, couplet 49; ♀ pp. 211, couplet 38]

Type species: *Nothodiplax dendrophila* Belle, 1984 [by original designation]
1 species:

dendrophila Belle, 1984*

References: Belle, 1984a.
Distribution: Surinam.

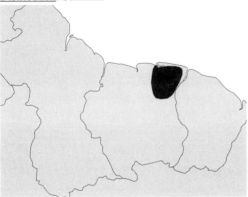

Map 107. Distribution of *Nothodiplax* sp.

Small libellulines (32 mm); face, frons and vertex brown, thorax marbled in greenish brown and black (Fig. 1555a); S1-7 yellow with distal portion black, S8-10 black (Fig. 1555b). Wings (Fig. 1550) hyaline;

last antenodal in FW incomplete; Mspl indistinct; 1 row of cells behind anal loop. Posterior lobe of prothorax constricted at base and upright (Fig. 1552); anterior lamina entire (Fig. 1556b); posterior hamule bifid with inner branch smaller than outer branch (Figs. 1556a-b); vulvar lamina shorter than 1/3 of S9 and not projected ventrally (Figs. 1559a-b). **Unique characters**: None known.

Status of classification: Poorly known species.

Potential for new species: Unknown. Thus far known only from the type series.

Habitat: Sand bottomed creeks within gallery forests of E coastal region of Surinam. Found on upper parts of trees, occasionally flying down to low bank vegetation of creeks (Belle, 1984a). Larva unknown.

1550   last antenodal incomplete   male wings

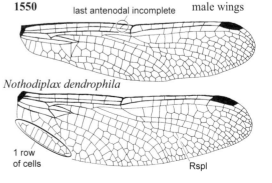

Nothodiplax dendrophila

1 row of cells   Rspl

head *Nothodiplax dendrophila*

1552 posterior lobe of prothorax
Nothodiplax dendrophila

dorsal view

1551a dorsal view **1551b** lateral view

lateral view

1553 male hind femur
Nothodiplax dendrophila

lateral view

1554 tarsal claw
Nothodiplax dendrophila

1555a male pterothorax and S1–4
Nothodiplax dendrophila
lateral view

lateral view

1555b male S5–10 – *Nothodiplax dendrophila*

hamule

genital fossa

male S10 – *Nothodiplax dendrophila*

anterior lamina

hamule
genital lobe

1557a lateral view
cercus

1557b detail

1556a
lateral view

Nothodiplax dendrophila

anterior lamina

hamule

genital lobe

epiproct

cercus

1557c medio-ventral view

1558a lateral view **1558b** ventral view

1556b ventral view

S10

1557d dorsal view

vesica spermalis

female S8–10
Nothodiplax dendrophila

vulvar lamina

S10

Nothodiplax dendrophila

cercus

S8 S9

S10

1559a lateral view

S8 S9

cercus

1559b ventral view

Oligoclada Karsch, 1890: 382.
[♂ pp. 175-176, 178, 186, 188-189, 198-201, couplets 9, 11, 17, 48-49, 54, 56, 59, 91, 93-94, 97-98, 101-103; ♀ pp. 205-206, 209, 211, 221-223, couplets 19-20, 32, 38, 76, 79-80, 84-85, 87]

Type species: *Oligoclada pachystigma* Karsch, 1890
[by monotypy]

  syn *Podothemis* Ris, 1909a: 27
    Type species: *Podothemis nemesis* Ris, 1909
    [by monotypy]

22 species:

abbreviata abbreviata (Rambur, 1842) [*Libellula*]*
– **L** [Fleck, 2003a]
    syn *raineyi* Ris, 1919
abbreviata limnophila Machado & Machado, 1993
– **L** [Machado and Machado, 1993]

amphinome Ris, 1919*
borrori Santos, 1945*
calverti Santos, 1951*
crocogaster Borror, 1931*
haywardi Fraser, 1947
heliophila Borror, 1931*
hypophane De Marmels, 1989*
laetitia Ris, 1911* – **L** [Souza, Costa, and Espindola, 2002]
leucotaenia De Marmels, 1989*
monosticha Borror, 1931
nemesis (Ris, 1909) [*Podothemis*]*
pachystigma Karsch, 1890* – **L** [Fleck, 2003a]
rhea Ris, 1911*
risi Geijskes, 1984*
stenoptera Borror, 1931*
sylvia (Kirby, 1889) [*Nannothemis*]
teretidentis Rehn, 2003*
umbricola Borror, 1931*
waikinimae De Marmels, 1992
walkeri Geijskes, 1931*
xanthopleura Borror, 1931*

References: Borror, 1931; Calvert, 1948b.
Distribution: Guatemala south to Bolivia, Paraguay and NE Argentina.

Map 108. Distribution of *Oligoclada* spp.

Small (20–35 mm), dark libellulines with hyaline wings (dark wing tips in *O. borrori*); pterothorax sometimes with metallic luster; abdomen usually becoming pruinose grayish-blue in males, and with venter of distal segments red in some species. Antefrons pale; postfrons and vertex brown to metallic blue in mature specimens; rear of occiput in some species bearing paired protuberances; posterior lobe of prothorax constricted at base, upright and wider than other prothoracic lobes. Wings (Figs. 1560-1562) with sectors of arculus usually stalked (rarely not stalked); last antenodal complete or incomplete; costal side of FW triangle bent at distal portion or straight; Mspl indistinct; anal loop with toe well developed or not; Aspl bent at heel level of anal loop at an obtuse angle or more or less straight. Anterior lamina (Fig. 1563b) entire or slightly bifid, shorter to longer than posterior hamule; posterior hamule (Figs. 1563a-b) not bifid to bifid with inner branch longer than outer branch. **Unique characters**: None known.

Status of classification: Borror (1931) with its excellent figures and keys remains the best introductory work for this genus, but eight new species have been described since 1931; Calvert (1948b) provided a key for two species and Geijskes (1984) more recently determined the status of the name *O. abbreviata* (Rambur). Identification of remaining species must be done by reference to species descriptions.

Potential for new species: Very likely.

Habitat: Adults found along canals, meandering streams or at margins of ponds, especially common at oxbow lakes and swamps, where they alight on surfaces of leaves or on ground with wings and abdomen in same plane as body. Known larvae collected in artificial lentic environments (Machado and Machado, 1993), flooded areas linked to a river (Souza, Costa, and Espindola, 2002), and in an artificial reservoir with strongly fluctuating water levels (Fleck, 2003a).

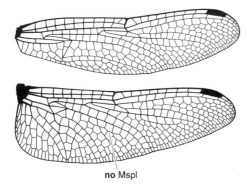

1560 male wings – *Oligoclada abbreviata*

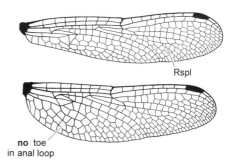

1561 male wings – *Oligoclada rhea*

1562 male wings – *Oligoclada xanthopleura*

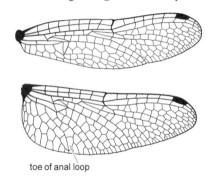

toe of anal loop

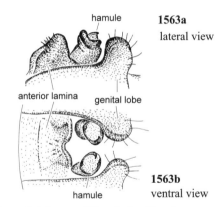

hamule

1563a
lateral view

anterior lamina   genital lobe

1563b
hamule   ventral view

male genital fossa – *Oligoclada abbreviata*

Orthemis Hagen, 1861: 160.
[♂ pp. 176, 179, 181, couplets 13, 22, 29, 31; ♀ pp. 203, couplets 6-8]
Type species: *Libellula ferruginea* Fabricius, 1775 [Kirby, 1889 by subsequent designation]
   syn *Neocysta* Kirby, 1889: 263, 300
      Type species: *Libellula attenuata* Erichson, 1848 [Kirby, 1889 by original designation]
17 species:

aequilibris Calvert, 1909* – **L** [Fleck, 2003b]
ambinigra Calvert, 1909*
ambirufa Calvert, 1909*
   syn *sibylla* Ris, 1919
anthracina De Marmels, 1989*
attenuata (Erichson, 1848) [*Libellula*]*
biolleyi Calvert, 1906* – **L** [Fleck, 2003b]
concolor Ris, 1919*
cultriformis Calvert, 1899*
discolor (Burmeister, 1839) [*Libellula*]*
ferruginea (Fabricius, 1775) [Libellula]* – **L?**
[Needham, 1904; Clavert, 1928; Klots, 1932; Geijskes, 1934]
   syn *macrostigma* (Rambur, 1842) [*Libellula*]
flavopicta Kirby, 1889*
levis Calvert, 1906* – **L** [De Marmels, 1990a]
nodiplaga Karsch, 1891* – **L** [Rodrigues Capítulo and Muzón, 1990]
plaumanni Buchholz, 1950
regalis Ris, 1910*
schmidti Buchholz, 1950*
sulphurata Hagen, 1868*

References: Calvert, 1906; Ris 1910; 1919; De Marmels, 1988.
Distribution: S United States to N Chile and central Argentina.

Large to very large (45–55 mm) orange, red, or dark libellulines, which often become pruinose blue or purple in mature males. Postfrons and vertex brown to metallic blue or violet; pterothorax pale with darker markings on side all of which can become obscured under dense coat of pruinosity. Wings (Figs. 1564-1565) hyaline sometimes with small brown patches at nodus and wing tip; sectors of arculus in FW stalked; last antenodal in FW complete; FW discoidal field parallel or narrowing distally; pterostigma long. Posterior lobe of prothorax widest at base, its posterior margin bent caudad; abdomen narrow to wide, triangular in cross section; anterior lamina entire (Fig. 1566b); posterior hamule bifid with inner branch shorter than outer branch (Figs. 1566a-b); vulvar lamina shorter than length of S9, bilobed and projected ventrally (shared with most species of *Libellula*); female S8 with lateral flanges (Fig. 1568) shared with *Cannaphila*, most species of *Libellula* and *Misagria*). **Unique characters**: None known.
Status of classification: The genus is badly in need of revision. Several names have been applied to the *Orthemis ferruginea* group (*O. aequilibris, O. discolor, O. ferruginea, O. macrostigma, O. schmidti, O. sulphurata*), and despite recent publications

Map 109. Distribution of *Orthemis* spp.

on this group (De Marmels, 1988; Donnelly, 1995; Paulson, 1998b), the status of several names is still unclear, and the identity of the Antillean forms is still unresolved (*O. ferruginea, O. schmidti* or *O. macrostigma*). Several authors described the larva of "*O. ferruginea,*" but based on locality data it could also be *O. discolor* (Needham, 1904: Texas) or the Antillean species if it is different from *O. ferruginea* (Calvert, 1928: Antigua; Klots, 1932: Puerto Rico; Geijskes, 1934: Aruba and Bonaire).

Based on mitochondrial CO1 and 16S rRNA sequence data, Artiss, Schultz, Polhemus, and Simon (2001) suggested that *Orthemis* and *Plathemis* would constitute the sister group of *Libellula sensu stricto*. However, according to studies by Carle and Kjer (2002) based on wing venation and vesica spermalis morphology, *Orthemis* would be the sister group of *Libellula sensu lato* (including *Plathemis*).

Potential for new species: Likely; at least one new species from Peru known to the authors.

Habitat: Adults common at almost any pond, ditch, puddle, lake or slow moving stream in the neotropical region, where males defend territories from a perch along or on tips of reeds, branches and snags. Female oviposits guarded by the hovering male. Reproductive behavior of "*O. ferruginea*" was documented by Novelo-Gutiérrez (1981), Novelo-Gutiérrez and Gonzalez-Soriano (1984), and Harvey and Hubbard (1987); feeding aggregation behavior by Young (1980).

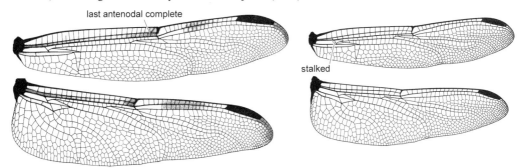

1564 male wings – *Orthemis regalis*

1565 male wings – *Orthemis attenuata*

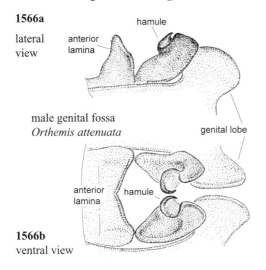

1566a

lateral view

hamule

anterior lamina

male genital fossa
Orthemis attenuata

genital lobe

anterior lamina   hamule

1566b
ventral view

1567 male S10
Orthemis ambirufa

lateral view

cercus

S10

epiproct

1568 female S8–10
Orthemis attenuata

cercus

S7   S8   S9

S10

lateral flange

lateral view

Pachydiplax Brauer, 1868: 368, 722.
[♂ pp. 173, couplet 3; ♀ pp. 205, 217, couplets 15, 63]
Type species: *Libellula longipennis* Burmeister, 1839 [by monotypy]
1 species:

longipennis (Burmeister, 1839) [*Libellula*]* – **L**
[Needham, 1901; Needham, Westfall, and May, 2000]
  syn *socia* (Rambur, 1842) [*Libellula*]

References: Needham, Westfall, and May, 2000.
Distribution: S Canada south through Mexico and Belize, the Bahamas, and W Cuba.

Medium libellulines (28–45 mm); face pale, postfrons and vertex brown to metallic blue in mature specimens; pterothorax pale with dark stripes which becomes wholly pruinose in males; abdomen triangular, dark with longitudinal pale stripes but becom-

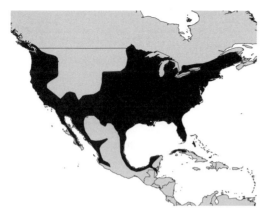

Map 110. Distribution of *Pachydiplax* sp.

ing pruinose blue in mature males. Wings (Fig. 1569) hyaline, sometimes with golden to dark basal spot; last antenodal complete; Mspl distinct; CuP in HW distinctly separated from anal angle of triangle. Pos-

terior lobe of prothorax constricted at base, upright and wider than other prothoracic lobes; anterior lamina entire (Fig. 1570b); posterior hamule bifid with inner branch smaller than outer branch (Figs. 1570a-b); vulvar lamina shorter than 1/3 of sternum S9, not projected ventrally. **Unique characters**: No crossveins under proximal portion of pterostigma, resulting in free triple space under second postnodal interspace (Fig. 1569).

Status of classification: Easily recognized species.
Potential for new species: Very unlikely.
Habitat: Adults at marshy or reedy ponds where they hover over water and perch on reed tips with wings fully set. One of the most common and best known nearctic libellulines which has been intensively studied as to foraging behavior (Baird and May, 1993; Dunham, 1994), reproduction (Robey, 1975; Sherman, 1983; McKinnon and May, 1994), territoriality (Johnson, 1962), and seasonal variation (Penn, 1951; Eller, 1963; Macklin, 1963).

1569	male wings – *Pachydiplax longipennis*

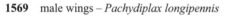

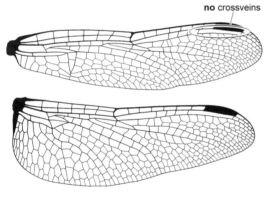

1570a	lateral view

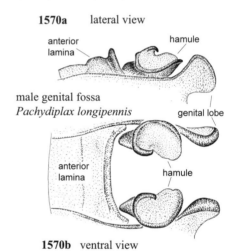

male genital fossa
Pachydiplax longipennis

1570b	ventral view

Paltothemis Karsch, 1890: 362.
[♂ pp. 189, 194, couplets 58, 76; ♀ pp. 207, 214, couplets 29, 49]
Type species: *Paltothemis lineatipes* Karsch, 1890 [by monotypy]
3 species:

cyanosoma Garrison, 1982*
lineatipes Karsch, 1890* – **L** [Needham, 1904]
	syn *russata* (Calvert, 1895) [*Dythemis*]
nicolae Hellebuyck, 2002*

References: Ris, 1913a; Garrison, 1982; Hellebuyck, 2002; González-Soriano, 2005.
Distribution: SW United States south to Panama [NOTE: Although *P. lineatipes* has been recorded from Brazil and Venezuela, De Marmels (1990b) provides compelling evidence for deleting records of this species from South America.]

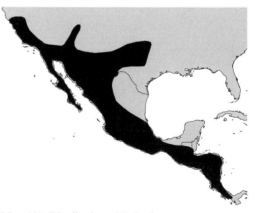

Map 111. Distribution of *Paltothemis* spp.

Large (47–54 mm) red or blue libellulines with pale face becoming red (males of *P. lineatipes*) to dark brown (*P. cyanosoma*); postfrons and vertex brown to red (males of *P. lineatipes*) or metallic blue (in mature males of *P. cyanosoma*); pterothorax pale with irregular dark markings becoming red in males of *P. lineatipes* or dark brown with blue pruinosity in males of *P. cyanosoma*; abdomen brown (*P. cyanosoma*) to gray or red (males of *P. lineatipes*) with irregular dark markings. Wings hyaline (*P. cyanosoma*) or infused with increasing intensity of orange toward wing base (in males of *P. lineatipes*); last antenodal in FW incomplete; Mspl and Rspl distinct; median and radial planates encompassing 2 rows of cells (Fig. 1571). Posterior lobe of prothorax widest at base, its posterior margin bent caudad; anterior lamina entire (Fig. 1573b); posterior hamule not bifid (Figs. 1573a-b); vulvar lamina shorter than 1/3 of sternum S9, not projected ventrally. **Unique characters**: Spines of male hind femur dimorphic; 2-3 long distal spines, the remaining short, stout, and quadrangular (Figs. 1572a-b).

Status of classification: Good. The three described species are easily recognized.

Potential for new species: Likely.

Habitat: Small rocky mountain streams, where they perch on rocks with wings set below horizontal plane of body. Males patrol portions of stream in the morning; copulation occurs in flight and females oviposit unattended (Alcock, 1990). Larvae live between rocks in areas of swift current (Dunkle, 1978).

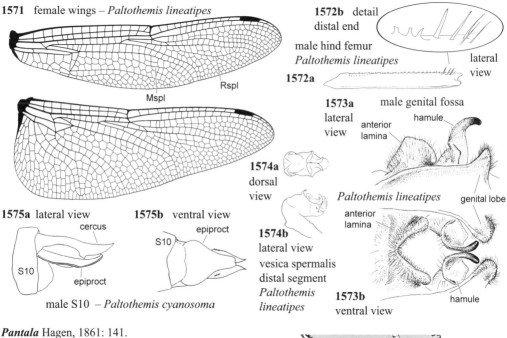

1571 female wings – *Paltothemis lineatipes*

Rspl

Mspl

1572b detail distal end

male hind femur
Paltothemis lineatipes

1572a

lateral view

1573a lateral view

male genital fossa

anterior lamina

hamule

1574a dorsal view

Paltothemis lineatipes

genital lobe

1574b lateral view vesica spermalis distal segment *Paltothemis lineatipes*

anterior lamina

1573b ventral view

hamule

1575a lateral view

cercus

S10

epiproct

1575b ventral view

epiproct

S10

male S10 – *Paltothemis cyanosoma*

Pantala Hagen, 1861: 141.

[♂ pp. 183, couplet 39; ♀ pp. 204, couplet 12]

Type species: *Libellula flavescens* Fabricius, 1798 [Kirby, 1889 by subsequent designation]

2 species; both species in the New World:

flavescens (Fabricius, 1798) [*Libellula*]* – **L** [Cabot, 1890; Geijskes, 1934; Byers, 1941]

   syn *analis* (Burmeister, 1839) [*Libellula*]

   syn *terminalis* (Burmeister, 1839) [*Libellula*]

   syn *viridula* (Beauvois, 1805) [*Libellula*]

hymenaea (Say, 1840) [*Libellula*]* – **L** [Kennedy, 1923b]

   syn *huanacina* (Förster, 1909) [*Tramea*]

References: Ris, 1913a; Needham, Westfall, and May, 2000.

Distribution: Breeds worldwide except for Europe; in

Map 112. Distribution of *Pantala* spp.

the New World from S Canada to Chile and central Argentina.

Large (45–50 mm), yellow to brown libellulines; frons yellow to reddish orange in mature males; pterothorax yellow or gray; abdomen yellow to pale brown with black dorsal markings. HW broadly widened at base, hyaline with wash of yellow or a round brown spot at HW base; last antenodal incomplete; FW subtriangle not defined; RP2 strongly undulated (Fig. 1576). Posterior lobe of prothorax widest at base, its posterior margin bent caudad; abdomen robust, tapering caudally; anterior lamina strongly bifid and higher than posterior hamule (Figs. 1578a-b); posterior hamule not bifid and directed caudally (Figs. 1578a-b); vulvar lamina shorter than 1/3 of S9 and not projecting ventrally. **Unique characters**: MA strongly elbowed at distal end of the median planate (Fig. 1576); S3-4 with a supplementary transverse carina, and S5 with a strong transverse carina (Fig.

1579); distal segment of vesica spermalis divided into a basal and a distal portion separated by a membranous area, the distal portion consisting of a pair of sclerotized, diamond-shaped lateral plates (Figs. 1577a-b).
Status of classification: Good.
Potential for new species: Unlikely.
Habitat: Permanent and temporary ponds and puddles with or without marginal vegetation, also brackish water and occasionally slow moving water. Adults engage in sustained gliding flight and are known migrants being found on oceanic islands, at sea, and on land from sea level to mountainous areas. They perch vertically on twigs and low vegetation. Oviposition is in tandem in the temporary pools, where larvae undergo a very fast development.

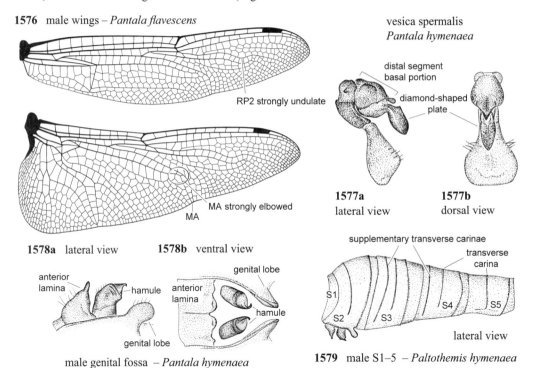

1576 male wings – *Pantala flavescens*

RP2 strongly undulate

MA strongly elbowed
MA

vesica spermalis
Pantala hymenaea

distal segment
basal portion

diamond-shaped
plate

1577a   **1577b**
lateral view   dorsal view

1578a lateral view   **1578b** ventral view

anterior
lamina   hamule
                 genital lobe
                           anterior
                           lamina
                                     genital lobe
                                              hamule

male genital fossa – *Pantala hymenaea*

supplementary transverse carinae
                              transverse
                              carina
S1
       S2   S3   S4   S5

lateral view

1579 male S1–5 – *Paltothemis hymenaea*

Perithemis Hagen, 1861: 185.
[♂ pp. 201, couplet 105; ♀ pp. 223, couplet 89]
Type species: *Libellula tenera* Say, 1840 [Kirby, 1889 by subsequent designation]
13 species:

bella Kirby, 1889*
  syn *austeni* Kirby, 1897
cornelia Ris, 1910*
domitia (Drury, 1773) [*Libellula*]* – **L** [Novelo-Gutiérrez, 2002b]
  syn *metella* (Selys *in* Sagra, 1857) [*Libellula*]
  syn *pocahontas* Kirby, 1889

  syn var. *iris* Hagen, 1861
electra Ris, 1930* – **L** [Santos, 1970e]
icteroptera (Selys *in* Sagra, 1857) [*Libellula*]* – **L** [von Ellenrieder and Muzón, 1999]
  syn *waltheri* Ris, 1910
intensa Kirby, 1889* – **L** [Novelo-Gutiérrez, 2002b]
  syn var. *californica* Martin, 1900
lais (Perty, 1834) [*Libellula*]* – **L** [Costa and Régis, 2005]
  syn *naias* Ris, 1910
mooma Kirby, 1889* – **L** [Santos, 1973d; von Ellenrieder and Muzón, 1999]

syn *cloe (nomen nudum)* Hagen, 1861
syn var. *octoxantha* Ris, 1910
parzefalli Hoffmann, 1991*
piperi Hoffmann, 1987
rubita Dunkle, 1982* – **L** [Dunkle, 1982]
tenera (Say, 1840) [*Libellula*]* – **L** [Needham, 1901]
　syn *chlora* (Rambur, 1842) [*Libellula*]
　syn *domitia seminole* Calvert, 1907
　syn *tenuicincta* (Say, 1840) [*Libellula*]
thais Kirby, 1889* – **L** [Spindola, Souza, and Costa, 2001]

References: Ris, 1930; von Ellenrieder and Muzón, 1999; Needham, Westfall, and May, 2000.
Distribution: SE Canada south through Uruguay and central Argentina.

Small (17–29 mm), yellow to orange libellulines; postfrons small and vertex pale to brown; pterothorax greenish or with a pair of dark lateral stripes; abdomen brown or with irregular series of chevrons and dark carinae. Wings in male yellow to orange to red, in female hyaline to yellow or with various bands of yellow and/or brown. HW broad in some species; FW triangle free or crossed, with costal side straight or broken near distal end; 2 or more bridge crossveins; Mspl and Rspl distinct and with 1 row of cells throughout (Figs. 1580-1581). Posterior lobe of prothorax constricted at base (Fig. 1582), upright and wider than other prothoracic lobes; anterior lamina entire (Fig. 1585b); posterior hamule not bifid, broadly flattened and armed with a small usually outwardly directed tooth (Figs. 1585a-b); vulvar lamina bilobed and about 1/5 to 1/2 length of sternum S9 (Figs. 1583-1584). **Unique characters**: Costal side of FW triangle almost as long as proximal side (Fig. 1580).

Map 113. Distribution of *Perithemis* spp.

Status of classification: Despite the revision by Ris (1930), the genus is in need of revision, as the number of species involved is uncertain. *Perithemis piperi*, based on one specimen from Mexico, is apparently a synonym of *P. mooma*, although this synonymy has not yet been published, and it is possible that *P. mooma* constitutes a synonym of *P. tenera*.
Potential for new species: Likely. At least one new species from Costa Rica known to the authors.
Habitat: Ponds, marshes and occasionally streams where adults fly just over water in sunshine or in partially shaded areas. They perch on small branches or reeds usually close to water's surface often with fore and hind wings slowly moving alternatively. Larvae crawl among submerged vegetation. Behavior of some species has been extensively studied (*i.e.* Montgomery, 1937; Jacobs, 1955; Wildermuth, 1991; 1992; Switzer, 1997a; 1997b; Switzer and Walter, 1999; Switzer and Eason, 2000; Switzer and Schultz 2000; Schultz and Switzer, 2001; Switzer, 2002a; 2002b). Species of *Perithemis* are highly sympatric; seven species were collected in a small area around Pakitza, Peru (Louton, Garrison, and Flint, 1996).

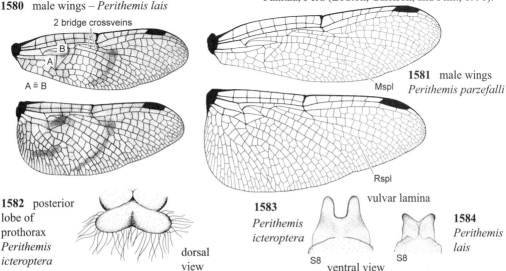

1580 male wings – *Perithemis lais*

2 bridge crossveins

A ≅ B

1581 male wings
Mspl   *Perithemis parzefalli*

Rspl

1582 posterior lobe of prothorax *Perithemis icteroptera*

dorsal view

1583 *Perithemis icteroptera*

vulvar lamina

1584 *Perithemis lais*

S8　　ventral view　　S8

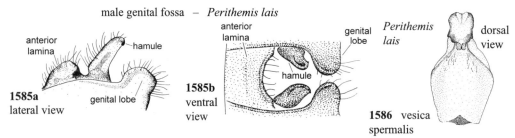

male genital fossa – *Perithemis lais*

1585a
lateral view

anterior
lamina

hamule

genital lobe

1585b
ventral
view

anterior
lamina

genital
lobe

hamule

*Perithemis
lais*

dorsal
view

1586 vesica
spermalis

Planiplax Muttkowski, 1910: 169.

[♂ pp. 198-199, 201, couplets 91, 97, 105; ♀ pp. 221, 223, couplets 76, 84, 89]

Type species: *Platyplax erythropyga* Karsch, 1891
[Muttkowski, 1910 by original designation]
    syn *Platyplax* Karsch, 1891: 268
        Type species: *Platyplax erythropyga* Karsch,
        1889 [by original designation]
[NOTE: *Planiplax* is a replacement name for *Platyplax*
Karsch, 1889, a junior primary homonym of *Platyplax*
Fieber, 1860 in Heteroptera.]
5 species:

arachne Ris, 1912*
erythropyga (Karsch, 1891) [*Platyplax*]*
machadoi Santos, 1949*
phoenicura Ris, 1912* – **L** [Souza, Costa, and
Santos, 1999a]
sanguiniventris (Calvert, 1907) [*Platyplax*]*

References: Ris, 1912; Santos, 1949.
Distribution: Mexico south to Uruguay and central-
eastern Argentina.

Medium (35–40 mm) libellulines; postfrons pale to
metallic blue in mature males; vertex pale to brown;
pterothorax brown to dark with varying patterns
of pruinosity on mesepisterna and sides; abdomen
brown to red or with a combination of dark blue
to brown on basal S1-2 (*P. sanguiniventris*) to S5
(*P. erythropyga*) in males. Wings hyaline or with
a basal dark reddish brown spot (in *P. machadoi*
and *P. sanguiniventris*); sectors of arculus in FW
not stalked; 1 or (in some *P. erythropyga*) 2 bridge
crossveins; Aspl almost straight; Mspl and Rspl

Map 114. Distribution of *Planiplax* spp.

distinct; median sector with 1 row of cells, radial
sector with 1 or 2 (*P. erythropyga*) rows of cells
(Fig. 1588). Posterior lobe of prothorax constricted
at base, upright and wider than other prothoracic
lobes; anterior lamina entire (Fig. 1587b); posterior
hamule not bifid (Figs. 1587a-b); vulvar lamina
small and bilobed, less than 1/5 of sternum S9.
Unique characters: None known.
Status of classification: Poorly known genus.
Specimens not numerous in collections.
Potential for new species: Likely.
Habitat: Flying over ponds with marginal vegetation,
and along banks of large rivers, rarely perching; hard
to approach. Only known larvae found on a river
outlet covered with aquatic vegetation (*Salvinia*
sp. and *Eichornia* sp.), connected to river during
flooding (Souza, Costa, and Santos, 1999a).

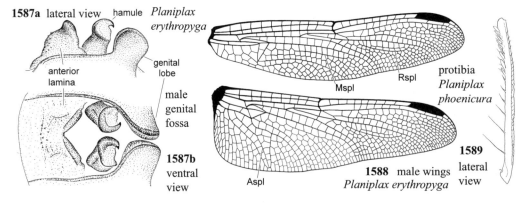

1587a lateral view   hamule   *Planiplax
erythropyga*

anterior
lamina

genital
lobe

male
genital
fossa

1587b
ventral
view

Aspl

Mspl

Rspl

1588 male wings
Planiplax erythropyga

protibia
*Planiplax
phoenicura*

1589
lateral
view

Pseudoleon Kirby, 1889: 261, 274.

[♂ pp. 183, couplet 40; ♀ pp. 204, couplet 13]

Type species: *Celithemis superba* Hagen, 1861 [by original designation]

1 species:

superbus (Hagen, 1861) [*Celithemis*]* – **L** [Needham, 1937]

References: Needham, Westfall, and May, 2000.

Distribution: SW United States south to Costa Rica.

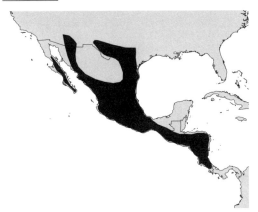

Map 115. Distribution of *Pseudoleon* sp.

Medium (35–46 mm) libellulines; brown and black filigree pattern on wings and body; eyes with alternate bands of dark and pale brown. Last antenodal in FW incomplete; Mspl and Rspl distinct; median and radial sectors encompassing 2 or more row of cells (Fig. 1590). Posterior lobe of prothorax widest at base, its posterior margin bent caudad; abdomen triangular in cross-section; anterior lamina entire (Fig. 1592b); posterior hamule bifid with inner branch smaller than outer branch (Figs. 1592a-b; distal segment of vesica spermalis long and cylindrical (Fig. 1591, shared with *Erythrodiplax*, *Nannothemis*, *Uracis* and *Ypirangathemis*). Vulvar lamina longer than 1/3 of S9, scoop shaped, and projecting ventrally (shared with *Crocothemis*, *Erythemis*, *Nannothemis* and some *Erythrodiplax*). **Unique characters**: Filigree pattern on wings, always with spots along antenodal subcostal space (Fig. 1590); acutely angulated tip of long and cylindrical distal segment of vesica spermalis (Fig. 1591).

Status of classification: Well diagnosed species.

Potential for new species: Unlikely.

Habitat: Rocky, clear water streams. Adults perch on banks or stones often in bright sunshine, where they adopt the obelisk position on hot days (Dunkle, 2000). Córdoba-Aguilar (2003) provided some notes on their reproductive behavior.

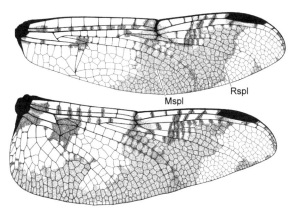

1590   male wings – *Pseudoleon superbus*

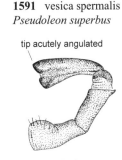

1591   vesica spermalis
Pseudoleon superbus

tip acutely angulated

lateral view

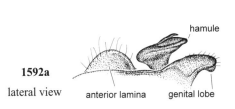

1592a

lateral view

male genital fossa
Pseudoleon superbus

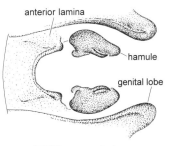

1592b   ventral view

Rhodopygia Kirby, 1889: 265, 299.

[♂ pp. 192, couplet 69; ♀ pp. 213, couplet 45]

Type species: *Libellula cardinalis* Erichson, 1848 [Kirby, 1889 by original designation]

5 species:

cardinalis (Erichson, 1848) [*Libellula*]*
　syn var. *colorata* Navás, 1923
geijskesi Belle, 1964* – **L** [Limongi, 1983]
hinei Calvert, 1907*
hollandi Calvert, 1907* – **L** [De Marmels, 1990a]
　syn *chloris* Ris, 1911
pruinosa Buchholz, 1953*

References: Ris, 1911b; Belle, 1998.
Distribution: Guatemala and Belize south to Bolivia and Brazil.

Large (44–50 mm), red libellulines; postfrons and vertex pale to red in mature males; pterothorax and abdomen pale brown to red, sometimes becoming pruinose in mature males (in *R. pruinosa*). Wings (Fig. 1593) hyaline or with a reddish-brown spot at base of HW (*R. cardinalis*); arculus in HW opposite or beyond antenodal 2; FW last antenodal incomplete; Mspl distinct with 1 cell of rows throughout. Posterior lobe of prothorax constricted at base; anterior lamina entire (Fig. 1594b); posterior hamule bifid with inner branch smaller than outer branch (Fig. 1594a-b); vulvar lamina shorter than 1/3 of S9 and not projected

Map 116. Distribution of *Rhodopygia* spp.

ventrally. **Unique characters**: None known.
Status of classification: Belle (1998) provides key for both sexes of all species.
Potential for new species: Likely. Some species bear superficial resemblance to common *Erythemis haematogastra* (Burmeister), a species likely to be ignored by collectors.
Habitat: Adults fly along forest trails and over fields and meadows. According to Belle (1998), adults occur at partially shaded pools and ditches. Known larvae found in large road-side ditches devoid of vegetation.

1593　female wings – *Rhodopygia hinei*

male genital fossa　– *Rhodopygia hinei*

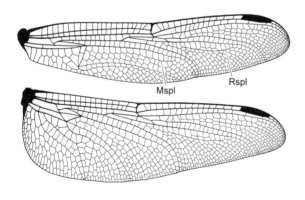

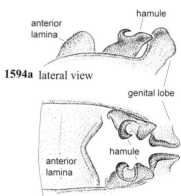

1594a lateral view

1594b　ventral view

Scapanea Kirby, 1889: 264, 298.

[♂ pp. 185, 193, couplets 44, 70; ♀ pp. 208, 215, couplets 31, 51]

Type species: *Libellula frontalis* Burmeister, 1839 [Kirby, 1889 by original designation]

1 species:

frontalis (Burmeister, 1839) [*Libellula*]* – **L** [Klots, 1932; García-Diaz, 1938]

References: Ris, 1913a; Donnelly, 1970; Garrison

and von Ellenrieder, 2006.
Distribution: Antilles (Cuba, Jamaica, Haiti, Dominica, Dominican Republic and Puerto Rico).

Large libellulines (40-47 mm); postfrons and vertex brown to metallic blue in mature males; pterothorax brown with yellow markings; mature males becoming pruinose on thorax and posterior abdominal segments. Wings hyaline (with slight opalescent band in

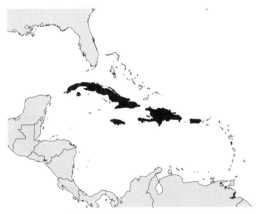

Map 117. Distribution of *Scapanea* sp.

populations of *S. frontalis* from Jamaica and Cuba); last antenodal in FW incomplete; FW discoidal field narrowing distally; Mspl distinct; median sector with 1 cell of rows throughout (Fig. 1595). Male S7-9 greatly widened and flattened (Fig. 1600) shared with

Brechmorhoga, *Nannothemis*, *Nephepeltia*, and some species of *Macrothemis* and *Micrathyria* (although widest point of abdomen less than 2 1/2 times as wide as base of S7 in these genera); anterior lamina entire (Fig. 1602b); posterior hamule not bifid (Figs. 1602a-b). **Unique characters**: Male S7-9 with widest point about 3-3 1/2 times as wide as base of S7 (Fig. 1600); distal segment of vesica spermalis with a ventral, horizontally flattened lobe at base (Figs. 1603a-b).

Status of classification: Good. *Scapanea frontalis* shares with *Brechmorhoga* and *Macrothemis* the short, stout, posteriorly directed hind femoral spines in the male; it differs from *Brechmorhoga* by the distally narrowed discoidal field and from most *Macrothemis* (except *M. griseofrons*) by the well-developed Mspl sector in FW.

Potential for new species: Unlikely.

Habitat: Males fly along mountain streams, preferring shaded areas, and females land near water (Trapero Quintana and Naranjo López, 2003).

1595 female wings – *Scapanea frontalis*

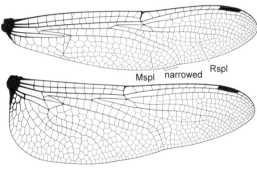

Mspl narrowed Rspl

1596 tarsal claw

obtuse angle

lateral view

Scapanea frontalis

1597 male hind femur

1598 female hind femur

lateral view

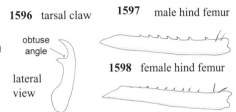

obtuse angle

S3   S4

ventral view   obtuse angle

1599 female ventral terga of S3–4 *Scapanea frontalis*

1600 male S6–10 – *Scapanea frontalis* dorsal view

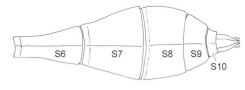

S6   S7   S8   S9

S10

1601 female S8–10 *Scapanea frontalis* ventral view

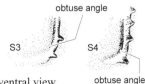

vulvar lamina   cercus

S8

S9   S10

male genital fossa – *Scapanea frontalis*

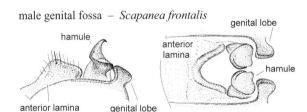

hamule

anterior lamina   genital lobe

1602a lateral view

genital lobe

anterior lamina

hamule

1602b ventral view

1603b dorsal view

flattened lobe

1603a lateral view

vesica spermalis – *Scapanea frontalis*

Sympetrum Newman, 1833: 511.

[♂ pp. 186, 190, couplets 50, 63; ♀ pp. 209, 211-212, couplets 32, 39-40]

Type species: *Libellula vulgata* Linnaeus, 1758 [Newman, 1833 by original designation]

    syn *Diplax* Burmeister, 1839: 848

        Type species: *Libellula pedemontana* Allioni, 1776 [Steinmann, 1997 by subsequent designation]

    syn *Thecadiplax* Selys, 1883a: 140

        Type species: *Diplax erotica* Selys, 1883 [Kirby, 1889 by subsequent designation]

    syn *Tarnetrum* Needham & Fisher, 1936: 114

        Type species: *Mesothemis illota* Hagen, 1861 [Needham and Fisher, 1936 by original designation]

    syn *Kalosympetrum* Carle, 1993: 10, 12

        Type species: *Libellula rubicundula* Say, 1840 [Carle, 1993 by original designation]

60 species; 22 New World species:

ambiguum (Rambur, 1842) [*Libellula*]* – **L** [Wright, 1946a]

    syn *albifrons* (Charpentier (*nec* Burmeister, 1839), 1840) [*Libellula*]

chaconi De Marmels, 1994

corruptum (Hagen, 1861) [*Mesothemis*]* – **L** [Musser, 1962]

costiferum (Hagen, 1861) [*Diplax*]* – **L** [Needham, 1901; Musser, 1962]

    syn *atripes* (Hagen, 1874) [*Diplax*]

danae (Sulzer, 1776) [*Libellula*]* – **L** [Walker, 1917; Walker and Corbet, 1975]

    syn *verum* Bartenef, 1915

evanescens De Marmels, 1992* – **L** [De Marmels, 1992a]

gilvum (Selys, 1884) [*Diplax*]* – **L** [Limongi, 1991]

illotum illotum (Hagen, 1861) [*Mesothemis*]* – **L** [Needham *in* Byers, 1927b]

illotum virgulum (Selys, 1884) [*Diplax*]*

internum Montgomery, 1943* – **L** [Walker and Corbet, 1975]

janeae Carle, 1993* – **L** [Carle, 1993]

madidum (Hagen, 1861) [*Diplax*]* – **L** [Cannings, 1981]

    syn *chrysoptera* (Selys, 1883) [*Diplax*]

    syn *flavicosta* (Hagen, 1877) [*Diplax*]

nigrocreatum Calvert, 1920* – **L** [Cannings, 1982b]

obtrusum (Hagen, 1867) [*Diplax*]* – **L** [Walker, 1917; Musser, 1962]

    syn *decisum* (Hagen, 1874) [*Diplax*]

occidentale californicum Walker, 1951* – **L** [Pritchard and Smith, 1956]

occidentale fasciatum Walker, 1951* – **L** [Musser, 1962]

occidentale occidentale Bartenef, 1915*

pallipes (Hagen, 1874) [*Diplax*]* – **L** [Walker, 1917;

Musser, 1962]

    syn *obtrusum morrisoni* Ris, 1911

paramo De Marmels, 2001*

roraimae De Marmels, 1988

rubicundulum (Say, 1840) [*Libellula*]* – **L** [Musser, 1962]

    syn *assimilatum* (Uhler, 1857) [*Libellula*]

semicinctum (Say, 1840) [*Libellula*]* – **L** [Needham, 1901]

signiferum Cannings & Garrison, 1991*

vicinum (Hagen, 1861) [*Diplax*]* – **L** [Needham, 1901]

villosum Ris, 1911* – **L** [Muzón and von Ellenrieder, 1997]

References: Ris, 1911b; Needham, Westfall, and May, 2000; De Marmels, 2001a.

Distribution: Holarctic and neotropical regions, with two species in NE Africa and one (extinct) in St. Helena; in the New World from Canada to Venezuela and south along the Andes to S Chile and Argentina.

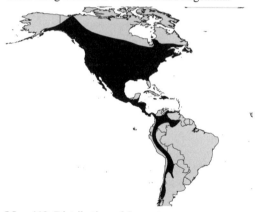

Map 118. Distribution of *Sympetrum* spp.

Small to medium (21–45 mm), red, brown to black libellulines; postfrons and vertex pale to dark, not metallic, pterothorax concolorous or with pale lateral stripes or spots, abdomen concolorous or with series of triangular dark or pale spots. Wings hyaline, with basal orange spots or bands, or with brown bands in some species (*S. occidentale* and *S. semicinctum*); last antenodal incomplete; FW triangle always crossed with costal side straight; Mspl distinct (Fig. 1604). Posterior lobe of prothorax constricted at base, upright and wider than other prothoracic lobes (Figs. 1605a-b); anterior lamina entire (Fig. 1606b); posterior hamule bifid with inner branch smaller than outer branch (Figs. 1606a-b); vulvar lamina (Fig. 1609) short, bilobed and horizontal to scoop shaped and ventrally directed. **Unique characters**: Distal segment of vesica spermalis with a pair of dorsoapical flagellae (Figs. 1607-1608).

Status of classification: Carle (1993) divided *Sympetrum* into four subgenera (*Sympetrum*, *Tarnetrum*, *Nesogonia* and *Kalosympetrum*), based on the morphology of the vesica spermalis, vulvar lamina, S4 and rear of the head. We have not adopted these names here because Carle (1993) did not include a phylogenetic analysis, nor did he adequately address relationships among the Old World species of *Sympetrum*. The status of some New World species (*i.e. S. janeae*, *S. internum*, and *S. occidentale*, *S. semicinctum*) and relationships of some species (*S. madidum* to *Tarnetrum* complex of species) are also controversial. A revision of the entire genus will be necessary in order to resolve these questions.

Potential for new species: Likely in the neotropical region.

Habitat: Marshes, bogs, swamps, lakes and ponds, including semipermanent ones, and slow streams. Adults are commonly found foraging on grassy meadows where adults perch on stems and twigs with wings set. Eggs are laid in tandem or by the female alone, by either tapping the water surface or wet vegetation or dropping them in flight onto mud of dry temporary pools. Mass migration of one species (*S. corruptum*) has been observed (Arnaud, 1972). Larvae lie among the detritus in the bottom of the waterbodies, or climb the submerged vegetation. Numerous studies on behavior and ecology of Old World species, of New World species scarce (Cannings, 1980; Van Buskirk, 1986; Schmidt, 1987; Upson, 2000).

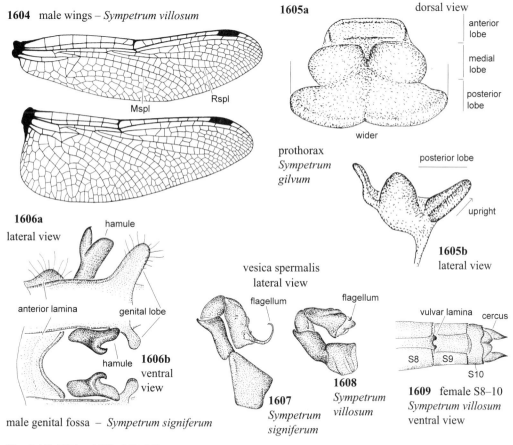

1604 male wings – *Sympetrum villosum*

Rspl
Mspl

1605a dorsal view

anterior lobe
medial lobe
posterior lobe

wider

prothorax *Sympetrum gilvum*

posterior lobe

upright

1605b lateral view

1606a lateral view

hamule

anterior lamina

genital lobe

hamule **1606b** ventral view

vesica spermalis lateral view

flagellum

flagellum

vulvar lamina cercus

S8 S9

S10

1607 *Sympetrum signiferum*

1608 *Sympetrum villosum*

1609 female S8–10 *Sympetrum villosum* ventral view

male genital fossa – *Sympetrum signiferum*

Tauriphila Kirby, 1889: 258, 268.
[♂ pp. 183, couplet 36; ♀ pp. 207-208, couplets 29-30]
Type species: *Tramea iphigenia* Hagen, 1867 [Kirby, 1889 by original designation]
5 species:

argo (Hagen, 1869) [*Tramea*]* – **L** [Costa and Assis, 1994?; Fleck, Brenk, and Misof, 2006]
 syn ?*nycteris* Karsch, 1890
australis (Hagen, 1867) [*Tramea*]* – **L** [Westfall, 1998]
 syn *iphigenia* (Hagen, 1867) [*Tramea*]
azteca Calvert, 1906*
risi Martin, 1896* – **L** [Rodrigues Capítulo, 1996]
xiphea Ris, 1913

References: Ris, 1913b; Ris, 1919; Needham, Westfall, and May, 2000.

Distribution: Southernmost United States to Uruguay and central Argentina.

Map 119. Distribution of *Tauriphila* spp.

Medium sized (40–50 mm); reddish to dark brown libellulines. Postfrons and vertex pale to dark, becoming metallic in mature males; pterothorax red to brown or black; abdomen red to black, or (*T. azteca*) brownish yellow with dark bands on intersegmental areas of S3-8. Wings hyaline, with basal brown spot on HW; last antenodal incomplete; FW triangle al-

ways crossed with costal side straight; Mspl distinct; median planate with 1 row of cells throughout (Fig. 1610). Posterior lobe of prothorax widest at base; anterior lamina entire, about 1 1/2-2 times as long as wide (Fig. 1611b, shared with *Miathyria marcella* and *Macrodiplax balteata*); posterior hamule not bifid (Figs. 1611a-b); vulvar lamina shorter than 1/3 of S9. **Unique characters**: None known.

Status of classification: Fair. Key to three of five known species in Needham, Westfall, and May (2000); relationship of *T. xiphea* Ris to *T. argo* (Hagen) as keyed and described in Ris (1913b) unclear.

Potential for new species: Possible. All known species (with exception of *T. xiphea* Ris) occupy broad geographic ranges.

Habitat: Adults engage in sustained flights over open fields often in company with individuals of *Miathyria*, *Pantala* or *Tramea*. *Tauriphila* males in sexual patrol flight over floating macrophytes (Paulson *pers. comm.*). Adults of *T. azteca* are reported as perching obliquely on twigs (Dunkle, 2000). Population ecology of *T. risi* studied by Rodrigues Capítulo (2000).

1610 male wings – *Tauriphila australis*

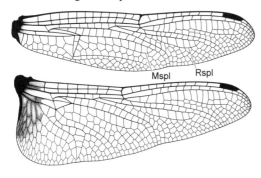

male genital fossa – *Tauriphila australis*

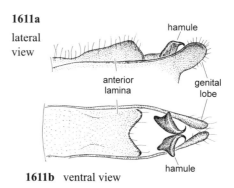

1611a lateral view

1611b ventral view

Tholymis Hagen, 1867: 221.

[♂ pp. 187, 196, couplets 51, 80; ♀ pp. 219, couplet 67]

Type species: *Libellula tillarga* Fabricius, 1798 [Kirby, 1889 by subsequent designation]

2 species; 1 New World species:

citrina Hagen, 1867 – **L** [Fleck, De Marmels, and Grand, 2004]*

References: Ris, 1913a; Needham, Westfall, and May, 2000.

Distribution: Pantropical; in the New World from southernmost United States to northernmost Argentina and Chile.

Large (48–53 mm) libellulines; body and abdomen brown. Wings hyaline, except for a yellow spot at the nodus; last antenodal incomplete; Mspl and Rspl

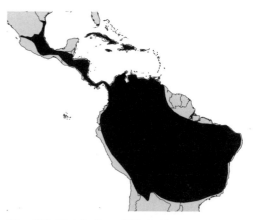

Map 120. Distribution of *Tholymis* sp.

distinct (Fig. 1612). Posterior lobe of prothorax widest at base, its posterior margin bent caudad; anterior lamina entire; posterior hamule short and not bifid (Figs. 1614a-b). **Unique characters**: HW anal loop open posteriorly (Fig. 1612); distal segment of vesica spermalis with a pair of very large, flat, vertical, sclerotized plates (Figs. 1613a-b); female sternum of S9 projected distally to about posterior margin of S10, rectangular, concave with a medial carina and lateromarginal hairs directed medially

1612 female wings – *Tholymis citrina*

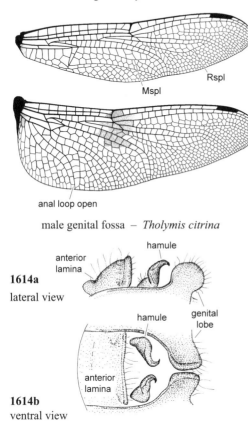

Rspl

Mspl

anal loop open

male genital fossa – *Tholymis citrina*

hamule

anterior lamina

1614a
lateral view

hamule

genital lobe

anterior lamina

1614b
ventral view

(Figs. 1615a-b).
Status of classification: Good.
Potential for new species: Unlikely.
Habitat: Adults crepuscular. They fly over marshes or fields during day and rapidly in evening hawking for prey. They rest obliquely on twigs in deep understory. Paulson (2001) described behavior of this species in Florida. Larva described from specimens collected at a rectangular water tank of concrete with less than half a meter of water and abundant floating vegetation (De Marmels, *pers. comm.*).

vesica spermalis – *Tholymis citrina*

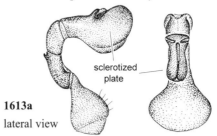

sclerotized plate

1613a
lateral view

1613b dorsal view

female S8–10 – *Tholymis citrina*

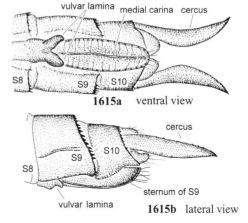

vulvar lamina   medial carina   cercus

S8   S9   S10

1615a   ventral view

cercus

S10

S9

S8

vulvar lamina

sternum of S9

1615b   lateral view

Tramea Hagen, 1861: 143.
[♂ pp. 183, 189, couplets 36, 58; ♀ pp. 211, couplet 39]
Type species: *Libellula carolina* Linnaeus, 1763 [Kirby, 1889 by subsequent designation]
[NOTE: Gloyd (1972) provided compelling reasons for using *Tramea* Hagen, 1861 over *Trapezostigma* Hagen, 1849, and this has been made a case to the ICZN (Dijkstra, van Tol, Legrand and Theischinger, 2005) in order to conserve the name usage for *Tramea* Hagen, 1861, and we follow this usage here.]
  syn *Trapezostigma* Hagen, 1849: 174
    Type species: *Libellula carolina* Linnaeus, 1763 [Cowley, 1935 by subsequent designation]
24 species; 10 New World species:

abdominalis (Rambur, 1842) [*Libellula*]* – **L** [Cabot, 1890; Klots, 1932]
  syn *basalis* (Burmeister (*nec* Stephens 1835), 1839) [*Libellula*]
binotata (Rambur, 1842) [*Libellula*]* – **L** [Needham, Westfall, and May, 2000]
  syn *brasiliana* Brauer, 1867
  syn *longicauda* Brauer, 1867
  syn *paulina* Förster, 1910
  syn *subbinotata* Brauer, 1867
  syn *walkeri* Whitehouse, 1943
calverti Muttkowski, 1910* – **L** [Souza, Costa, and Santos, 1999b]

carolina (Linnaeus, 1763) [*Libellula*]* – **L** [Cabot, 1890; Needham, Westfall, and May, 2000]
cophysa Hagen, 1867* – **L** [Santos, 1968c]
  syn *darwini* Kirby, 1889
insularis Hagen, 1861*
lacerata Hagen, 1861* – **L** [Cabot, 1890; Bick, 1951b]
minuta De Marmels & Rácenis, 1982
onusta Hagen, 1861* – **L** [Byers, 1927b; Needham, Westfall, and May, 2000]
rustica De Marmels & Rácenis, 1982*

Map 121. Distribution of *Tramea* spp.

References: Ris, 1913b; De Marmels and Rácenis, 1982; Needham, Westfall, and May, 2000.

Distribution: Worldwide except in most of the palearctic region (present in Japan); in the New World from southernmost Canada to Uruguay and central Argentina.

[NOTE: Belle (1988a) recorded the capture of a male of the Old World species *Tramea basilaris burmeisteri* Kirby from the international airport at Zanderij, Surinam. The species has not been listed from the New World again and we consider the specimen only an accidental introduction.]

Large (41–60 mm), orange or red to black libellulines; postfrons and vertex pale to red to metallic blue in adult males; pterothorax concolorous or, in *T. cophysa* group, with a pair of pale lateral stripes; abdomen concolorous with contrasting dark markings on S8-10 or, in *T. lacerata*, with vestigial (S4-7) to prominent (S8) pale spots. Wings hyaline, with broad HW usually with a dark anal spot; FW with 1 cubito-anal crossvein; FW subtriangle not defined; pterostigma short, surmounting 1-2 cells in FW; arculus before antenodal 2 in HW; Mspl and Rspl distinct and with 2 or more cell rows (Fig. 1616). Posterior lobe of prothorax widest at base, its posterior margin bent caudad; anterior lamina entire (Fig. 1618b); posterior hamule not bifid, broad at base, broadly flattened and bent ventroposteriorly toward tip, which is armed with a small usually outwardly directed tooth (Figs. 1617- 1618a-b); cercus long with basal 1/4 or more concave in lateral view; vulvar lamina forming a pair of large oval lobes at 1/2 or more of sternum S9. **Unique characters**: None known.

Status of classification: Good. Keys in Needham, Westfall, and May (2000) and De Marmels and Rácenis (1982) will allow for identification of all described species.

Potential for new species: Unlikely. All described species occupy broad geographic ranges. The most recently described species (*T. minuta, T. rustica*) are cryptic species earlier confused with *T. cophysa* (De Marmels and Rácenis, 1982).

Habitat: Adults fly over open pastures or exposed ponds, lakes and pools and are seemingly always on the wing. They perch on tips of branches or reeds with wings held at same plane as body or slightly raised. Male holds female and releases her during oviposition and again grabs her after releasing eggs; tandem pairs will defend oviposition sites (e.g. openings in floating vegetation). Further details of behavior for Floridian species provided by Paulson (1999). Larvae found on bottom of ponds or lakes.

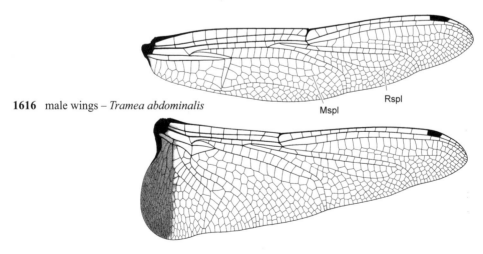

1616 male wings – *Tramea abdominalis*

Rspl

Mspl

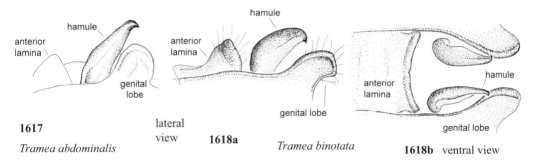

male genital fossa

1617

Tramea abdominalis

lateral view **1618a**

Tramea binotata

1618b ventral view

Uracis Rambur, 1842: 31.

[♂ pp. 184, 192, couplets 43, 68; ♀ pp. 220, couplet 69]

Type species: *Libellula imbuta* Burmeister, 1839 [Kirby, 1889 by subsequent designation]

   syn *Urothemis* Hagen (*nec Urothemis* Brauer, 1868), 1877: 94

      Type species: *Libellula infumata* Rambur, 1842 [Cowley, 1935 by subsequent designation]

   syn *Pronomaja* Förster, 1909: 225

      Type species: *Pronomaja mimetica* Förster, 1909 [by original designation]

7 species:

fastigiata (Burmeister, 1839) [*Libellula*]*
   syn *fastigiata* aberration *pura* Förster, 1909
   syn *fastigiata* forma *machadina* Förster, 1910
imbuta (Burmeister, 1839) [*Libellula*]*
   syn *quadra* Rambur, 1842
infumata (Rambur, 1842) [*Libellula*]*
   syn *ovata* Calvert, 1909
ovipositrix Calvert, 1909*
   syn *mimetica* (Förster, 1909) [*Pronomaja*]
reducta Fraser, 1946
siemensi Kirby, 1897*
turrialba Ris, 1919*

Map 122. Distribution of *Uracis* spp.

References: Ris, 1911a; Ris, 1919; Costa and Santos, 1997.

Distribution: S Mexico south to SE Brazil.

[NOTE: Rambur (1842) cited *Uracis quadra* Rambur (= *U. imbuta*) from "Buénos-Ayres" but no species of *Uracis* has subsequently been collected in Argentina.]

Medium (28–40 mm) libellulines; postfrons and vertex pale to black, not metallic; pterothorax gray often with reticulate cross-stripings across mesepisternum but this and abdomen becoming mostly pruinose blue in mature males. Wings hyaline, with brown wing tips (which can be extensive) or with broad wing bands (males can be polymorphic for wing banding); FW with last antenodal incomplete; FW with 1 or more cubito-anal crossveins; FW supratriangle free or crossed; Mspl and Rspl distinct, Mspl rarely indistinct; arculus in HW opposite to antenodal 2 to distal to antenodal 3 (Fig. 1620, latter condition shared with *Misagria*). Vertex with tubercles (Figs. 1619a-b); posterior lobe of prothorax widest at base, its dorsal margin bent caudad; anterior lamina entire or slightly bifid (Fig. 1621); posterior hamule bifid with inner branch smaller than outer branch (Figs 1621a-b); distal segment of vesica spermalis long and cylindrical (shared with *Erythrodiplax, Nannothemis, Pseudoleon* and *Ypirangathemis*), gradually widening toward bluntly rounded tip (shared with *Erythrodiplax* and *Ypirangathemis*); vulvar lamina and sternum of S9 projected distally beyond tip of cerci (projected vulvar lamina shared with *Argyrothemis* and *Ypirangathemis*, and projected sternum S9 with *Ypirangathemis*). Very similar to *Ypirangathemis*, from which it can be distinguished by the vertex with tubercles (vertex smoothly rounded in *Ypirangathemis*), the spines of male hind femur gradually increasing in length (dimorphic in *Ypirangathemis*) and details of the complex membranous distal lobes of vesica spermalis (see Borror, 1947). **Unique characters**: None known.

Status of classification: Fair. Key for all known species provided by Costa and Santos (1997). Individual variation in wing spotting/banding occurs in some species.

Potential for new species: Likely. One known to the authors from Venezuela.

Habitat: Adults can be common in shaded under story or along partially shaded forest trails. They perch frequently on small twigs and vegetation near the ground. RWG observed a female *Uracis* ovipositing by repeatedly thrusting her abdomen in the muddy bottom of a small rain puddle in Rondônia State, Brazil; and Williamson noted (in Ris, 1919) for *U. ovipositrix*: "female…oviposits in damp earth, male hovering near, and thrusts her abdomen (ovipositor) into soil but does not alight." De Marmels (1992d) stated the following for *U. siemensi*: "This species behaves quite differently from its congeners. The males hold territories at small water holes in the forest, close to the creek. They rarely settle, but spend considerable intervals hovering close to the water surface. A female was laying eggs stabbing its ovipositor into the humid bank some thirty centimeters above water." Larvae unknown.

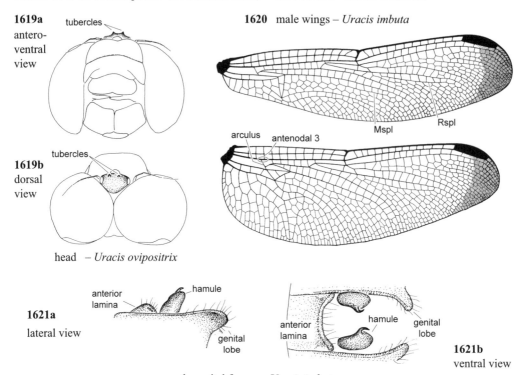

1619a antero-ventral view — tubercles

1619b dorsal view — tubercles

head — *Uracis ovipositrix*

1620 male wings — *Uracis imbuta*

Mspl   Rspl

arculus   antenodal 3

1621a lateral view

anterior lamina   hamule   genital lobe

1621b ventral view

anterior lamina   hamule   genital lobe

male genital fossa — *Uracis imbuta*

Ypirangathemis Santos, 1945: 457.

[♂ pp. 191, couplet 66; ♀ pp. 220, couplet 70]

Type species: *Ypirangathemis calverti* Santos, 1945 [by original designation]

1 species:

calverti Santos, 1945*

References: Borror, 1947.

Distribution: S Venezuela and Guiana to SE Brazil.

Small to medium (27–34 mm), brown libellulines; postfrons and vertex brown to metallic blue in mature specimens, pterothorax pale brown with black transverse reticulation on dorsum, fading laterally; abdomen brown to dark brown. Wings hyaline or

Map 123. Distribution of *Ypirangathemis* sp.

with broad dark brown band between triangles and pterostigma; pterostigma yellowish; last antenodal crossvein incomplete in FW; arculus between antenodals 2-3 in FW; FW with 1 cubito-anal crossvein; FW supratriangle free; Mspl and Rspl very poorly developed or indistinct (Fig. 1622). Vertex rounded; posterior lobe of prothorax widest at base, its posterior margin bend caudad. Anterior lamina entire (Fig. 1623b); posterior hamule bifid, with inner branch smaller than outer branch (Figs. 1623b); distal segment of vesica spermalis long and cylindrical (shared with *Erythrodiplax, Nannothemis, Pseudoleon* and *Uracis*), gradually widening to rounded or bluntly angulated tip (shared with *Erythrodiplax* and *Uracis*); vulvar lamina and sternum of S9 (Fig. 1624) projected distally beyond tip of cerci (projected vulvar lamina shared with *Argyrothemis* and *Uracis*, and projected sternum S9 with *Uracis*). Very similar to *Uracis*,

from which it can be distinguished by the rounded vertex (vertex with tubercles in *Uracis*), the dimorphic spines on male hind femur (gradually increasing in length in *Uracis*), and details of the complex membranous distal lobes of vesica spermalis (see Borror, 1947). **Unique characters**: None known.

Status of classification: Good. Borror (1947) considered the genus related to *Erythrodiplax* based on wing venation and vesica spermalis morphology, and to *Uracis* based on vulvar lamina and vesica spermalis morphology.

Potential for new species: Likely. One known to the authors from Venezuela and Amazonian areas Brazil.

Habitat: Adults perch on grass stems along swampy trails in the open and along forest edges, often together with *Zenithoptera americana* (De Marmels, *pers. comm.*). Larva unknown.

1622 female wings – *Ypirangathemis calverti*

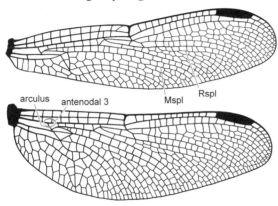

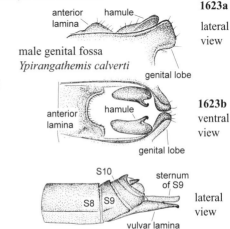

1623a lateral view

male genital fossa
Ypirangathemis calverti

1623b ventral view

1624 female S8–10 – *Ypirangathemis calverti*

Zenithoptera Selys, 1869b: 16.
[♂ pp. 197, couplet 86; ♀ pp. 220, couplet 72]

Type species: *Libellula fasciata* Linnaeus, 1758 [Kirby, 1890 by subsequent designation]
[NOTE: Pujol-Luz (1995) credits the genus *Zenithoptera* to Bates *in* Selys, 1869, based on a later statement by Selys (1881) crediting the genus to Bates. However, the original brief description of *Zenithoptera* is entirely by Selys ("Il convient d'adapter pour elles ["*P. americana* L. and *P. fasciata* F [*sic*] (*violacea* De Geer)"] le genre *Zenithoptera*, propose pour elles par célèbre voyageur dans les notes manuscrites qu'il m'a gracieusement adressées, lorsqu'il m'a cede sa riche collection d'Odonates de l'Amazone"] and we credit the name to Selys. Pujol-Luz (1995) cites *L. americana* L. as senior synonym based on first reviewer status by Ris (1910) as stated by De Marmels (1985). However, we agree with Calvert (1948a), and Jurzitza (1982) in choosing *L. fasciata* as type species since Erichson (1848) and Rambur (1842) recognized both names as synonyms previously, and both choose *L. fasciata* over *L. americana*. Pujol-Luz (1995) designated the type species of *Zenithoptera* as follows: "*Libellula americana* Linnaeus, 1758: 545, based on ichnotaxon of Edwards (1751), by present designation." However, Kirby

(1889: 272) had already designated *Libellula fasciata* Linnaeus as type of his new genus *Potamothemis* which later (1890: 178) he synonymized with *Zenithoptera* Selys.]
    syn *Potamothemis* Kirby, 1889: 257, 272
        Type species: *Libellula fasciata* Linnaeus, 1758 [Kirby 1889 by original designation]
4 species:

anceps Pujol-Luz, 1993* – **L** [Costa, Pujol-Luz, and Regis, 2004]
fasciata (Linnaeus, 1758) [*Libellula*]*
    syn *americana* (Linnaeus, 1758) [*Libellula*]
    syn *fasciata* (Linnaeus, 1758) [*Libellula*]
    syn *violacea* (De Geer, 1773) [*Libellula*]
lanei Santos, 1941*
viola Ris, 1910*

References: Jurzitza, 1982; Pujol-Luz and Rodrigues da Fonseca, 1997.
Distribution: Nicaragua to Paraguay and NE Argentina.

Map 124. Distribution of *Zenithoptera* spp.

Small (21–29 mm), dark brown libellulines; postfrons and vertex brown to metallic black; thorax brown or with lateral yellow stripes; abdomen black or with pale longitudinal stripes above lateral carinae. Wings (Fig. 1625) largely black with metallic blue to violet luster, and a dim to well defined transverse white bands after nodus and at wing tips; FW costa undulate (shared with *Diastatops*); sectors of arculus in FW not stalked; 2 or more bridge crossveins; Mspl and Rspl distinct and with 2 or more rows of cells; Aspl

in HW more or less straight; anal loop closed posteriorly, but in some specimens of *Z. viola* open on basal side (Fig. 1625). Posterior lobe of prothorax constricted at base; anterior lamina entire (Fig. 1626b); posterior hamule bifid with inner branch smaller than outer branch (Fig. 1626b); vulvar lamina shorter than 1/3 of S9 and not projected ventrally; sternum of S9 projected distally to about posterior margin of S10, triangular and smoothly convex (shared with *Elga*, *Fylgia*, *Nephepeltia* and some *Micrathyria*). **Unique characters**: None known.

Status of classification: Fair. The most recently described species, *Z. anceps* Pujol-Luz, is a cryptic species separable from *Z. fasciata* apparently only by morphology of the vesica spermalis.

Potential for new species: Possible given that some species approach one another closely.

Habitat: Adults occur at grassy marshes or in clearings. When at rest on tips of twigs or on grass, they often hold their wings over the back (unique field character for New World libellulines), which they start opening slowly to then rapidly lower to display a brief flash of metallic blue (as in Plate 8), as seen in some butterflies. Pujol-Luz and dias Viera (1998) studied the territorial behavior of *Z. anceps*.

1625 male wings – *Zenithoptera viola*

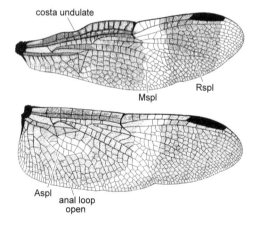

male genital fossa – *Zenithoptera fasciata*

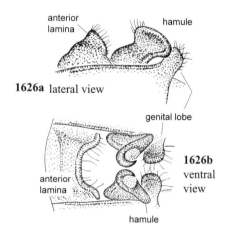

Literature Cited

Aguesse, P. 1959. Notes sur l'accouplement et la ponte chez *Crocothemis erythraea* Brulle (Odonata, Libellulidae) (I). *Vie et Milieu* 10(2): 176-184.

Abbott, J.C. 2005. Dragonflies and damselflies of Texas and the south-central United States. Princeton University Press, Princeton, NJ. viii + 344 pp.

Alcock, J. 1989. The mating system of *Libellula saturata* Uhler (Anisoptera: Libellulidae). *Odonatologica* 18(1): 89-93.

———. 1990. Oviposition resources, territoriality and male reproductive tactics in the dragonfly *Paltothemis lineatipes* (Odonata: Libellulidae). *Behaviour* 113(3-4): 252-263.

Arango, M.C., and G. Roldán. 1983. Odonatos inmaduros del Departamento de Antioquía en diferentes pisos altitudinales. *Actualidades biológicas* 12(46): 91-105.

Arnaud, P.H. 1972. Mass movement of *Sympetrum corruotum* [*sic*] (Hagen) (Odonata: Libellulidae) in central California. *Pan-Pacific Entomologist* 48(1): 75-76.

Artiss, T. 2001. Structure and function of male genitalia in *Libellula*, *Ladona* and *Plathemis* (Anisoptera: Libellulidae). *Odonatologica* 30(1): 13-27.

Artiss, T., T. Schultz, D. Polhemus, and C. Simon. 2001. Molecular phylogenetic analysis of the dragonfly genera *Libellula*, *Ladona* and *Plathemis* (Odonata; Libellulidae) based on mitochondrial COI and 16S rRNA sequence data. *Molecular Phylogenetics and Evolution* 18: 348-361.

Assis, J.C.F., A.L. Carvalho, and L.F.M. Dorvillé. 2000. Aspects of larval development of *Limnetron debile* (Karsch), in a mountain stream of Rio de Janeiro State, Brazil (Anisoptera: Aeshnidae). *Odonatologica* 29(2): 151-155.

Assis, C.V., and J.M. Costa. 1994. Seis novas larvas do gênero *Micrathyria* Kirby e notas sobre a distribuição no Brasil (Odonata: Libellulidae). *Revista Brasileira de Zoología* 11(2): 195-209.

Bachmann, A.O. 1963. La ninfa de *Staurophlebia bosqi* Navás, 1927 (Odonata – Aeshnidae). *Revista de la Sociedad Entomológica Argentina* 26: 71-73.

Baird, J.M., and M.L. May. 1993. Behavioral ecology of foraging in *Pachydiplax longipennis* (Odonata: Libellulidae). *Journal of Insect Behavior*. 10: 655-678.

Beatty, G.H., and A.F. Beatty. 1963. Gregarious roosting behavior of *Mecistogaster ornatus* in Mexico. *Proceedings North Central Branch, Entomological Society of America* 18: 153-155.

Bechly, G. 1996. Morphologische Untersuchungen am Flügelgeäder der rezenten Libellen und deren Stammgruppenvertreter (Insecta; Pterygota; Odonata), unter besonderer Berücksichtigung der Phylogenetischen Systematik und des Grundplanes der *Odonata. *Petalura* (special volume) 2: 1-402.

———. 1999. Phylogeny and systematics of fossil dragonflies (Insecta: Odonatoptera) with special reference to some Mesozoic outcrops. PhD thesis, Eberhard-Karls-University Tübingen. x + 755 pp.

Bechly, G., A. Nel, X. Martínez-Delclôs, E.A. Jarzembowski, R. Coram, D. Martill, G. Fleck, F. Escuillié, M.M. Wisshak, and M. Maisch. 2001. A revision and phylogenetic study of Mesozoic Aeshnoptera, with description of numerous new taxa (Insecta: Odonata: Anisoptera). *Neue Paläontologische Abhandlungen* 4: 1-219.

Bede, L.C., W. Piper, G. Peters, and A.B.M. Machado. 2000. Phenology and oviposition behaviour of *Gynacantha bifida* Rambur (Anisoptera: Aeshnidae). *Odonatologica* 29(4): 317-324.

Belle, J. 1963. Dragon flies of the genus *Zonophora* with special reference to its Surinam representatives. *Studies on the Fauna of Suriname and other Guyanas* 5(16): 60-69.

———. 1964. Surinam dragon flies of the genus *Aphylla*, with a description of a new species. *Studies on the Fauna of Suriname and other Guyanas* 7(23): 22-34.

———. 1966a. Surinam dragon-flies of the genus *Progomphus*. *Studies on the Fauna of Suriname and other Guyanas* 8(28): 1-28.

———. 1966b. Surinam dragon-flies of the *Agriogomphus* complex of genera. *Studies on the Fauna of Suriname and other Guyanas* 8(29): 29-60.

———. 1966c. Additional notes on some dragon-flies of the genus *Zonophora*. *Studies on the Fauna of Suriname and other Guyanas* 8(30): 61-64.

———. 1970. Studies on South American Gomphidae (Odonata) with special reference to the species from Surinam. *Studies on the Fauna of Suriname and other Guyanas* 11(43): 1-158.

———. 1972a. Further studies on South American

Gomphidae (Odonata). *Tijdschrift voor Entomologie* 115(5): 217-240.

———. 1972b. On *Diaphlebia* Selys, 1854 from Central America (Odonata: Gomphidae). *Odonatologica* 1(2): 63-71; errata p. 279-280.

———. 1973. A revision of the New World genus *Progomphus* Selys, 1854 (Anisoptera: Gomphidae). *Odonatologica* 2(4): 191-308.

———. 1975. On *Agriogomphus tumens* (Calvert, 1905) with a description of its male (Anisoptera: Gomphidae). *Odonatologica* 4(4): 237-242.

———. 1977a. Revisional notes on *Diaphlebia* Selys, 1854 (Anisoptera: Gomphidae). *Odonatologica* 6(2): 111-117.

———. 1977b. Some gomphine material from Surinam, preserved in the Leyden Museum of Natural History, with a note on the larva of *Desmogomphus tigrivensis* Williamson (Anisoptera: Gomphidae). *Odonatologica* 6(4): 289-292.

———. 1979. A new species of *Peruviogomphus* Klots, 1944 from Ecuador (Anisoptera: Gomphidae). *Odonatologica* 8(2): 111-114.

———. 1980. Notes on three species of *Cyanogomphus* Selys, 1873 (Odonata: Gomphidae). *Entomologische Berichten* 40: 151-155.

———. 1982. A review of the genus *Archaeogomphus* Williamson (Odonata, Gomphidae). *Tijdschrift voor Entomologie* 125(3): 37-56.

———. 1983a. On the species of the *polygonus* group of *Progomphus* with a description of a new species (Odonata, Gomphidae). *Tijdschrift voor Entomologie* 126(7-8): 137-144.

———. 1983b. A review of the genus *Zonophora* Selys (Odonata, Gomphidae). *Tijdschrift voor Entomologie* 126(7-8): 145-173.

———. 1984a. *Nothodiplax dendrophila*, a new genus and a new species from Surinam (Odonata: Libellulidae). *Entomologische Berichten* 44: 6-8.

———. 1984b. *Idiogomphoides*, a new genus from Brazil (Odonata: Gomphidae). *Entomologische Berichten* 44: 106-109.

———. 1984c. A synopsis of the South American species of *Phyllogomphoides*, with a key and descriptions of three new taxa (Odonata, Gomphidae). *Tijdschrift voor Entomologie* 127(4): 79-100.

———. 1986a. New World Lindeniinae, with *Melanocacus interioris* gen. nov. et spec. nov. (Odonata: Gomphidae). *Entomologische Berichten* 46: 97-102.

———. 1986b. *Cyanogomphus pumilus*, a new species from Venezuela (Odonata: Gomphidae). *Entomologische Berichten* 46: 111-112.

———. 1988a. A record of the Old World Species *Tramea basilaris burmeisteri* Kirby from Suriname (Odonata: Libellulidae). *Zoologische Mededelingen*, Leiden 62(1): 1-3.

———. 1988b. A synopsis of the species of *Phyllocycla* Calvert, with descriptions of four new taxa and a key to the genera of neotropical Gomphidae (Odonata, Gomphidae). *Tijdschrift voor Entomologie* 131: 73-102.

———. 1988c. *Epigomphus gibberosus*, a new species from Peru, with lectotype designations for the eligible species of the genus *Epigomphus* (Odonata: Gomphidae). *Tijdschrift voor Entomologie* 131: 135-140.

———. 1989. A revision of the New World genus *Neuraeschna* Hagen, 1867 (Odonata: Aeshnidae). *Tijdschrift voor Entomolgie* 132: 259-284.

———. 1990. *Progomphus nigellus* and *Phyllocycla hamata*, two new dragonflies from Brazil (Odonata: Gomphidae). *Tijdschrift voor Entomolgie* 133: 27-30.

———. 1991. The ultimate instar larvae of the Central American species of *Progomphus* Selys, with a description of *P. belyshevi* spec. nov. from Mexico (Anisoptera: Gomphidae) *Odonatologica* 20(1): 9-27.

———. 1992a. Studies on ultimate instar larvae of neotropical Gomphidae, with the description of *Tibiagomphus* gen. nov. (Anisoptera). *Odonatologica* 21(1): 1-24.

———. 1992b. A revision of the South American species of *Aphylla* Selys, 1854 (Odonata: Gomphidae). *Zoologische Mededelingen*, Leiden 66(12): 239-264.

———. 1994a. Three new neotropical Gomphidae from the genera *Archaeogomphus* Williamson, *Cyanogomphus* Selys and *Epigomphus* Hagen (Anisoptera). *Odonatologica* 23(1): 45-50.

———. 1994b. Four new species of *Aphylla* from Brazil (Odonata: Gomphidae). *Entomologische Berichten* 54(7): 138-144.

———. 1995. On the female sex of some elusive South-American Gomphidae with the descriptions of three new genera and four new species (Odonata). *Zoologische Mededelingen*, Leiden 69: 19-36.

———. 1996. Higher classification of the South-American Gomphidae (Odonata). *Zoologische Mededelingen*, Leiden 70: 297-324.

———. 1998. Synopsis of the genus *Rhodopygia* Kirby, 1889 (Odonata: Libellulidae). *Zoologische Mededelingen*, Leiden 72(1): 1-13.

Bennefield, B.L. 1965. A taxonomic study of the subgenus *Ladona* (Odonata: Libellulidae). *University of Kansas Scientific Bulletin* 45(4): 361-396.

Bick, G.H. 1941. Life-history of the dragonfly, *Erythemis simplicicollis* (Say). *Annals of the Entomological Society of America* 34(1): 215-230.

———. 1951a. The nymph of *Libellula semifasciata* Burmeister (Odonata, Libellulidae). *Proceedings of the Entomological Society of Washington* 53(5): 247-250.

———. 1951b. The early nymphal stages of *Tramea lacerata* Hagen (Odonata: Libellulidae). *Entomological News* 62(10): 293-303.

———. 1953. The nymph of *Miathyria marcella* (Selys) (Odonata, Libellulidae). *Proceedings of the Entomological Society of Washington* 55(1): 30-36.

———. 1955. The nymph of *Macrodiplax balteata* (Hagen). *Proceedings of the Entomological Society of Washington* 57(4): 191-196.

Bick, G.H., and D. Sulzbach. 1966. Reproductive behaviour of the damselfly, *Hetaerina americana* (Fabricius) (Odonata: Calopterygidae). *Animal Behaviour* 14: 156-158.

Biggs, K.R. 2000. Common dragonflies of California. A beginner's Pocket Guide. Second edition. Azalea Creek Publishing 96 pp.

———. 2004. Common dragonflies of the Southwest. A beginner's Pocket Guide. Azalea Creek Publishing 160 pp.

Bird, R.D. 1934. The emergence and nymph of *Gomphus militaris* (Odonata: Gomphinae). *Entomological News* 45: 44-46.

Borror, D.J. 1931. The Genus *Oligoclada* (Odonata). *Miscellaneous Publications of the Museum of Zoology University of Michigan* 22: 42 pp.

———. 1942. A revision of Libelluline genus *Erythrodiplax* (Odonata). Ohio State University Graduate Studies, Contributions in Zoology and Entomology 4, Biological Series xv +286 pp.

———. 1947. Notes on *Ypirangathemis calverti* Santos (Odonata: Libellulidae) with a description of the female of *Y. calverti* Santos. *Annals of the Entomological Society of America* 40(2): 247-256.

———. 1957. New *Erythrodiplax* from Venezuela (Odonata: Libellulidae). *Acta Biologica Venezuelica* 2(5): 32-42.

Brauer, F. 1864. Erster Bericht über die auf der Weltfahrt der kais. Fregatte Novara gesammelten Neuropteren. *Verhandlungen der zoologisch-botanischen Gesellschaft in Wien* 14: 159-164.

———. 1865. Vierter Bericht über die auf der Weltfahrt der kais. Fregatte Novara gesammelten Neuropteren. *Verhandlungen der zoologisch-botanischen Gesellschaft in Wien* 15: 903-908.

———. 1868. Verzeichniss der bis jetzt bekannten Neuropteren im Sinne Linne's. *Verhandlungen der zoologisch-botanischen Gesellschaft in Wien* 18: 359-416, 711-742.

Bridges, C.A. 1994. Catalogue of the family-group, genus-group, and species-group names of the Odonata of the world (Third Edition). Urbana, xlvi+905 pp.

Brittinger, C.C. 1850. Die Libelluliden des Kaiserreichs Oesterreich. *Sitzungsberichte der Akademie Wissenschaften Wien* 4: 328-336.

Broughton, E. 1928. Some new Odonata nymphs. *Canadian Entomologist* 60: 34.

Brunelle, P.-M. 2000. A new species of *Neurocordulia* (Odonata: Anisoptera: Corduliidae) from eastern North America. *Canadian Entomologist* 132(1): 39-48.

Buchecker, H. 1876. Systema entomologiae sistens insectorum classes, genera, species. Pars 1. Odonata (Fabric) europ. München: 16 + 4 pp, 47 pl.

Buchholz, K.-F. 1952. Eine neue *Antidythemis*-Art (Libellulinae, Odonata) mit bemerkungen über das genus *Antidythemis* Kirby und *A. trameiformis* Kirby. *Bulletin of the Institut Royal des Sciences naturelles de Belgique* 28(35): 1-11.

Burmeister, H.C.C. 1839. Neuroptera. *In*: Handbuch der Entomologie, T.C.F. Enslin, Berlin, 2(2): 757-1050.

Byers, C.F. 1927a. The nymph of *Libellula incesta* and a key for the separation of the known nymphs of the genus *Libellula* (Odonata). *Entomological News* 38: 113-115.

———. 1927b. Notes on some American dragonfly nymphs (Odonata, Anisoptera). *Journal of the New York Entomological Society* 35: 65-74.

———. 1930. A contribution to the knowledge of Florida Odonata. *University of Florida Publica-*

tions, *Biological Science Series* 1(1): 1-327.

———. 1936. The immature form of *Brachymesia gravida*, with notes on the taxonomy of the group (Odonata: Libellulidae). *Entomological News* 47: 35-37, 60-64.

———. 1937. A review of the dragonflies of genera *Neurocordulia* and *Platycordulia*. *Miscellaneous Publications, Museum of Zoology, University of Michigan* 36: 36.

———. 1941. Notes on the emergence and life history of the dragonfly *Pantala flavescens*. *Proceedings of the Florida Academy of Science* 6: 14-25.

Cabot, L. 1890. Immature State of the Odonata. *Memoirs of the Museum of Comparative Zoology* 17(1): 1-52.

Carpenter, T. 1991. Dragonflies and damselflies of Cape Cod. The Cape Cod Museum of Natural History, Natural History Series 479 pp.

Calil, E.R., and A.L. Carvalho. 1999. Descrições da larva de ultimo estádio e do adulto de *Triacanthagyna septima* (Selys, 1857) (Odonata, Aeshnidae), com notas sobre a biologia da espécie. *Revista Brasiliera de Entomologia* 43(1/2): 73-83.

Calvert, P.P. 1890. Additional notes on some North American Odonata. *Entomological News* 1: 73-74.

———. 1895. The Odonata of Baja California, Mexico. *Proceedings of the California Academy of Science*, Series 2, 4: 463-558.

———. 1898. The Odonate genus *Macrothemis* and its allies. *Proceedings of the Boston Society of Natural History* 28(12): 301-332.

———. 1899. A contribution to knowledge of the Odonata of Paraguay. *Anales del Museo Nacional de Buenos Aires* 7: 25-35.

———. 1901. Odonata, pp. 17-72. *In*: Biologia Centrali Americana: Insecta Neuroptera, R.H. Porter and Dulau Co., London.

———. 1902. Odonata, pp. 73-128. *In*: Biologia Centrali Americana: Insecta Neuroptera, R.H. Porter and Dulau Co., London.

———. 1903. On some American Gomphinae. *Entomological News* 183-192.

———. 1904. On a nymph of *Micrathyria* [sic!] *berenice* Drury. *Entomological News* 15: 174.

———. 1905. Odonata, pp. 145-212. *In*: Biologia Centrali Americana: Insecta Neuroptera, R.H. Porter and Dulau Co., London.

———. 1906. Odonata, pp. 213-308. *In*: Biologia Centrali Americana: Insecta Neuroptera, R.H. Porter and Dulau Co., London.

———. 1909. Contributions to a knowledge of the Odonata of the Neotropical region, exclusive of Mexico and Central America. *Annals of the Carnegie Museum* 6(3): 73-280.

———. 1920. The Costa Rican species of *Epigomphus* and their mutual adaptations (Odonata). *Transactions of the American Entomological Society* 46: 323-354.

———. 1928. Report on Odonata, including notes on some internal organs of the larvae. *University of Iowa Studies* 12(2): 1-44.

———. 1934. The rates of growth, larval development and seasonal distribution of dragonflies of the genus *Anax* (Odonata: Aeshnidae). *Proceedings of the American Philosophical Society* 73(1): 1-70.

———. 1948a. Odonata (dragonflies) of Kartabo, Bartica District, British Guiana. Zoologica. *Scientific Contributions of the New York Zoological Society* 33(2): 47-87.

———. 1948b. Odonata from Pirassununga (Emas), state of São Paulo, Brazil: ecological and taxonomic data. *Boletim do Museu Nacional Nova Série* 87: 1-34.

———. 1952. New taxonomic entities in neotropical Aeshnas (Odonata: Aeshnidae). *Entomological News* 63(10): 253-264.

———. 1956. The neotropical species of the "subgenus *Aeshna*" *sensu* Selysii 1883 (Odonata). *Memoirs of the American Entomological Society* 15: 1-251.

———. 1958. Resultados Zoológicos de la expedición de la Universidad Central de Venezuela a la Región del Auyantepui en la Guayana Venezolana, April de 1956. 4. Genus *Racenaeschna* new genus (Odonata: Aeshnidae). *Acta Biologica Venezuelica* 2(20): 227-234.

Campanella, P.J. 1975. The evolution of mating systems in temperate zone dragonflies (Odonata: Anisoptera) II: *Libellula luctuosa* (Burmeister). *Behaviour* 54(2-4): 278-310.

Campanella, P.J., and L.L. Wolf. 1974. Temporal leks as a mating system in a temperate zone dragonfly (Odonata: Anisoptera) I: *Plathemis lydia* (Drury). *Behaviour* 51(1-2): 49-87.

Cannings, R.A. 1980. Ecological notes on *Sympetrum madidum madidum* (Hagen) in British Columbia,

Canada (Anisoptera: Libellulidae). *Notulae odonatologicae* 1(6): 97-99.

———. 1981. The larva of *Sympetrum madidum* (Hagen) (Odonata: Libellulidae). *Pan-Pacific Entomologist* 57(2): 341-346.

———. 1982a. Notes on the biology of *Aeshna sitchensis* Hagen (Anisoptera: Aeshnidae). *Odonatologica* 11(3): 219-223.

———. 1982b. The larvae of the *Tarnetrum* subgenus of *Sympetrum*, with a description of the larva of *Sympetrum nigrocreatum* Calvert (Odonata: Libellulidae). *Advances in Odonatology* 1: 9-14.

———. 2002. Introducing dragonflies of British Columbia and the Yukon. Royal British Columbia Museum, Victoria, BC. 96 pp.

Cannings, S.G., and R.A. Cannings, 1985. The larva of *Somatochlora sahlbergi* Trybom, with notes on the species in the Yukon Territory (Anisoptera: Corduliidae). *Odonatologica* 14(4): 319-330.

Carle, F.L. 1979. Two new *Gomphus* (Odonata: Gomphidae) from eastern North America with adult keys to the subgenus *Hylogomphus*. *Annals of the Entomological Society of America* 72(3): 418-426.

———. 1980. A new *Lanthus* (Odonata: Gomphidae) from Eastern North America with adult and nymphal keys to American octogomphines. *Annals of the Entomological Society of America* 73(2): 172-179.

———. 1981. A new species of *Ophiogomphus* from eastern North America, with a key to the regional species (Anisoptera: Gomphidae). *Odonatologica* 10(4): 271-278.

———. 1982a. The wing vein homologies and phylogeny of the Odonata: a continuing debate. *Societas Internationalis Odonatologica Rapid Communications* 4: 1-66.

———. 1982b. *Ophiogomphus incurvatus*: a new name for *Ophiogomphus carolinus* Hagen (Odonata: Gomphidae). *Annals of the Entomological Society of America* 75(3): 335-339.

———. 1983. A new *Zoraena* (Odonata: Cordulegastridae) from Eastern North America, with a key to the adult Cordulegastridae of America. *Annals of the Entomological Society of America* 76(1): 61-68.

———. 1986. The classification, phylogeny and biogeography of the Gomphidae (Anisoptera). I. Classification. *Odonatologica* 15(3): 275-326.

———. 1992. *Ophiogomphus (Ophionurus) australis* spec. nov. from the Gulf Coast of Louisiana, with larval and adult keys to American *Ophiogomphus* (Anisoptera: Gomphidae). *Odonatologica* 21(2): 141-152.

———. 1993. *Sympetrum janeae* spec. nov. from eastern North America, with a key to nearctic *Sympetrum* (Anisoptera: Libellulidae). *Odonatologica* 22(1): 1-128.

———. 1995. Evolution, taxonomy, and biogeography of ancient Gondwanian libelluloides, with comments on anisopteroid evolution and phylogenetic systematics (Anisoptera: Libellulidae). *Odonatologica* 24(4): 383-506.

———. 1996. Revision of Austropetaliidae (Anisoptera: Aeshnoidea). *Odonatologica* 25(3): 231-259.

Carle, F.L., and C. Cook. 1984. A new *Neogomphus* from South America, with extended comments on the phylogeny and biogeography of the Octogomphini trib. nov. (Anisoptera: Gomphidae). *Odonatologica* 13(1): 55-70.

Carle, F.L., and K.M. Kjer. 2002. Phylogeny of *Libellula* Linnaeus (Odonata: Insecta). *Zootaxa* 87: 1-18.

Carle, F.L., and J.A. Louton. 1994. The larva of *Neopetalia punctata* and establishment of Austropetaliidae fam. nov. (Odonata). *Proceedings of the Entomological Society of Washington* 96(1): 147-155.

Carpenter, V. 1991. Dragonflies and damselflies of Cape Cod. The Cape Cod Museum of Natural History, Natural History Series 4: 1-79.

Carvalho, A.L. 1987. Description of the larva of *Gynacantha bifida* Rambur (Anisoptera: Aeshnidae). *Odonatologica* 16(3): 281-284.

———. 1988. Descrição da larva de *Triacanthagyna ditzleri* Williamson, 1923 (Odonata, Aeshnidae, Gynacanthini). *Revista brasiliera de Entomologia* 32(2): 223-226.

———. 1989. Description of the larva of *Neuraeschna costalis* (Burmeister), with notes on its biology, and a key to the genera of Brazilian Aeshnidae larvae (Anisoptera). *Odonatologica* 18(4): 325-332.

———. 1992a. Revalidation of the genus *Remartinia* Navas, 1911, with the description of a new species and a key to the genera of Neotropical Aeshnidae (Anisoptera). *Odonatologica* 21(3): 289-298.

———. 1992b. Aspectos da biologia de *Coryphaeschna perrensi* (McLachlan, 1887) (Odonata,

Aeshnidae), com enfase no periodo larval. *Revista brasiliera de Entomologia* 36(4): 791-802.

———. 1993. A morfologia externa da larva de ultimo estadio de *Coryphaeschna perrensi* (McLachlan, 1887) (Odonata, Aeshnidae). *Revista brasiliera de Entomologia* 37(1): 167-179.

———. 1995. Revisão de *Coryphaeschna* Williamson, 1903 *sensu* Calvert, 1956 (Insecta, Odonata, Aeshnidae). Tese de Doutorado. São Paulo, Univeridade de São Paulo, Instituto de Biociências, 1995, 191 pp.

———. 2000. Descriptions of the last instar larva and some structures in the pharate male adult of *Praeviogomphus proprius* Belle, 1995, with notes on the occurrence and taxonomic status of the species (Anisoptera: Gomphidae, Octogomphinae). *Odonatologica* 29(3): 239-246.

Carvalho, A.L., and N. Ferreira. 1989. Descrição da larva de *Gynacantha mexicana* e notas sobre sua Biologia. *Revista brasiliera de Entomologia* 33(3/4): 413-419.

Carvalho, A.L., and J.L. Nessimian. 1998. Odonata do estado do Rio de Janeiro, Brasil: hábitats e hábitos das larvas. *In*: J.L. Nessimian, and A.L. Carvalho eds., Ecologia de Insectos Acuáticos, pp. 3-28. Oecologia Brasiliensis, Rio de Janeiro Vol. V: xvii + 309 pp.

Carvalho, A.L., N. Ferreira, and J.L. Nessiminian. 1991. Descrição das larvas de três espécies do gênero *Erythrodiplax* Brauer (Odonata: Libellulidae). *Revista brasileira de Entomologia* 35(1): 165-171.

Carvalho A.L., P.C. Werneck-de-Carvalho, and E.R. Calil. 2002. Description of the larvae of two species of *Dasythemis* Karsch, with a key to the genera of Libellulidae occurring in the states of Rio de Janeiro and São Paulo, Brazil (Anisoptera). *Odonatologica* 31(1): 23-33.

Carvalho, A.L., and L.G.V. Salgado. 2004. Two new species of *Aeshna* in the *punctata* group from southeastern Brazil (Anisoptera: Aeshnidae). Odonatologica 33(1): 25-39.

Carvalho A.L., L.G.V. Salgado, and P.C. Werneck-de-Carvalho. 2004. Description of a new species of *Lauromacromia* Geijskes, 1970 (Odonata: Corduliidae) from Southeastern Brazil. *Zootaxa* 666: 1-111.

Cashatt, E.D., and T.E. Vogt. 2001. Description of the larva of *Somatochlora hineana* with a key to the larvae of the North American species of *Somatochlora* (Odonata: Corduliidae). *International*

Journal of Odonatology 4(2): 93-105.

Chao, H.-f. 1984. Reclassification of Chinese gomphid dragonflies, with the establishment of a new subfamily and the descriptions of a new genus and species (Anisoptera: Gomphidae). *Odonatologica* 13(1): 71-80.

Charlton, R.E., and R.A. Cannings. 1993. The larva of *Williamsonia fletcheri* Williamson (Anisoptera: Corduliidae). *Odonatologica* 22(3): 335-343.

Charpentier, T. de. 1825. *Horae entomologicae, adjectis tabulis novem coloratis*. Wratislaviae, xvi + 255 pp.

———. 1840. *Libellulinae Europaeae descriptae ac depictae*. Lipsiae, Leopold Voss 180 pp.

Cockerell, T.D.A. 1913. Two fossil insects from Florissant, Colorado, with a discussion of the venation of the aeshnine dragon flies. *Proceedings of the United States Natural History Museum* 45: 577-583.

Comstock, J.H., and J.G. Needham. 1898-1899. The wings of insects. *The American Naturalist* 32: 43-48, 81-89, 231-257, 335-340, 413-424, 561-565, 769-777, 903-911; 33: 117-126, 573-582, 845-860.

Convey, P. 1992. Predation risks associated with mating and oviposition for female *Crocothemis erythraea* (Brulle) (Anisoptera: Libellulidae). *Odonatologica* 21(3): 343-350.

Cook, C., and J.J. Daigle. 1985. *Ophiogomphus westfalli* spec. nov. from the Ozark region of Arkansas and Missouri, with a key to the *Ophiogomphus* species of eastern North America (Anisoptera: Gomphidae). *Odonatologica* 14(2): 89-99.

Cook, C., and E.L. Laudermilk. 2004. *Stylogomphus sigmastylus* sp. nov., a new North American dragonfly previously confused with *S. albistylus* (Odonata: Gomphidae). *International Journal of Odonatology* 7(1): 3-24.

Corbet, P.S. 1999. Dragonflies - Behavior and ecology of Odonata. Comstock Publishing Associates, Cornell University Press, Ithaca, New York, xxxii + 829 pp.

———. 2000. The first recorded arrival of *Anax junius* Drury (Anisoptera: Aeshnidae) in Europe: A scientist's perspective. *International Journal of Odonatology* 3(2): 153-162.

Córdoba-Aguilar, A. 1995. Changes from territorial to feeding activity in adult *Anax walsinghami* McL. before sunset (Anisoptera: Aeshnidae). *Notulae odonatologicae* 4(5): 90-91.

tropical *Entomology* 31(3): 377-389.

————. 2003. A note on the territorial and mating behaviour in the dragonfly *Pseudoleon superbus* (Hag.) (Anisoptera: Libellulidae). *Notulae odonatologicae* 6(1): 10.

Costa, J.M. 1986. Contribução ao conhecimento da fauna da Guanabara. 59 - Notas sobre a ninfa de *Epigomphus paludosus* Hagen in Selys, 1854 (Odonata, Gomphidae). *Atas da Sociedade de Biologia de Rio de Janeiro* 11(4): 157-158.

————. 1991. *Macrothemis absimile* spec. nov., a remarkable new species of Libellulidae from Brazil (Anisoptera). *Odonatologica* 20(2): 233-237.

Costa, J.M., and C.V. de Assis. 1994. Description of the larva of *Tauriphila argo* Hagen, 1869 from São Paulo, Brazil (Anisoptera: Libellulidae). *Odonatologica* 23(1): 51-54.

Costa, J.M., and B. Mascareñas. 1998. Catálogo do material-tipo de Odonata (Insecta) do Museu Nacional. *Publicações Avulsas do Museu Nacional* 76: 1-30.

Costa, J.M., and J.R. Pujol-Luz. 1993. Descrição da larva de *Erythemis mithroides* (Brauer) e notas sobre outras larvas conhecidas do gênero (Odonata, Libellulidae). *Revista brasileira de Zoologia* 10(3): 443-448.

Costa, J.M., and L.P.R.B. Régis. 2005. Description of the last larval instar of *Perihemis thais* (Perty) and comparison with other species of the genus (Anisoptera: Libellulidae). *Odonatologica* 34(1): 51-57.

Costa, J.M., and T.C. Santos. 1992. *Santosia marshalli* gen. nov., spec. nov. - a new genus and species of Corduliinae from Brazil (Anisoptera: Corduliidae). *Odonatologica* 21(2): 235-239.

————. 1997. Intra- and interspecific variation in the genus *Uracis* Rambur, 1842, with a key to the known species (Anisoptera: Libellulidae). *Odonatologica* 26(1): 1-7.

————. 2000. Two new species of *Santosia* Costa, and Santos, 1992, with a description of five new corduliid larvae (Anisoptera: Corduliidae). *Odonatologica* 29(2): 95-111.

Costa, J.M., J.R. Pujol-Luz, and L.P.B. Regis. 2004. Descrição da larva de *Zenithoptera anceps* (Odonata, Libellulidae). *Iheringia*, Série Zoologia 94(4): 421-424.

Costa, J.M., A.N. Lourenço, and L.P. Vieira. 2002. *Micrathyria pseudohypodidyma* sp. n. (Odonata, Libellulidae), com chave das espécies do gênero que ocorrem no estado do Rio de Janeiro. *Neo-*

Costa, J.M., T.C. Santos, and A.M. Telles. 1999a. *Phyllogomphoides annectens* (Selys): description of the last instar, with a key to the South American species (Anisoptera: Gomphidae). *Odonatologica* 28(1): 79-82.

Costa, J.M., G.M. Souza-Franco, and A.M. Takeda. 1999b. Descrição da larva de *Diastatops intensa* Montgomery, 1940 e morfologia dos diferentes estádios de desenvolvimiento (Odonata: Libellulidae). *Boletim do Museu Nacional*, N.S. 410: 1-14.

Costa, J.M., L.P. Vieira, and A.N. Lourenço. 2001. Descrição de três larvas de *Erythrodiplax* Brauer, 1868, e redescrição das larvas de *E. pallida* (Needham, 1904) e *E. umbrata* (Linnaeus, 1758), com chave para identificação das larvas conhecidas das espécies Brasileiras (Odonata, Libellulidae). *Boletim do Museu Nacional*, N.S. 465: 1-16.

Cowley, J. 1934a. Changes in the generic names of the Odonata. *The Entomologist* 67: 200-205.

————. 1934b. Notes on some generic names of Odonata. *Entomologist's Monthly Magazine* 70: 240-247.

————. 1934c. The types of some genera of Odonata. *The Entomologist* 67: 247-253.

————. 1934d. The types of some genera of Gomphidae (Odonata). *The Entomologist* 67:273-276.

————. 1935. Nomenclature of Odonata: three generic names of Hagen. *The Entomologist* 68: 283-284.

Currie, N. 1963. Mating behavior and local dispersal in *Erythemis simplicicollis*. *Proceedings North Central Branch, Entomological Society of America* 18: 112-115.

Curry, G. 2001. Dragonflies of Indiana. The Indiana Academy of Science xiv + 303 pp.

Daigle, J.J. 1994. The larva and adult male of *Somatochlora georgiana* Walker (Odonata: Corduliidae). *Bulletin of American Odonatology* 2(2): 21-26.

Dalchetti, E., and C. Utzeri. 1974. Preliminary observations on the territorial behaviour of *Crocothemis erythraea* (Brulle) (Odonata: Libellulidae). *Fragmenta Entomologica*, Rome 10(3): 295-300.

Dallwitz, M. J., Paine, T. A. and Zurcher, E. J. 2000. Principles of interactive keys. http://biodiversity.uno.edu/delta/

Davies, D.A.L., and P. Tobin. 1985. The dragonflies of the world: a systematic list of the extant species of Odonata. Vol. 2 Anisoptera. *Societas Internationalis Odonatologica Rapid Communications* (Supplements) 5: ix + 151 pp.

Davis, W.T. 1913. *Williamsonia*, a new genus of dragonflies from North America. *Bulletin of the Brooklyn Entomological Society* 8: 93-96.

DeBano, S.J. 1993. Territoriality in the dragonfly *Libellula saturata* Uhler: mutual avoidance or resource defence? (Anisoptera: Libellulidae). *Odonatologica* 22(4): 411-429.

———. 1996. Male mate searching and female availability in the dragonfly *Libellula saturata*: Relationships in time and space. *The Southwestern Naturalist* 41(3): 293-298.

De Marco, P., A.O. Latini, and P.H.E. Ribeiro. 2002. Behavioural ecology of *Erythemis plebeja* (Burmeister) at a small pond in southeastern Brazil (Anisoptera: Libellulidae). *Odonatologica* 31(3): 305-312.

De Marmels, J. 1981a. The larva of *Progomphus abbreviatus* Belle, 1973 from Venezuela (Anisoptera: Gomphidae). *Odonatologica* 10(2): 147-149.

———. 1981b. A new *Misagria* Kirby, 1889 from southern Venezuela. *Odonatologica* 10(4): 319-322.

———. 1982a. Cuatro náyades nuevas de la familia Libellulidae (Odonata: Anisoptera). *Boletin de Entomología de Venezuela*, Nueva Serie 2(11): 94-101.

———. 1982b. Dos náyades nuevas de la familia Aeshnidae (Odonata: Anisoptera). *Boletin de Entomología de Venezuela*, Nueva Serie 2(12): 102-106.

———. 1983. Hallazgo de Odonata nuevos para Venezuela o poco conocidos. *Boletin de Entomología de Venezuela*, Nueva Serie 3 2(19): 155-156.

———. 1985. Hallazgo de Odonata nuevos para Venezuela o poco conocidos. 4. *Boletin de Entomología de Venezuela*, Nueva Serie 4(11): 85-91.

———. 1988. Odonata del Estado Táchira. *Revista Científica Unet* 2(1): 91-111.

———. 1989. Odonata or dragonflies from Cerro de la Neblina. *Academia de las Ciencias Físicas, Matemáticas y Naturales, Caracas, Venezuela* 25: 1-78.

———. 1990a. Nine new Anisoptera larvae from Venezuela (Gomphidae, Aeshnidae, Corduliidae, Libelulidae). *Odonatologica* 19(1): 1-15.

———. 1990b. An updated list of the Odonata of Venezuela. *Odonatologica* 19(4): 333-345.

———. 1991a. *Progomphus incurvatus bivittatus* subspec. nov. from Venezuela (Odonata: Gomphidae). *Opuscula zoologica fluminensia* 71: 1-7.

———. 1991b. *Dorocordulia nitens* sp. n., eine neue Smaragdlibelle aus Venezuela (Odonata: Corduliidae). *Mitteilungen der Entomologische Gesellschaft Basel*, Neue Folge 41(4): 106-111.

———. 1992a. *Sympetrum evanescens* spec. nov., a hitherto overlooked dragonfly from the central Andes of Venezuela (Odonata: Libellulidae). *Opuscula zoologica fluminensia* 79: 1-7.

———. 1992b. Odonata del Cerro Guaiquinima (Edo. Bolivar) y zonas aledañas. *Boletín de Entomología Venezolana* 7(1): 37-47.

———. 1992c. The female and larva of *Aeshna andresi* Racenis, 1958 (Anisoptera: Aeshnidae). *Odonatologica* 21(3): 351-355.

———. 1992d. Caballitos del Diablo (Odonata) de las Sierras de Tapirapecó y Unturán, en el extremo sur de Venezuela [Dragonflies (Odonata) from the Sierras of the Tapirapeco and Unturan, in the extreme south of Venezuela]. *Acta Biológica Venezuélica* 14(1): 57-78.

———. 1993. Hallazgo de Odonata nuevos para Venezuela o poco conocidos. 8. *Boletín de Entomología Venezolana*, Nueva Serie 8(2): 156-158.

———. 1994. A new genus of Aeshnini (Odonata: Aeshnidae) from the Andes, with description of a new species. *Entomologica Scandinavica* 25: 427-438.

———. 1999. Rare Venezuelan dragonflies (Odonata) evaluated for their possible inclusions in the National Red Data Book. *International Journal of Odonatology* 2(1): 55-67.

———. 2000. The larva of *Allopetalia pustulosa* Selys, 1873 (Anisoptera: Aeshnidae), with notes on Aeshnoid evolution and biogeography. *Odonatologica* 29: 113-128.

———. 2001a. *Sympetrum paramo* sp. n. (Odonata: Libellulidae) from the Venezuelan high Andes, with a key to the species of *Sympetrum* Newman, 1833 found in Venezuela. *Entomotropica* 16(1): 15-19.

———. 2001b. *Aeshna (Hesperaeschna) condor* sp. nov. from the Venezuelan Andes, with a redescription of *A (R.) joannisi*, comments on other species and descriptions of larvae (Odonata, Aeshnidae). *International Journal of Odonatology* 4(2): 119-134.

De Marmels, J., and J. Rácenis. 1982. An analysis of the *cophysa*-group of *Tramea* Hagen, with descriptions of two new species (Anisoptera: Libellulidae). *Odonatologica* 11(2): 109-128.

Dijkstra, K.-D. B., J. van Tol, J. Legrand, and G. Theischinger. 2005. Case 3324. *Tramea* Hagen, 1861 (Insecta, Odonata): proposed conservation. *Bulletin of Zoological Nomenclature* 62(2): 68-71.

Donnelly, T.W. 1970. The Odonata of Dominica British West Indies. *Smithsonian Contributions to Zoology* 37: 1-20.

———. 1979. The genus *Phyllogomphoides* in Middle America (Anisoptera: Gomphidae). *Odonatologica* 8(4): 245-265.

———. 1984. A new species of *Macrothemis* from Central America with notes on the distinction between *Brechmorhoga* and *Macrothemis* (Odonata: Libellulidae). *Florida Entomologist* 67(1): 169-174.

———. 1989.Three new species of *Epigomphus* from Belize and Mexico (Odonata: Gomphidae). *Florida Entomologist* 72(3): 428-435.

———. 1992a. The Odonata of Central Panama and their position in the neotropical odonate fauna, with a checklist, and descriptions of new species, pp. 52-90. *In*: D. Quintero, and A. Aiello (eds.), Insects of Panama and Mesoamerica: selected studies. Oxford University Press, xxii + 692 pp.

———. 1992b. Taxonomic problems (?) with *Tetragoneuria*. *Argia* 4(1): 11-14.

———. 1995. *Orthemis ferruginea* - An adventure in Caribbean biogeography. *Argia* 7(4): 9-12.

———. 2003. Problems with *Tetragoneuria*? *Argia* 14(4): 10-11.

Donnelly, T.W., and F.L. Carle. 2000. A new subspecies of *Gomphus (Gomphurus) septima* from the Delaware River of New Jersey, New York, and Pennsylvania (Odonata, Gomphidae). *International Journal of Odonatology* 3(2): 111-123.

DuBois, R.B. 2003. Unreliability of taxonomic keys to larval *Leucorrhinia*. *Argia* 15(1): 13-14.

Dunham, M., 1994. The effect of physical characters on foraging in *Pachydiplax longipennis* (Burmeister) (Anisoptera: Libellulidae). *Odonatologica* 23(1): 55-62.

Dunkle, S.W. 1976. Larva of the dragonfly: *Ophiogomphus arizonicus* (Odonata: Gomphidae). *Florida Entomologist* 59(3): 317-320.

———. 1977a. The larva of *Somatochlora filosa* (Odonata: Corduliidae). *Florida Entomologist* 60(3): 187-191.

———. 1977b. Larvae of the genus *Gomphaeschna* (Odonata: Aeshnidae). *Florida Entomologist* 60(3): 223-225.

———. 1978. Notes on adult behavior and emergence of *Paltothemis lineatipes* Karsch, 1890 (Anisoptera: Libellulidae). *Odonatologica* 7(3): 277-279.

———. 1982. *Perithemis rubita* spec. nov., a new dragonfly from Ecuador (Anisoptera: Libellulidae). *Odonatologica* 11(1): 33-39.

———. 1984. Novel features of reproduction in the dragonfly genus *Progomphus* (Anisoptera: Gomphidae). *Odonatologica* 13(3): 477-480.

———. 1985. *Phyllopetalia pudu* spec. nov., a new dragonfly from Chile, with a key to the family (Anisoptera: Neopetaliidae). *Odonatologica* 14(3): 191-199.

———. 1989. Dragonflies of the Florida peninsula, Bermuda and the Bahamas. Scientific Publishers Nature Guide No. 1, Gainesville. vii + 154 pp.

———. 2000. Dragonflies through binoculars. A field guide to dragonflies of North America. Oxford University Press, New York, 266 pp.

Edwards, G. 1751. A natural history of birds. The most of which have not been figured or described, and the rest, by reason of obscure, or too brief descriptions without figures, or of figures very ill designed, are hitherto but little known. Containing the representatives of thirty-nine birds, engraved on thirty-seven copper-plates after curious original drawings from life; together with a full and accurate description of each. To which are added, by way of appendix, sixteen copper-plates, representing the figures of many curious and undescribed animals, such a quadripedes (both land and amphibious) serpents, fishes and insects: the whole containing fifty-three copper plates, which is the full number given in each of the foregoing parts of this work. Every bird, beast, and c. is colour'd from the original painting, according to nature. College of Physicians Vol. 4: 158-248.

Eller, J.G. 1963. Seasonal regulation in *Pachydiplax longipennis*. *Proceedings North Central Branch, Entomological Society of America* 18: 135.

Erichson, W.F. 1848. Insekten, pp. 533-617. *In*: Schomburgk, Richard. Reisen in British Guiana in den Jahren 1840-1844, Dritter Teil, Versuch einer Fauna und Flora von British Guiana. Weber,

Leipzig.

Fabricius, I.C. 1775. *Systema Entomologiae, Sistens Insectorum Classes, Ordines, Genera, Species, Adiectis, Synonymis, Locis, Descriptionibus, Observationibus.* Flensburgi et Lipsiae in Officina Libraria Kortii xxx + 832 pp.

———. 1798. *Supplimentum Entomologiae Systematicae.* Christ. Gottl. Proft et Storch, Copenhagen, iv + 572 pp.

Fleck, G. 2002a. Une larve d'odonate remarquable de la Guyane française, probablement *Lauromacromia dubitalis* (Fraser, 1939) (Odonata, Anisoptera, 'Corduliidae'). *Bulletin de la Société entomologique de France* 107(3): 223-230.

———. 2002b. Contribution à la connaissance des Odonates de Guyane française: notes sur les genres *Epigomphus* Hagen, 1854, et *Phyllocycla* Calvert, 1948 (Anisoptera, Gomphidae). *Bulletin de la Société entomologique de France* 107(5): 493-501.

———. 2003a. Contribution à la connaissance des Odonates de Guyane française. Les larves des genres *Argyrothemis* Ris, 1911 et *Oligoclada* Karsch, 1889 (Insecta, Odonata, Anisoptera, Libellulidae). *Die Annalen des Naturhistorischen Museums in Wien*, Serie B 104: 341-352.

———. 2003b. Contribution à la connaissance des Odonates de Guyane française: Notes sur des larves des genres *Orthemis, Diastatops* et *Elga* (Anisoptera: Libellulidae). *Odonatologica* 32(4): 335-344.

———. 2004. Contribution à la connaissance des Odonates de Guyane française: les larves des *Macrothemis pumilla* Karsch, 1889 et de *Brechmorhoga praedatrix* Calvert, 1909. Notes biologiques et conséquences taxonomiques (Anisoptera: Libellulidae). *Annales de la Société entomologique de France* 40(2): 177-184.

Fleck, G., M. Brenk, and B. Misof. 2006. DNA taxonomy and the identification of immature insect stages: the true larva of *Tauriphila argo* (Hagen 1869) (Odonata: Anisoptera: Libellulidae). *Annales de la Société entomologique de France* 42(1): 91-98.

Fleck, G., J. De Marmels, and D. Grand. 2004. La larve de *Tholymis citrina* Hagen, 1867 (Odonata, Anisoptera, Libellulidae). *Bulletin de la Société entomologique de France* 109(5): 455-457.

Flint, Jr. O.S. 1991. The Odonata collection of the National Museum of Natural History, Washington, U.S.A. *Advances in Odonatology* 5: 49-58.

———. 1996. The Odonata of Cuba, with a report on a recent collection and checklist of the Cuban species. *Cocuyo* 5: 17-20.

Förster, F. 1900. Odonaten aus Neu-Guinea. II. *Természetrajzi Füzetek* 23: 81-108.

———. 1905. Neotropische Libellen. II. *Entomologische Zeitschrift* 22(19): 75-76.

———. 1907. Neotropische Libellen. V. *Entomologische Wochenblatt* 24: 153-154, 157, 163, 166-167.

———. 1909. Beiträge zu den Gattungen und Arten der Libellen. *Jahrbücher des nassauischen Vereins für Naturkunde, Wiesbaden* 62: 211-235.

———. 1914. Beiträge zu den Gattungen und Arten der Libellen. III. *Archiv für Naturgeschichte* 80: 59-83

Förster, S. 2001. The dragonflies of Central America exclusive of Mexico and the West Indies. A guide to their identification, 2nd edition. Gunnar Rehfeldt, Braunschweig. Odonatological Monographs, 141pp.

Fraser, F.C. 1922. New and rare Indian Odonata in the Pusa collection. *Memoirs of the Department of Agriculture in India*, Entomological Series 7: 39-77.

———. 1929. A revision of the Fissilabioidae (Cordulegasteridae, Petaliidae and Petaluridae) (Order Odonata). Part. I. Cordulegasteridae. *Memoirs of the Indian Museum* 9(3): 69-167.

———. 1933. A Revision of the Fissilabioidea (Cordulegasteridae, Petaliidae and Petaluridae) (Order Odonata). Part II. -Petaliidae and Petaluridae and appendix to Part I. *Memoirs of the Indian Museum* 9(6): 205-260.

———. 1936. The fauna of British India, including Ceylon and Burma. Odonata. Vol. III. Taylor and Francis, London, xi + 461 pp.

———. 1939. Additions to the family Corduliidae including descriptions of two new species and a new genus (Order--Odonata). *Proceedings of the Royal Entomological Society of London*, (B) 8(5): 91-94.

———. 1947. The Odonata of the Argentine Republic I. *Acta Zoológica Lilloana* 4: 427-462.

———. 1957. A reclassification of the order Odonata. Royal Zoological Society of New South Wales, 133 pp.

———. 1960. A Handbook of the Dragonflies of Australasia with keys for the identification of all species. Royal Zoological Society of New South Wales, 67 pp.

———. 1961. A note on the invalidity of the generic name *Acanthagyna* Kirby (Odonata, Aeshnoidea). *Entomologist's Monthly Magazine* 96: 119-120.

García-Diaz, J. 1938. An ecological survey of the fresh water insects of Puerto Rico. 1. The Odonata: with new life histories. *Journal of Agriculture of the University of Puerto Rico* 22(1): 43-97.

Garman, P. 1927. Guide to the insects of Connecticut. Part V. The Odonata or dragonflies of Connecticut. *State of Connecticut, State Geological and Natural History Survey Bulletin* 39: 1-331.

Garrison, R.W. 1982. *Paltothemis cyanosoma*, a new species of dragonfly from Mexico (Odonata: Libellulidae). *Pan-Pacific Entomologist* 58(2): 135-138.

———. 1983. Odonata collected in Canaima, Venezuela, in September 1980. *Notulae odonatologicae* 2(2): 24-25.

———. 1986a. *Diceratobasis melanogaster* spec. nov., a new damselfly from the Dominican Republic (Zygoptera: Coenagrionidae), with taxonomic and distributional notes on the Odonata of Hispaniola and Puerto Rico. *Odonatologica* 15(1): 61-76.

———. 1986b. The genus *Aphylla* in Mexico and Central America, with a description of a new species, *Aphylla angustifolia* (Odonata: Gomphidae). *Annals of the Entomological Society of America* 79(6): 938-944.

———. 1991. A synonymic list of the New World Odonata. *Argia* 3(2): 1-30.

———. 1994. A revision of the New World genus *Erpetogomphus* Hagen *in* Selys (Odonata: Gomphidae). *Tijdschrift voor Entomologie* 137(2): 173-269.

Garrison, R.W., and J. Muzón. 1995. Collecting down at the other "down under.*"Argia* 7(3): 23-26.

Garrison, R.W., and N. von Ellenrieder. 2006. Generic diagnoses within a confusing group of genera: *Brechmorhoga, Gynothemis, Macrothemis,* and *Scapanea* (Odonata: Libellulidae). *Canadian Entomologist* 138: 269-284.

Garrison, R.W., N. von Ellenrieder, and M.F. O'Brien. 2003. An annotated list of the name bearing types of species-group names in Odonata preserved in the University of Michigan, Museum of Zoology. *Occasional Papers of the Museum of Zoology, University of Michigan* 736: 1-73.

Geijskes, D.C. 1934. Notes on the Odonata fauna of the Dutch West Indian Islands Aruba, Curacao and Bonaire with an account on their nymphs. *Internationale Revue der Gesamten Hydrobiologie und Hydrographie* 31: 284-311.

———. 1943. Notes on Odonata of Surinam. III. The genus *Coryphaeschna*, with descriptions of a new species and the nymph of *C. virens*. *Entomological News* 54(3): 61-72.

———. 1946. Observations on the Odonata of Tobago, B.W.I. *Transactions of the Royal Entomological Society of London* 97(9): 213-235.

———. 1951. Notes on Odonata of Surinam. V. A new species of *Misagria* with a redescription of the genus (Odonata: Libellulidae). *Entomological News* 62(2): 70-76.

———. 1959. The aeschnine genus *Staurophlebia*. *Studies on the Fauna of Suriname and other Guyanas* 3(9): 147-172.

———. 1964a. The identity of *Staurophlebia gigantula* Martin (Odonata, Aeshnidae). *Zoologische Mededelingen*, Leiden 39: 67-70.

———. 1964b. The female sex of *Cacus mungo, Gomphoides undulatus, Planiplax phoenicura, Planiplax arachne* and *Dythemis williamsoni*. Notes on Odonata of Suriname IX. *Studies on the Fauna of Suriname and other Guyanas* 7: 36-47.

———. 1968. *Anax longipes versus Anax concolor.* Notes on Odonata of Suriname X. *Studies on the Fauna of Suriname and other Guyanas* 10: 67-100.

———. 1970. Generic characters of the South American Corduliidae with descriptions of the species found in the Guyanas. *Studies on the Fauna of Suriname and other Guyanas* 12: 1-42.

———. 1971. List of Odonata known from French Guyana, mainly based on a collection brought together by the mission of the "Museum National D'Histoire Naturelle," Paris. *Annales de la Société entomologique de France* (N.S.) 7(3): 655-677.

———. 1972. A new species of *Gynothemis* and its larva (Odonata, Libellulidae). Notes on Odonata of Surinam XII. *Zoologische Mededelingen*, Leiden 47: 401-409.

———. 1984. What is *Oligoclada abbreviata* (Rambur, 1942)? (Odonata: Libellulidae). *Zoologische Mededelingen*, Leiden 58(12): 175-185.

Gistl, J. 1848. Naturgeschichte des Thierreichs für höhere Schulen. Scheitlin, and Krais xvi + 216 pp.

Gloyd, L.K. 1959. Elevation of the *Macromia* group to family status (Odonata). *Entomological News* 70(8): 197-205.

———. 1972. *Tramea, Trapezostigma*, and time (Anisoptera: Libellulidae). A nomenclatural problem. *Odonatologica* 1(3): 131-136.

———. 1973. The status of the generic names *Gomphoides, Negomphoides, Progomphus*, and *Ammogomphus* (Odonata: Gomphidae). *Occasional Papers of the Museum of Zoology, University of Michigan* 668: 1-7.

———. 1974. A correction concerning the gender of the generic name *Gomphoides* (Anisoptera: Gomphidae). *Odonatologica* 3(3):179-180.

Glozthober, R.C., and D. McShaffrey, 2002. The dragonflies and damselflies of Ohio. *Bulletin of the Ohio Biological Survey New Series* 14(2): ix+364 pp.

González-Soriano, E. 1987. *Dythemis cannacrioides* Calvert, a libellulid with unusual ovipositing behaviour (Anisoptera). *Odonatologica* 16(2): 175-182.

González-Soriano, E. 2005. The female of *Paltothemis cyanosoma* Garrison (Odonata: Libellulidae). *Folia Entomológica Mexicana* 44 (Suplemento 1): 107-110.

González-Soriano, E., and R. Novelo-Gutiérrez. 1998. *Oplonaeschna magna* sp. nov. (Odonata: Aeshnidae), from Mexico with a description of its larva. *Revista de Biología Tropical* 46(3): 705-715.

Götz, H.J. 1923. Zur Nomenklature der Gattung *Aeshna* Fab. (=*Aeschna* auct.). *Mitteilungen der Münchener Entomologischen Gesellschaft* 13(1-5): 37-39.

Hagen, H.A. 1849. Übersicht der neuren Litteratur betreffend die Neuropteren Linn. *Entomologische Zeitung* 10(2): 55-61, 10(3): 66-74, 10(5):141-156, 10(6): 167-175.

———. 1861. A synopsis of the Neuroptera of North America. Smithsonian Miscellaneous Collections, Washington xx + 347 pp.

———. 1867. Notizen beim Studium von Brauer's Novara-Neuropteren. *Verhandlungen der zoologisch-botanischen Gesellschaft in Wien* 17: 31-62.

———. 1868. Odonaten Cubas (Fortsetzung). *Stettiner Entomologischer Zeitung* 29: 274-287.

———. 1877. Synopsis of the Odonata of America. *Proceedings of the Boston Society of Natural History* 18: 20-96.

[NOTE: Although this paper is usually cited 1875 as given in the running head, the title page of this volume states " Vol. XLIII/ 1875-1876" and "Boston:/Printed for the Society./1877. Hagen's paper was read before the society in 1875 but was published in 1877.]

———. 1885. Monograph of the earlier stages of the Odonata. Sub-families Gomphina and Cordulegastrina. *Transactions of the American Entomological Society* 12: 249-291.

———. 1889. Synopsis of the Odonata of North America. No. 1. *Psyche* 5(160-164): 241-250.

Harvey, F.L. 1898. Contributions to the Odonata of Maine. - III. *Entomological News* 9: 59-64.

Harvey, I.F., and S.F. Hubbard. 1987. Observations on the reproductive behaviour of *Orthemis ferruginea* (Fabricius) (Anisoptera: Libellulidae). *Odonatologica* 16(1): 1-8.

Hedge, T.A., and T.E. Crouch. 2000. A catalogue of the dragonflies and damselflies (Odonata) of South Africa with nomenclatoral clarification. *Durban Museum Novitates* 25: 40-55.

Hellebuyck, V.J. 2002. *Paltothemis nicolae*, spec. nov., a new dragonfly from El Salvador (Odonata: Libellulidae). *Revista Nicaraguense de Entomología* 59/62: 5-15.

Higashi, K. 1969. Territorality and dispersal in the population of the dragonfly *Crocothemis servilia* Drury (Odonata: Anisoptera). *Memoirs of the Faculty of Sciences of the Kyushu University*, Series Biology 5: 95-113.

Hilton, D.F.J. 1983a. Reproductive behavior of *Cordulia shurtleffi* [sic!] Scudder (Anisoptera: Corduliidae). *Odonatologica* 12(1): 15-23.

———. 1983b. Territoriality in *Libellula julia* Uhler (Anisoptera: Libellulidae). *Odonatologica* 12(2): 115-124.

———. 1985. Reproductive behavior of *Leucorrhinia hudsonica* (Selys) (Odonata: Libellulidae). *Journal of the Kansas Entomological Society* 57(4): 580-590.

Huggins, D.G., and G.L. Harp. 1985. The nymph of *Gomphus (Gomphurus) ozarkensis* Westfall (Odonata: Gomphidae). *Journal of the Kansas Entomological Society* 58(4): 656-661.

ICZN. 1999. International Code of Zoological Nomenclature. Fourth edition. The International Trust for Zoological Nomenclature, London, xxix + 306 pp.

———. 2005. Opinion 2110 (Case 3253). *Libellula aenea* Linnaeus, 1758 (currently *Cordulia aenea*) and *L. flavomaculata* Vander Linden, 1825 (currently *Somatochlora flavomaculata*; Insecta, Odonata): usage of the specific names conserved by the replacement of the lectotype of *L. aenea* with a newly designated lectotype. *Bulletin of Zoological Nomenclature* 62(2): 99-100.

Illiger, C. 1801. Namen der Insekten-Gattungen, ihr Genitiv, ihr grammatisches Geschlecht, ihr Silbenmass, ihr Herleitung; zugleich mit den deutschen Benennungen. *Magazine für Insektenkunde* 1: 125-155.

Jacobs, M.E. 1955. Studies on territorialism and sexual selection in dragonflies. *Ecology* 36: 566-586.

Jarzembowski, E.A., and A. Nel. 1996. New fossil dragonflies from the Lower Cretaceous of SE England and the phylogeny of the superfamily Libelluloidea. *Cretaceous Research* 17: 67-85.

Jödicke, R., P. Langhoff, and B. Misof. 2004. The species-group taxa in the Holarctic genus *Cordulia*: a study in nomenclature and genetic differentiation (Odonata: Corduliidae). *International Journal of Odonatology* 7(1): 37-52.

Jödicke, R. and J. van Tol, 2003. *Libellula aenea* Linnaeus, 1758 (currently *Cordulia aenea*) and *L. flavomaculata* Vander Linden, 1825 (currently *Somatochlora flavomaculata*; Insecta, Odonata): proposed conservation of usage of the specific names by the replacement of the lectotype of *L. aenea* with a newly designated lectotype. *Bulletin of Zoological Nomenclature* 60(4): 272-274.

Johnson, C. 1962. A study of territoriality and breeding behavior in *Pachydiplax longipennis* Burmeister (Odonata: Libellulidae). *Southwestern Naturalist* 7(3-4): 191-197.

———. 1968. Seasonal ecology of the dragonfly *Oplonaeschna armata* Hagen (Odonata: Aeshnidae). *American Midland Naturalist* 80(2): 449-457.

Joseph, H.C. 1928. Observaciones sobre el *Phenes raptor* Rambur. *Revista Chilena de Historia Natural* 32: 8-10.

Jurzitza, G. 1975. Ein Beitrag zur Faunistik und Biologie der Odonaten von Chile. *Stuttgarter Beiträge zur Naturkunde*, Ser. A 280: 1-20.

———. 1981. Identificación de los representantes chilenos del género *Gomphomacromia* (Corduliidae: Odonata). *Revista Chilena de Entomología* 11: 31-36.

———. 1982. Die Unterscheidung der Mannchen von *Zenithoptera fasciata* (Linnaeus, 1758), *Z. viola* Ris, 1910, und *Z. lanei* Santos, 1941 (Anisoptera: Libellulidae). *Odonatologica* 11(4): 331-338.

———. 1989. Versuch einer Zusammenfassung unserer Kenntnisse über die Odonatenfauna Chiles. *Societas Internationalis Odonatologica Rapid Communications* (Supplement) 9: 1-32.

Kambhampati, S., and R.E. Charlton. 1999. Phylogenetic relationship among *Libellula*, *Ladona* and *Plathemis* (Odonata: Libelllulidae) based on DNA sequence of mitochondrial 16S rRNA gene. *Systematic Entomology* 24: 37-49.

Karsch, F. 1889. Beitrag zur Kenntniss der Libellulinen mit vierseitiger cellula cardinalis (*Nannophya* Rambur). *Entomologische Nachrichten* 15(16): 245-264.

———. 1890. Beiträge zur Kenntniss der Arten und Gattungen der Libellulinen. *Berliner entomologische Zeitschrift* 33: 347-392.

———. 1891. Kritik des Systems der Aeschniden. *Entomologische Nachrichten* 17(18): 273-290.

Karube, H., and W.C. Yeh. 2001. *Sarasaeschna* gen. nov., with descriptions of female *S. minuta* (Asahina) and male penile structures in *Linaeschna* (Anisoptera: Aeshnidae). *Tombo* 43: 1-8.

Kennedy, C.H. 1915. Notes on the life history and ecology of the Dragonflies (Odonata) of Washington and Oregon. *Proceedings of the United States National Museum* 49: 259-345.

———. 1917. Notes on the life history and ecology of the Dragonflies (Odonata) of Central California and Nevada. *Proceedings of the United States National Museum* 52: 483-635.

———. 1920. Forty-two hitherto unrecognized genera and subgenera of Zygoptera. *Ohio Journal of Science* 21(2):83-88

———. 1922a. The morphology of the penis in the genus *Libellula* (Odonata). *Entomological News* 33(2): 33-40.

———. 1922b. The phylogeny and the geographical distribution of the genus *Libellula* (Odonata). *Entomological News* 33(3): 65-71, 105-111.

———. 1923a. The phylogeny and the distribution of the genus *Erythemis* (Odonata). *Miscellaneus Publications Museum of Zoology, University of Michigan* 11: 19-21.

———. 1923b. The naiad of *Pantala hymenea*. *Canadian Entomologist* 54: 36-38.

————. 1924. Notes and descriptions of naiads belonging to the dragonfly genus *Helocordulia*. *Proceedings of the United States National Museum* 64(12): 1-4.

————. 1946. *Epigomphus subquadrices*, a new dragonfly (Odonata: Gomphidae) from Panama, with notes on *E. quadrices* [sic!] and *Eugomphus* n. subgen. *Annals of the Entomological Society of America* 39(4): 662-666.

Kennedy, C.H., and H.B. White III. 1979. Description of the nymph of *Ophiogomphus howei* (Odonata: Gomphidae). *Proceedings of the Entomological Society of Washington* 81(1): 64-69.

Kenner, R.D., R.A. Cannings, and S.G. Cannings. 2000. The larva of *Leucorrhinia patricia* Walker (Odonata: Libellulidae). *International Journal of Odonatology* 3(1): 1-10.

Kenner, R.D. 2001. Redescription of the larva of *Leucorrhinia glacialis* Hagen with a key to the nearctic *Leucorrhinia* species (Anisoptera: Libellulidae). *Odonatologica* 30(3): 281-288.

Kimmins, D.E. 1936. VIII. Odonata, Ephemeroptera, and Neuroptera of the New Hebrides and Banks Island. *Annals and Magazine of Natural History* 18: 68-88.

————. 1966. A list of the Odonata types described by F.C. Fraser, now in the British Museum (Natural History). *Bulletin of the British Museum, Natural History* 18(6): 173-227.

————. 1969a. A list of the type-specimens of Libellulidae and Corduliidae (Odonata). *Bulletin of the British Museum, Natural History* 22(6): 277-305.

————. 1969b. A list of the type-specimens of Odonata in the British Museum (Natural History) Part II. *Bulletin of the British Museum, Natural History* 23(7): 287-314.

————. 1970. A list of the type-specimens of Odonata in the British Museum (Natural History) Part III. *Bulletin of the British Museum, Natural History* 24(6): 171-205.

Kirby, W.F. 1889. A revision of the subfamily Libellulinae, with descriptions of new genera and species. *Transactions of the Zoological Society of London* 12(9): 249-348.

————. 1890. A synonymic catalogue of Neuroptera Odonata, or dragonflies, with an appendix of fossil species. Gurney, and Jackson, London, ix + 202 pp.

————. 1894. On some small collections of Odonata (dragonflies) recently received from the West Indies. *Annals and Magazine of Natural History* 6(14): 261-269.

————. 1897. List of the Neuroptera collected by Mr. E. E. Austen on the Amazons etc. during the recent expedition of Messrs. Siemens Bros. Cable S. S. 'Faraday', with descriptions of several new species of Odonata. *Annals and Magazine of Natural History* 6(19): 598-617.

————. 1900. Report on the Neuroptera Odonata collected by Mr. E.E. Austen at Sierra Leone during August and September 1899. *Annals and Magazine of Natural History* 7(6): 67-79.

Klots, E.B. 1932. Insects of Porto Rico and the Virgin Islands, Odonata or dragonflies. *Scientific Survey of Porto Rico and the Virgin Islands, New York Academy of Sciences* 16: 107 pp.

————. 1944. Notes on the Gomphinae (Odonata), with descriptions of new species. *American Museum Novitates* 1259: 1-11.

Knopf, K.W. 1977. Dragonfly collecting in Trinidad. *Selysia* 7(2): 6-7.

König, W.D, and S.S. Albano. 1987. Breeding site fidelity in *Plathemis lydia* (Drury) (Anisoptera: Libellulidae). *Odonatologica* 16(3): 249-259.

Landwer, B.H.P., and R.W. Sites. 2003. Redescription of the larva of *Gomphus militaris* Hagen (Odonata: Gomphidae), with distributional and life history notes. *Proceedings of the Entomological Society of Washington* 105(2): 304-311.

Latreille, P.A. 1810. Considérations générales sur l'ordre natural des animaux composant les classes des Crustacés, des Arachnides, et des Insectes; avec un tableau méthodique de leurs genres, dispsés en famillas. Chez F. Schoell, Paris, 444 pp.

Leach, W.E. 1815. Entomology. *In*: Brewester D., The Edinburgh Encyclopedia. Vol. IX, Part I. Edinburgh: 57-172.

Legler, K., D. Legler and D. Westover. 2003. Color guide to dragonflies of Wisconsin. Edition 4.0. Amberwing Publishing 68 pp.

Leonard, J.W. 1934. The Naiad of *Celithemis monomelaena* Williamson (Odonata: Libellulidae). *Occasional Papers of the Museum of Zoology, University of Michigan* 297: 1-5.

Levine, H.R. 1957. Anatomy and taxonomy of the mature naiads of the dragonfly genus *Plathemis* (family Libellulidae). *Smithsonian Miscellaneous Collections* 134(11): 1-28.

Lieftinck, M.A. 1954. Handlist of Malaysian Odonata. A catalogue of the dragonflies of the Malay Peninsula, Sumatra, Java and Borneo, including the adjacent small islands. *Treubia* 22, Supplement, xii + 202 pp.

———. 1971. A catalog of the type-specimens of Odonata preserved in the Netherlands, with a supplementary list of the Odonata types described by Dutch scientists deposited in foreign institutional collections. *Tijdschrift voor Entomologie* 114(2): 65-139.

Limongi, J. 1983. Estudio morfo-taxonómico de nayades en algunas especies de Odonata (Insecta) en Venezuela. *Memorias de la Sociedad de ciencias naturales "La Salle"* 43(119): 95-117.

———. 1991. Estudio morfo-taxonómico de náyades de algunas especies de Odonata (Insecta) en Venezuela (II). *Memorias de la Sociedad de ciencias naturales "La Salle"* 49(131-132): 405-420.

Lincoln, E. 1940. Growth in *Aeshna tuberculifera* (Odonata). *Proceedings of the American Philosophical Society* 83(5): 589-605.

Linnaeus, C. 1758. Systema naturae per regna tria naturae, secundum Classes, Ordines, Genera, Species, cum Characteribus, Differentiis, Synonymis, Locis. Holmiae, Laurentii Salvii, (Edition 10). 1 (Animalia), iv + 824 pp.

Lohmann, H. 1981. Zur Taxonomie einiger *Crocothemis*-Arten, nebst Beschreibung einer neuen Art von Madagaskar (Anisoptera: Libellulidae). *Odonatologica* 10(2): 109-116.

———. 1992. Revision der Cordulegastridae. 1. Entwurf einer neuen Klassifizierung der Familie (Odonata: Anisoptera). *Opuscula zoologica fluminensia* 96: 1-18.

———. 1996. Das Phylogenetische System der Anisoptera (Odonata). *Entomologische Zeitschrift* 106: 209-296.

Louton, J.A. 1982a. A new species of *Ophiogomphus* (Insecta: Odonata: Gomphidae) from the western highland rim in Tennessee. *Proceedings of the Biological Society of Washington* 95(1): 198-202.

———. 1982b. Lotic dragonfly (Anisoptera: Odonata) nymphs of the Southeastern United States: Identification, Distribution and Historical Biogeography. Ph.D. Dissertation, University of Tennessee, Knoxville, 357 pp.

———. 1983. The larva of *Gomphurus ventricosus* (Walsh), and comments on relationships within the genus (Anisoptera: Gomphidae). *Odonatologica* 12(1): 83-86.

Louton, J.A., R.W. Garrison, and O.S. Flint. 1996. The Odonata of Parque Nacional Manu, Madre de Dios, Peru: natural history, species richness and comparisons with other Peruvian sites, pp. 431-449. *In*: Wilson, D.E. and A. Sandoval (eds.), Manu, the biodiversity of southeastern Peru, Smithsonian Institution.

Lutz, P.E. 1963a. Life cycle and seasonal differences in photoperiodic response by nymphs of *Tetragoneuria cynosura* (Odonata). *The ASB Bulletin* 10(2): 33.

———. 1963b. Seasonal regulation in nymphs of *Tetragoneuria cynosura* (Say). *Proceedings North Central Branch, Entomological Society of America* 18: 135-138.

———. 1970. Effects of temperature and photoperiod on seasonal development of *Tetragoneuria cynosura* larvae (Insecta: Odonata). *The ASB Bulletin* 17(2): 53.

Lutz, P.E., and C.E. Jenner. 1964. Life-history and photoperiodic responses of *Tetragoneuria cynosura* (Say). *Biological Bulletin* 127(2): 304-316.

Machado, A.B.M. 1954. *Elga santosi* sp. n. e redescrição de *Elga leptostyla* Ris, 1911 (Odonata, Libellulidae). *Revista brasileira de Biologia* 14(3): 303-312.

———. 1992. A taxonomic note on *Elga* Ris, with *E. newtonsantosi* nom. nov. for *E. leptostyla* Machado, 1954 (Anisoptera: Libellulidae). *Notulae odonatologica* 3(9): 153-154.

———. 2002. Description of *Lauromacromia flaviae* spec. nov., with notes on the holotype of *L. luismoojeni* (Santos) (Anisoptera: Corduliidae). *Odonatologica* 31(3): 313-318.

———. 2005a. *Peruviogomphus bellei* spec. nov. from the Amazonian region of Brazil (Anisoptera: Gomphidae). *Odonatologica* 34(1): 59-63.

———. 2005b. *Neocordulia matutuensis* spec. nov. from Brazil (Anisoptera: Corduliidae). *Odonatologica* 34(3): 299-302.

———. 2005c. *Schizocordulia* gen. nov. related to *Aeschnosom* [*sic*] Selys with description of the female and additional data on the male of *Schizocordulia rustica* (Selys) comb. nov. (Odonata, Corduliidae). *Revista Brasileira de Zoologia* 22 (3): 775-779.

———. 2005d. *Lauromacromia bedei* sp. nov. from the State of Minas Gerais, Brazil (Odonata, Corduliidae). *Revista Brasileira de Entomologia* 49 (4): 453-456.

Machado, A.B.M., and P.A.R. Machado. 1993. *Oligoclada abbreviata limnophila* ssp. nov., with notes on its ecology and distribution (Anisoptera: Libellulidae). *Odonatologica* 22(4): 479-486.

Machado, A.B.M., and J.M. Costa. 1995. *Navicordulia* gen. nov., a new genus of neotropical Corduliinae, with descriptions of seven new species (Anisoptera: Corduliidae). *Odonatologica* 24(2): 187-218.

Machet, P. 1990. Deux nouvelles especes d'Aeshnidae de la Guyane francaise: *Neuraeschna clavulata* et *Neuraeschna capillata* (Odonata: Anisoptera). *L'Entomologiste* 46(5): 209-218.

Machet, P., and M. Duquef. 2004. Un visiteur inattendu, et de taille! *Hemianax ephippiger* (Burmeister, 1839) capturé à la Guyane française. *Martinia* 20(3): 121-124.

Macklin, J.M. 1963. Growth rate of *Pachydiplax longipennis* as influenced by environmental factors. *Proceedings North Central Branch, Entomological Society of America* 18: 138-139.

Manolis, T. 2003. Dragonflies and damselflies of California. University of California Press, Berkeley x + 201pp.

Martin, R. 1907. Cordulines. *In*: Collections Zoologiques du Baron Edm. de Selys Longchamps, Catalogue Systématique et descriptif 17: 1-94.

⸻. 1908. Aeschnines. *In*: Collections zoologiques du Baron Edmund de Sélys-Longchamps, Catalogue Systématique et Descriptif 18: 1-84.

⸻. 1909. Aeschnines. *In*: Collections zoologiques du Baron Edmund de Sélys-Longchamps, Catalogue Systématique et Descriptif 20: 157-223.

⸻. 1921. Sur les odonates du Chili. *Revista Chilena de Historia Natural* 25: 19-25.

May, M.L. 1977. Thermoregulation and reproductive activity in tropical dragonflies of the genus *Micrathyria*. *Ecology* 58: 787-798.

⸻. 1980. Temporal activity patterns of *Micrathyria* in Central America (Anisoptera: Libellulidae). *Odonatologica* 9(1): 57-74.

⸻. 1986. A preliminary investigation of variation in temperature among body regions of *Anax junius* (Drury) (Anisoptera: Aeshnidae). *Odonatologica* 15(1): 119-128.

⸻. 1987. Body temperature regulation and responses to temperature by male *Tetragoneuria cynosura* (Anisoptera: Corduliidae). *Advances in Odonatology* 3: 103-119.

⸻. 1992a. Morphological and ecological differences among species of *Ladona* (Anisoptera: Libellulidae). *Bulletin of American Odonatology* 1(3): 51-56.

⸻. 1992b. A review of the genus *Neocordulia*, with a description of *Mesocordulia* subgen. nov. and of *Neocordulia griphus* spec. nov. from Central America, and a note on *Lauromacromia* (Odonata: Corduliidae). *Folia Entomológica Mexicana* 82: 17-67.

⸻. 1995. The subgenus *Tetragoneuria* (Anisoptera: Corduliidae: *Epitheca*) in New Jersey. *Bulletin of American Odonatology* 2(4): 63-74.

⸻. 1997. Reconsideration of the status of the genera *Phyllomacromia* and *Macromia* (Anisoptera: Corduliidae). *Odonatologica* 26(4): 375-506.

⸻. 1998. *Macrothemis fallax*, a new species of dragonfly from Central America (Anisoptera: Libellulidae), with a key of male *Macrothemis*. *International Journal of Odonatology* 1(2): 137-153.

May, M.L., and F.L. Carle. 1996. An annotated list of the Odonata of New Jersey. *Bulletin of American Odonatology* 4(1): 1-35.

McKinnon, B.I., and M.L. May. 1994. Mating habitat choice and reproductive success of *Pachydiplax longipennis* (Burmeister) (Anisoptera: Libellulidae). *Advances in Odonatology* 6: 59-77.

McLachlan, R. 1896. LX. On some Odonata of the subfamily Aeschnina. *Annals and Magazine of Natural History*, Ser. 6, 17(102): 409-425.

⸻. 1870. Descriptions of a new genus and four new species of Calopterygidae, and of a new genus and species of Gomphidae. *Transactions of the Entomological Society of London* 18: 165-172.

McVey, M.E. 1985. Rates of color maturation in relation to age, diet, and temperature in male *Erythemis simplicicollis* (Say) (Anisoptera: Libellulidae). *Odonatologica* 14(2): 101-114.

Mead, K. 2003. Dragonflies of the North woods. Kollath-Stensaas Publishing x + 203 pages.

Miller, P.L. 1995. Dragonflies. Second Edition (revised). Naturalists' Handbooks 7, The Richmond Publishing Co. Ltd, Slough, England, 118 pp.

Miller, K.B., and D.L. Gustafson, 1996. Distribution records of the Odonata of Montana. *Bulletin of*

American Odonatology, 3(4): 75-88.

Misof, B., A.M. Rickert, T.R. Buckley, G. Fleck, and K.P. Sauer. 2001. Phylogenetic signal and its decay in mitochondrial SSU and LSU rRNA gene fragments of Anisoptera. *Molecular Biology and Evolution* 18(1): 27-37.

Montgomery, B.E. 1937. Oviposition of *Perithemis* (Odonata, Libellulidae). *Entomological News* 48(3): 61-63.

———. 1940. A revision of the genus *Diastatops* (Libellulidae, Odonata) and a study of the leg characters of related genera. *Lloydia* 3(4): 213-280.

Moore, A.J. 1987. Behavioral ecology of *Libellula luctuosa* Burmeister (Anisoptera: Libellulidae). 2. Proposed functions for territorial behaviors. *Odonatologica* 16(4): 385-391.

Moore, A.J., and A.B.M. Machado, 1992. A note on *Cacoides latro* (Erichson), a territorial lacustrine gomphid (Anisoptera: Gomphidae). *Odonatologica* 21(4): 499-503.

Müller, O., and F. Suhling. 2001. *Phyllogomphoides litoralis* Belle: description of the final instar larva (Anisoptera: Gomphidae). *Odonatologica* 30(4): 451-474.

Munz, P.A. 1919. A venational study of the suborder Zygoptera (Odonata) with keys for the identification of genera. *Memoirs of the Entomololical Society of America* 3: 1-76.

Musser, R.J., 1962. Dragonfly nymphs of Utah (Odonata: Anisoptera). University of Utah Biological Series 12(6): vii + 74 pp.

Muttkowski, R.A. 1910. Catalogue of the Odonata of North America. *Bulletin of the Public Museum, Milwaukee* 1: 1-207.

Muttkowski, R.A., and A.D. Whedon, 1915. On *Gomphus cornutus* Tough (Odonata). *Bulletin of the Wisconsin Natural History Society* 13(2): 88-101.

Muzón, J., and A. Garré. 2005. Description of the last instar larva of *Erythrodiplax paraguayensis* (Förster) (Anisoptera: Libellulidae). *Revista de la Sociedad Entomologica Argentina* 64(1-2): 85-91.

Muzón, J., P. Pessacq, and N. von Ellenrieder. 2006. Description of the female and larva of *Phyllogomphoides joaquini* Rodrigues Capítulo 1992 (Anisoptera, Gomphidae). *Odonatologica* 35(1): 47-52.

Muzón, J., and N. von Ellenrieder. 1996. Estadios lar-

vales de Odonata de la Patagonia. I. Descripción de *Aeshna variegata* Fabricius (Odonata: Aeshnidae). *Revista de la Sociedad Entomológica Argentina* 56(1-4): 143-146.

———. 1997. Description of the last larval instar of *Sympetrum villosum* Ris (Odonata: Libellulidae). *Neotropica* 43(109-110): 41-43.

———. 2001. Revision of the subgenus *Marmaraeschna* (Anisoptera, Aeshnidae). *International Journal of Odonatology* 4(2): 93-124.

Navás, L. 1911. Neurópteros del Brasil. *Revista del Museo Paulista* 8: 476-481.

———. 1915. Neue Neuropteren. *Entomologische Mitteilungen* 4(4/6): 146-153.

———. 1916a. Neuoptera nova americana. II Series. *Memorie dell' Accademia pontificia dei Nuovi Lincei*, Ser. 2, 2: 71-80.

———. 1916b. Neuropteres sudamericanos. Tercera serie. Neuropteres del Brasil recogidos por el R.P. Joaquin da Silva Tavares S.J. *Broteria*, Ser. zool. 14(1): 14-35.

Needham, J.G. 1897a. *Libellula deplanata* of Rambur. *Canadian Entomologist* 29: 144-146.

———. 1897b. Preliminary studies of N. American Gomphinae. *Canadian Entomologist* 29: 164-186.

———. 1901. Aquatic insects in the Adirondacks. *New York State Museum Bulletin* 47: 383-612.

———. 1904. New dragonfly nymphs in the United States National Museum. *Proceedings of the United States National Museum* 27: 685-720.

———. 1905a. Two elusive dragonflies. *Entomological News* 16: 3-6.

———. 1905b. A new genus and species of Libellulinae from Brazil. *Proceedings of the Biological Society of Washington* 18: 113-116.

———. 1937. The nymph of *Pseudoleon superbus* Hagen. *Pomona College Journal of Entomology and Zoology* 29: 107-109.

———. 1940. Studies on Neotropical Gomphine dragonflies (Odonata). *Transactions of the American Entomological Society* 65: 363-394.

———. 1941. Life history studies on *Progomphus* and its nearest allies (Odonata: Aeshnidae). *Transactions of the American Entomological Society* 67: 221-245.

———. 1943a. Life history notes on *Micrathyria*. *Annals of the Entomological Society of America*

36: 185-189.

———. 1943b. New species of North American Gomphine dragonflies and life-history notes on some of them. *Bulletin of the Brooklyn Entomological Society* 38(5): 143-152.

———. 1944. Further Studies on Neotropical Gomphinae (Odonata). *Transactions of the American Entomological Society* 69: 171-224.

———. 1950. Three new species of North American dragonflies with notes on related species (Odonata). *Transactions of the American Entomological Society* 76: 1-12.

———. 1951. Prodrome for a manual of the dragonflies of North America, with extended comments on wing venation systems. *Transactions of the American Entomological Society* 77: 21-62.

Needham, J.G., and D.S. Bullock. 1943. The Odonata of Chile. *Zoological Series of the Field Museum* 24(32): 357-373.

Needham, J.G., and E. Broughton, 1927. The venation of the Libellulinae (Odonata). *Transactions of the American Entomological Society* 53: 157-190.

Needham, J.G., and E. Fisher. 1936. The nymphs of North American Libellulinae dragonflies (Odonata). *Transactions of the American Entomological Society* 62: 107-116.

———. 1940. Two Neotropical damselflies (Odonata) from Mts. Duida and Roraima. *American Museum Novitates* 1081: 1-3.

Needham, J.G., and C.A. Hart. 1901. The dragonflies (Odonata) of Illinois. Part I. Petaluridae, Aeschnidae, and Gomphidae. *Bulletin of the Illinois State Laboratory of Natural History* 6: 1-94.

Needham, J.G., and H.B. Heywood. 1929. A handbook of the dragonflies of North America, Charles C Thomas Pub., Springfield, viii + 372 pp.

Needham, J.G., and M.J. Westfall. 1955. A manual of the Dragonflies of North America (Anisoptera). University of California Press/Berkeley xii + 615 pp.

Needham, J.C., M.J. Westfall, and M.L. May. 2000. Dragonflies of North America. Revised edition. Scientific Publishers, Gainesville, FL. xv + 939 pp.

Newman, E. 1833. Entomological Notes. *Entomological Monthly Magazine* 1(5): 505-514.

Nikula, B., J. Sones, D. Stokes, and L.Stokes. 2002. Beginner's guide to dragonflies. Little, Brown,

and Company, Boston 159 pp.

Nikula, B., J.L. Loose, M.R. Burne. 2003. A field guide to the dragonflies and damselflies of Massachusetts. Natural Heritage and Endangered Species Program, Westborough, MA. vii + 197 pp.

Novelo-Gutiérrez, R. 1981. Comportamiento sexual y territorial en *Orthemis ferruginea* (Fab.) (Odonata: Libellulidae). Thesis, Universidad Nacional Autónoma de Mexico, Fac. de Ciencias, Mex. D.F. 1-63.

———. 1989. The larva of *Agriogomphus tumens* Calvert (Anisoptera: Gomphidae). *Odonatologica* 18(2): 203-207.

———. 1991. Los odonatos de la Reserva de la Biosfera de Sian Ka'an, Quintana Roo, Mexico (Insecta: Odonata). Diversidad Biológica en la Reserva de la Biosfera de Sian Ka'an Quintana Roo, Mexico (Insecta: Odonata) 18: 257-274.

———. 1993. Four new larvae of *Phyllogomphoides* Belle from Mexico (Anisoptera: Gomphidae). *Odonatologica* 22(1): 17-26.

———. 1995a. La náyade de *Brechmorhoga praecox* (Hagen, 1861), y notas sobre las náyades de *B. rapax* Calvert, 1898, *B. vivax* Calvert, 1906 y *B. mendax* (Hagen, 1861) (Odonata: Libellulidae). *Folia Entomológica Mexicana* 94: 33-40.

———. 1995b. Náyade de *Brechmorhoga pertinax* (Odonata: Libellulidae). *Anales del Instituto de Biología de la Universidad Nacional Autónoma de México*, Ser. Zool. 66(2): 181-187.

———. 1998. Description of the larva of *Remartinia secreta* and notes on the larva of *Remartinia luteipennis florida* (Odonata: Aeshnidae). *Canadian Entomologist* 130: 893-897.

———. 2002a. Larvae of the *ophibolus*-species group of *Erpetogomphus* Hagen *in* Selys from Mexico and Central America (Anisoptera: Gomphidae). *Odonatologica* 31(1): 35-46.

———. 2002b. Descripción de las larvas de *Perithemis intensa* Kirby, 1889 y *P. domitia* (Drury, 1773), con notas sobre otras larvas del género en México (Odonata: Anisoptera: Libellulidae). *Folia Entomológica Mexicana* 41(3): 321-327.

———. 2002c. Two new Mexican larvae of the genus *Erpetogomphus* Hagen *in* Selys (Odonata: Gomphidae). *Journal of the New York Entomological Society* 110(3-4): 370-375.

———. 2005. Five new *Erpetogomphus* Hagen in Selys larvae from Mexico, with a key to the known species (Anisoptera: Gomphidae). *Odona-*

tologica 34(3): 243-257.

Novelo-Gutiérrez, R., and R.W. Garrison. 1999. *Erpetogomphus erici* spec. nov. from Mexico, and description of the male of *E. agkistrodon* Garrison (Anisoptera: Gomphidae). *Odonatologica* 28(2): 171-179.

Novelo-Gutiérrez, R., and E. González-Soriano. 1984. Reproductive behavior in *Orthemis ferruginea* (Fab.) (Odonata: Libellulidae). *Folia Entomológica Mexicana* 59: 11-24.

―――. 1991. Odonata de la Reserva de la Biósfera la Michilia, Durango, Mexico. Parte II. Náyades. *Folia Entomológica Mexicana* 81: 107-164.

―――. 2004. The larva of *Dythemis maya* Calvert, 1906 and a redescription of the larva of *Dythemis sterilis* Hagen, 1861 with a key to the larvae of the genus (Anisoptera: Libellulidae). *Odonatologica* 33(3): 279-289.

Novelo-Gutiérrez, R., and A. Ramírez. 1995. The larva of *Neocordulia batesi longipollex* Calvert, 1909 (Odonata: Corduliidae). *Journal of the New York Entomological Society* 103(2):180-184.

―――. 1998. The larva of *Macrothemis inacuta* (Odonata: Libellulidae). *Entomological News* 109(5): 301-306.

Paulson, D.R. 1969. Oviposition in the tropical dragonfly genus *Micrathyria* (Odonata, Libellulidae). *Tombo* 12(1/4): 12-16.

―――. 1973. Temporal isolation in two species of dragonflies, *Epitheca sepia* (Gloyd, 1933) and *E. stella* (Williamson, 1911) (Anisoptera Corduliidae). *Odonatologica* 2(2): 115-119.

―――. 1978. An Asiatic dragonfly, *Crocothemis servilia* (Drury), established in Florida (Anisoptera: Aeshnidae). *Notulae odonatologicae* 1(1): 9-10.

―――. 1983. A new species of dragonfly, *Gomphus (Gomphurus) lynnae* spec. nov., from the Yakima River, Washington, with notes on the pruinosity in Gomphidae (Anisoptera). *Odonatologica* 12(1): 59-70.

―――. 1994. Two new species of *Coryphaeschna* from Middle America, and a discussion of the red species of the genus (Anisoptera: Aeshnidae). *Odonatologica* 23(4): 379-398.

―――. 1998a. Possible morphological and behavioral male mimicry in a libellulid dragonfly, *Erythrodiplax umbrata* (L.) (Anisoptera). *Odonatologica* 27(2): 249-252.

―――. 1998b. The distribution and relative abundance of the sibling species *Orthemis ferruginea* (Fabricius, 1775) and *O. discolor* (Burmeister, 1839) in North and Middle America (Anisoptera: Libellulidae). *International Journal of Odonatology* 1(1): 89-93.

―――. 1999. Dragonflies (Odonata: Anisoptera) of south Florida. Slater Museum of Natural History, University of Puget Sound, Occasional Paper 57, 139 pp.

―――. 2001. Recent Odonata records from southern Florida – effects of global warming? *International Journal of Odonatology* 4(1): 57-69.

―――. 2003. Comments on the *Erythrodiplax connata* (Burmeister, 1839) group, with the elevation of *E. fusca* (Rambur, 1842), *E. minuscula* (Rambur, 1842), and *E. basifusca* (Calvert, 1895) to full species (Anisoptera: Libellulidae). *Bulletin of American Odonatology* 6(4): 101-110.

Paulson, D.R., and N. von Ellenrieder. 2005. The synonymy of *Subaeschna* Martin, 1909 with *Gynacantha* Rambur, 1842, and a new species of *Gynacantha* from Peru (Anisoptera, Aeshnidae). *Odonatologica* 34(1): 65-72.

Penn, G.H. 1951. Seasonal variation in the adult size of *Pachydiplax longipennis* (Burmeister) (Odonata, Libellulidae). *Proceedings of the Entomological Society of America* 44: 193-197.

Peña, G. L.E. Distribución geográfica de *Hypopetalia pestilens* Mc.L. (Odonata). *Revista chilena de Entomología* 6: 6.

Peters, G. 1998. Taxonomic and population studies of British Columbia *Aeshna* species. *Bulletin of American Odonatology* 5(2): 33-42.

Pilgrim, E.M., S.A. Roush, and D.E. Krane. 2002. Combining DNA sequences and morphology in systematics: testing the validity of the dragonfly species *Cordulegaster bilineata*. *Heredity* 89: 184-190.

Pilon, J.-G., and J. Desforges. 1989. Morphologie larvaire de *Libellula julia* Uhler (Anisoptera: Libellulidae). *Odonatologica* 18(1): 51-64.

Pinhey, E.C.G. 1962. A descriptive catalogue of the Odonata of the African Continent (up to December 1959). *Publicações Culturais, Companhia de Diamantes de Angola, Museu do Dundo*, Lisboa, 59: 1-321.

Pritchard, A.E. 1936. Notes on *Somatochlora ozarkensis* Bird. (Odonata, Libellulidae, Corduliinae). *Entomological News* 47: 99-101.

Pritchard A.E., and R.F. Smith. 1956. Odonata, pp. 106-153. *In*: Usinger, R.L. (ed.), Aquatic insects of California. University of California Press, Berkeley, ix + 508 pp.

[NOTE: The sequence of authors is Pritchard and Smith in the table of contents and on the dust jacket, but Smith and Pritchard on page 106. According to L.K. Gloyd (*pers. comm.* 17 Aug. 1981), who asked R.F. Smith about the sequence of authorship, the senior author should be A.E. Pritchard.]

Provonsha, A.V., and W.P. McCafferty. 1973. Previously unknown nymphs of Western Odonata (Zygoptera: Calopterygidae, Coenagrionidae). *Proceedings of the Entomological Society of Washington* 75(4): 449-454.

Pujol-Luz, J.R. 1990. Descrição da larva de *Elasmothemis constricta* (Calvert, 1898) (Odonata: Libellulidae). *Revista brasileira de Biologia* 50(2): 487-490.

―――. 1995. Nomenclatural notes on the genus *Zenithoptera* Bates *in* Selys, 1869 and on *Z. fasciata* (L., 1758) versus *Z. americana* (L., 1758) (Anisoptera: Libellulidae). *Odonatologica* 24(2): 229-235.

Pujol-Luz, J.R., and R. Rodrigues da Fonseca. 1997. Variação da coloração das asas e distribuição geográfica do gênero *Zenithoptera* Bates in Selys (Odonata: Libellulidae). *Revista da Universidade Rural, Série Ciência Vida, Rio de Janeiro* 19(1-2): 13-26.

Pujol-Luz, J.R., and F. dias Viera. 1998. Observações sobre o comportamento territorial de machos de *Zenithoptera anceps* Pujol-Luz, 1993 (Odonata: Libellulidae). *Revista da Universidade Rural, Série Ciência Vida, Rio de Janeiro* 20(1-2): 97-102.

Rácenis, J. 1953. Algunas notas sobre las especies venezolanas del género *Nephepeltia* (Odonata: Libellulidae). *Acta Biológica Venezuélica* 1(7): 133-140.

―――. 1969. Las especies del género *Idiataphe* (Odonata: Libellulidae). *Publicaciones Ocasionales del Museo de Ciencias Naturales* 14: 1-15.

―――. 1970. Los Odonatos de la región del Auyatepui y de la sierra de Lema, en la Guayana Venezolana. 2. Las Familias Gomphidae, Aeshnidae y Corduliidae. *Acta Biológica Venezuélica* 7(1): 23-39.

Rambur, P. 1842. Histoire Naturelle des Insectes: Neuropteres. Libraire Encyclopedique de Roret, Paris xvii + 534 pp.

Ramírez, A. 1994. Descripción e historia natural de las larvas de odonatos de Costa Rica. III: *Gynacantha tibiata* (Karsch 1891) (Anisoptera, Aeshnidae). *Bulletin of American Odonatology* 2(1): 9-14.

―――. 1996. Six new dragonfly larvae of the family Gomphidae in Costa Rica, with a key to the Central American genera (Anisoptera). *Odonatologica* 25(2): 143-156.

Ramírez, A., and R. Novelo-Gutiérrez. 1999. The Neotropical genus *Macrothemis*: new larval descriptions and an evaluation of its generic status based on larval stages (Odonata, Libellulidae). *Journal of the North American Benthological Society* 18(1): 67-73.

Rehn, A.C. 2003. Phylogenetic analysis of higher-level relationships of Odonata. *Systematic Entomology* (London) 28: 181-239.

Riek, E.F., and J. Kukalová-Peck. 1984. A new interpretation of dragonfly wing venation based upon Early Upper Carboniferous fossils from Argentina (Insecta: Odonatoidea) and basic character states in pterygote wings. *Canadian Journal of Zoology* 62(6): 1150-1166.

Ris, F. 1904. Odonaten, 44 pp. *In*: Friederichsen, and Co.: Hamburg Magallaenischen Sammelreise 1892/93. Hamburg.

―――. 1909a. Collections zoologiques du Baron de Selys Longchamps; catalogue systematique et descriptif. Fasc. IX. Libellulinen 1: 1-120.

―――. 1909b. Collections zoologiques du Baron de Selys Longchamps; catalogue systematique et descriptif. Fasc. X. Libellulinen 2: 121-244.

―――. 1910. Collections zoologiques du Baron de Selys Longchamps; catalogue systématique et descriptif. Fasc. XI. Libellulinen 3: 245-384.

―――. 1911a. Collections zoologiques du Baron de Selys Longchamps; catalogue systématique et descriptif. Fasc. XII. Libellulinen 4: 385-528.

―――. 1911b. Collections zoologiques du Baron de Selys Longchamps; catalogue systématique et descriptif. Fasc. XIII. Libellulinen 5: 529-700.

―――. 1912. Collections zoologiques du Baron de Selys Longchamps; catalogue systématique et descriptif. Fasc. XIV. Libellulinen 6: 701-836.

―――. 1913a. Collections zoologiques du Baron de Selys Longchamps; catalogue systématique et descriptif. Fasc. XV. Libellulinen 7: 837-964.

―――. 1913b. Collections zoologiques du Baron de

Selys Longchamps; catalogue systématique et descriptif. Fasc. XVI (Première partie). Libellulinen 8: 965-1042.

———. 1913c. Neuer Beitrag zur Kenntnis der Odonatenfauna von Argentina. *Mémoires de la Société de Belgique* 22: 55-102 (1-48 separate).

———. 1919. Collections zoologiques du Baron de Selys Longchamps; catalogue systématique et descriptif. Fasc. XVI (Deuxième partie). Libellulinen 9: 1043-1278.

———. 1930. A revision of the Libelluline genus *Perithemis* (Odonata). Misc. Pub. Mus. Zool. Univ. Mich. 21: 50 pp.

Robey, C.W. 1975. Observations on breeding behavior of *Pachydiplax longipennis* (Odonata: Libellulidae). *Psyche* 82(1): 89-96.

Rodrigues Capítulo, A. 1980. Contribución al conocimiento de los Anisoptera de la República Argentina. I. Descripción de los estadios preimaginales de *Aeshna bonariensis* Rambur (Insecta Odonata). *Limnobios* 2(1): 1-21.

———. 1981. Presencia de *Anax amazili* Burmeister (Odonata Anactinae) en la República Argentina. Algunos datos acerca del comportamiento y determinación del metabolismo energético de las ninfas. *Limnobios* 2(4): 207-214.

———. 1983a. La ninfa de *Phyllocycla argentina* (Hagen in Selys) 1878 (Odonata, Gomphidae). *Revista de la Sociedad Entomológica Argentina* 42(1-4): 267-271.

———. 1983b. Descripción de los estadios preimaginales de *Erythemis attala* Selys (Odonata Libellulidae). *Limnobios* 2(7): 533-548.

———. 1985. Una nueva especie del género *Cyanogomphus* Selys, incluyendo la descripción del último estadio preimaginal (Odonata Gomphidae). *Revista de la Sociedad Entomológica Argentina* 43(1-4): 329-336.

———. 1996. Description of the last instar larva of *Tauriphila risi* Martin (Anisoptera: Libellulidae). *Odonatologica* 25(4): 391-395.

———. 2000. Populations dynamics of larval stages of *Tauriphila risi* Martin and *Erythemis attala* (Selys) in Punta Lara gallery forest, Buenos Aires, Argentina (Anisoptera: Libellulidae). *Odonatologica* 29(4): 333-340.

Rodrigues Capítulo, A., and G. Jurzitza. 1989. Erstbeschreibung der Larve von *Castoraeschna decurvata* Dunkle & Cook 1984 (Odonata: Aeshnidae). *Entomologische Zeitschrift* 99(21): 312-317.

Rodrigues Capítulo, A., and J. Muzón. 1989. Nuevas citas y localidades para los Odonata de la Argentina. *Revista de la Sociedad Entomológica Argentina* 47(1-4): 143-156.

———. 1990. The larval instars of *Orthemis nodiplaga* Karsch 1891 from Argentina (Anisoptera: Libellulidae). *Odonatologica* 19(3): 283-291.

Santos, N.D. 1945. *Ypirangathemis calverti*, novo genero e nova especie (Odonata, Libellulidac). *Revista de Entomologia* 16(3):457-462.

———. 1949. *Planiplax machadoi* n. sp. e notas sôbre outras espécies (Odonata, Libellulidae). *Revista brasileira de Biologia* 9(4): 427-432.

———. 1950. A especiação no gênero *Nephepeltia* (Libellulidae: Odonata). Tese de Doutoramento, Universidade do Brasil, 66 pp.

———. 1966a. Notas sobre *Aeshna (Hesperaeschna) punctata* Martin, 1908 e sua ninfa (Odonata, Aeshnidae). *Atas da Sociedade de Biologia do Rio de Janeiro* 10(4): 97-100.

———. 1966b. Notas sobre *Aeshna (Hesperaeschna) peralta* Ris, 1918 e sua ninfa (Odonata, Aeshnidae). *Atas da Sociedade de Biologia do Rio de Janeiro* 10(5): 123-12.

———. 1967a. Notas sôbre a ninfa de *Erythrodiplax connata fusca* (Rambur, 1842) Brauer, 1868 (Odonata, Libellulidae). *Atas da Sociedade de Biologia do Rio de Janeiro* 10(6): 145-147.

———. 1967b. *Neocordulia luis-moojeni* sp. n. (Odonata, Corduliidae). *Atas da Sociedade de Biologia do Rio de Janeiro* 11(3): 113-115.

———. 1968a. Contribuição ao conhecimento da fauna do estado da Guanabara. 61 - Notas sobre a ninfa de *Progomphus complicatus* (?) Selys 1854 e seu imago (Gomphidae, Odonata). *Atas da Sociedade de Biologia do Rio de Janeiro* 11(5): 171-174.

———. 1968b. Contribuição ao conhecimento da fauna do estado da Guanabara. 62 - Notas sobre a ninfa e o imago de *Micrathyria hypodidyma* Calvert, 1906. *Atas da Sociedade de Biologia do Rio de Janeiro* 11(5): 195-197.

———. 1968c. Contribuição ao conhecimento da fauna do estado da Guanabara. 66 - Descrição da ninfa de *Trapezostigma cophysa* (Selys, 1857) Cowley, 1934 e notas sobre a emergencia (Odonata, Libellulidae). *Atas da Sociedade de Biologia do Rio de Janeiro* 12(3): 169-171.

———. 1969a. Notas sobre a ninfa e o imago de *Coryphaeschna perrensi* (MacLachlan, 1887) Ris,

1913 (Odonata, Aeshnidae). *Atas da Sociedade de Biologia do Rio de Janeiro* 12(4): 173-174.

———. 1969b. Contribuição ao conhecimento da fauna do estado da Guanabara. 67 - Descrição da ninfa e emergencia de *Brechmorhoga nubecula* (?) (Rambur, 1842) Calvert, 1898 (Odonata - Libellulidae). *Atas da Sociedade de Biologia do Rio de Janeiro* 12(4): 221-223.

———. 1969c. Contribuição ao conhecimento da fauna do estado da Guanabara. 69 - Descrição da ninfa de *Erythemis credula* (Hagen, 1861) Calvert 1907 (Odonata: Libellulidae). *Atas da Sociedade de Biologia do Rio de Janeiro* 12(5,6): 287-288.

———. 1970a. Contribuição ao conhecimento da fauna do estado da Guanabara. 71 - Notas sobre a ninfa e o imago de *Limnetron debile* (Karsch, 1891) Forster, 1914 (Odonata, Aeshnidae). *Atas da Sociedade de Biologia do Rio de Janeiro* 13(1,2): 15-17.

———. 1970b. Contribuição ao conhecimento da fauna do estado da Guanabara. 72 - Descrição da ninfa de *Castoraeschna castor* (Brauer, 1865) Calvert, 1952 (Odonata: Aeshnidae). Atas da Sociedade de Biologia do Rio de Janeiro 13(1,2): 47-48.

———. 1970c. Contribuição ao conhecimento da fauna do estado da Guanabara. 73 - Notas sobre a ninfa, o imago e a emergencia de *Coryphaeschna adnexa* (Hagen, 1861) Calvert, 1903 (Odonata, Aeshnidae). *Atas da Sociedade de Biologia do Rio de Janeiro* 13(1,2): 75-77.

———. 1970d. Descrição da ninfa de *Macrothemis musiva* (Hagen, 1861) Calvert, 1898 (Odonata: Libellulidae). *Atas da Sociedade de Biologia do Rio de Janeiro* 13(5,6): 157-158.

———. 1970e. Contribuição ao conhecimento da fauna do estado da Guanabara. 76 - Descrição da ninfa de *Perithemis electra* Ris, 1930 e notas sobre o macho (Odonata: Libellulidae). *Atas da Sociedade de Biologia do Rio de Janeiro* 14(3,4): 49-50.

———. 1972. Contribuição ao conhecimento da fauna do estado da Guanabara e arredores. 80 – Descrição da ninfa de *Micrathyria artemis* (Selys ms.) Ris, 1911 (Odonata: Libellulidae). *Atas da Sociedade de Biologia do Rio de Janeiro* 15(3): 141-143.

———. 1973a. Contribuição ao conhecimento da fauna da Guanabara e arredores. 81 - Descrição da ninfa de *Triacanthagyna caribbea* Williamson, 1923 (Odonata: Aeshnidae). *Atas da Sociedade de*

Biologia do Rio de Janeiro 16(2,3): 53-54.

———. 1973b. Contribuição ao conhecimento da fauna da Guanabara e arredores 82 - Descrição da ninfa de *Gynacantha gracilis* (Burmeister, 1839) Kolbe, 1888 (Aeshnidae: Odonata). *Atas da Sociedade de Biologia do Rio de Janeiro* 16(2,3): 55-57.

———. 1973c. Contribuição ao conhecimento da fuana do estado da Guanabara e arredores. 83 - Descrição da ninfa de *Anatya januaria* Ris, 1911 (Odonata: Libellulidae). *Atas da Sociedade de Biologia do Rio de Janeiro* 16(2,3): 67-69.

———. 1973d. Contribuição ao conhecimento da fauna do estado da Guanabara e arredores. 84 - Descrição da ninfa de *Perithemis mooma* Kirby, 1889 (Odonata: Libellulidae). *Atas da Sociedade de Biologia do Rio de Janeiro* 16(2,3): 71-72.

———. 1978. Contribuição ao conhecimento da fauna do Municipio do Rio de Janeiro, RJ e arredores 85 - Descrição da ninfa de *Micrathyria atra* (Martin, 1897) Calvert, 1906 (Odonata: Libellulidae). *Atas da Sociedade de Biologia do Rio de Janeiro* 19: 17-18.

———. 1981. A new species of *Aeschnosoma* Selys, 1871 from Brazil, with new distributional records and notes on *A. forcipula* Selys, 1871 (Anisoptera: Corduliidae). *Odonatologica* 10(1): 43-47.

Santos, T.C., and J.M. Costa. 1999. Description of the last instar larva of *Brechmorhoga travassosi* Santos and comparison with other *Brechmorhoga* species (Anisoptera: Libellulidae). *Odonatologica* 28(4): 425-428.

Santos, N.D., J.M. Costa, and J.R. Pujol-Luz. 1987. Descrição da ninfa de *Gynacantha membranalis* Karsch, 1891 (Odonata: Gynacanthini) e notas sobre o imago. *Anais da Sociedade Entomologica do Brasil* 16(2): 437-443.

———. 1993. Descrição da larva de *Diastatops obscura* (Fabricius) (Odonata, Libellulidae). *Revista Brasileira de Zoologia* 10(3): 467-472.

Say, T. 1840. Descriptions of new North American neuropterous insects, and observations on some already described. *Proceedings of the Academy of Natural Sciences of Philadelphia* 8: 9-46.

Schmidt, E. 1941a. Revision der Gattung *Zonophora* Selys (Odonata Gomphidae neotrop.). *Deutsche Entomologische Zeitschrift* 1-2: 76-96.

———. 1941b. Petaluridae, Gomphidae und Petaliidae der Schonemannschen Sammlung aus Chile (Ordnung Odonata). *Archiv für Naturgeschichte*,

N.F. 10(2): 231-258.

———. 1987. Notes on a peculiar reproductive behaviour and on habitat recognition in *Sympetrum internum* Montgomery (Anisoptera: Libellulidae). *Notulae odonatologicae* 2(9): 144-147.

Schultz, J.K., and P.V. Switzer. 2001. Pursuit of heterospecific targets by territorial amberwing dragonflies (*Perithemis tenera* Say): A case of mistaken identity. *Journal of Insect Behavior* 14(5): 607-620.

Selys-Longchamps, E. de. 1850. Revue des Odonates ou Libellules d'Europe. *Mémoires de la Société Royale des Sciences de Liége* 6: xxii + 408 pp.

———. 1854. Synopsis des Gomphines. *Bulletin de l'Académie royale de Belgique* 21(2): 23-112 (1-93 reprint).

———. 1858. Monographie de Gomphines. *Mémoires de la Société royale des Sciences de Liége* 11: 257-720 (xiii + 460 reprint).

———. 1859. Additions au Synopsis des Gomphines. *Bulletin de l'Académie royale de Belgique* (2)7: 530-552 (1-26 reprint).

———. 1869a. Secondes additions au synopsis des Gomphines. *Bulletin de l'Académie royale de Belgique* (2)28: 168-208 (1-45 reprint).

———. 1869b. Odonates recuellis a Madagascar, aux iles Mascareignes et Comores déterminés et décrits. pp. 15-25. *In*: Van Vollenhoven, S. C., and Edm. de Selys-Longchamps. Recherches sur la faune de Madagascar et de ses dépendances, d'apres les découvertes de François P. L. Pollen et D. C. Van Dam. 5ᵐᵉ Partie, 1ʳᵉ Livraison, Insectes. Leyde, J. K. Steenhoff.

———. 1870. Resume d'une nouvelle classification des Cordulines. *Bulletin de l'Académie royale de Belgique* 14: 4-7.

———. 1871a. Synopsis des Cordulines. *Bulletin de l'Académie royale de Belgique* (2)31: 238-316 (1-128 reprint).

———. 1871b. Aperçu statistique sur les Néuroptères Odonates. *Transactions of the Entomological Society of London* 19: 409-416.

———. 1873. Appendices aux troisiemès additions et liste des Gomphines, décrites dans le synopsis et ses trois additions. *Bulletin de l'Académie royale de Belgique* (2)36: 492-531 (47-87 reprint).

———. 1874. Additons au synopsis des Cordulines. *Bulletin de l'Académie royale de Belgique* (2)37: 16-34 (1-24 reprint).

———. 1878. Quatrièmes additions au synopsis des Gomphines. *Bulletin de l'Académie royale de Belgique* (2)46: 408-698 (1-106 reprint).

———. 1879. Revision des *Ophiogomphus* et descriptions de quatre nouvelles Gomphines Americaines. *Comptes rendus de la Societe Entomologique de Belgique* 22: lxvii-lxx.

———. 1881. Sur la distribution des insects Odonates en Afrique. *Association Française pour l'Avancement des Sciences Congrès d'Algerie - 1891*: 663-669 (1-7 reprint).

———. 1882. Note sur le genre *Gomphomacromia* Brauer. *Annales de la Societe Entomologique de Belgique* 26: 166-169.

———. 1883a. Les Odonates du Japon. *Annales de la Societe Entomologique de Belgique* 27: 82-143.

———. 1883b. Synopsis des Aeschnines. Premiere partie: Classification. *Bulletin de l'Académie royal de Belgique* 3(5): 712-748.

———. 1900. *In*: Förster: Odonaten aus Neu-Guinea. *Termeszetrajzi Füzetek* 23: 81-108.

Sherman, K.J. 1983. The adaptive significance of postcopulatory mate guarding in a dragonfly *Pachydiplax longipennis*. *Animal Behaviour* 31: 1107-1115.

Silsby, J. 2001. Dragonflies of the world. CSIRO Publishing, vii + 216 pp.

Siva-Jothy, M.T. 1984. Sperm competition in the family Libellulidae (Anisoptera) with special reference to *Crocothemis erythraea* (Brulle) and *Orthetrum cancellatum* (L.). *Advances in Odonatology* 2: 195-207.

Souza, L.O.I., and J.M. Costa. 2002. Descrição de tres larvas de *Micrathyria* Kirby, 1889, com chave para identificação das larvas conhecidas das espécies Brasileiras (Odonata, Libellulidae). *Arquivos do Museu Nacional, Rio de Janeiro* 60(4): 321-331.

Souza, L.O.I., J.M. Costa, and L.A. Espindola. 2002. Description of the last instar larva of *Oligoclada laetitia* Ris, 1911 and comparison with other Libellulidae (Anisoptera). *Odonatologica* 31(4): 403-407.

Souza, L.O.I., J.M. Costa, and T.C. Santos. 1999a. Description of the larva of *Planiplax phoenicura* Ris, from Pantanal sul-Matogrossense, Brazil (Anisoptera, Libellulidae). *Odonatologica* 28(2): 159-163.

———. 1999b. Redescrição da larva de *Tramea cal-*

verti Muttkowski, 1910, com chave para identificação das larvas conhecidas do gênero (Odonata, Libellulidae). *Boletim do Museu Nacional, Rio de Janeiro*, N.S. 409: 1-7.

Spindola, L.A., L.O.I. Souza, and J.M. Costa. 2001. Descrição da larva de *Perithemis thais* Kirby, 1889, com chave para identificação das larvas conhecidas do gênero citadas para o Brasil (Odonata, Libellulidae). *Boletim do Museu Nacional, Rio de Janeiro*, N.S. 442: 1-8.

Steinmann, H. 1997. World catalogue of Odonata. Volume II. Anisoptera. *In*: Wermuth, H., and M. Fischer (eds.). Das Tierreich. The Animal Kingdom. Eine zusammenstellung und Kennzeichnung der rezenten Tierformen, Walter de Gruyter, Berlin, 111: xiv + 636 pp.

Stout, A.L. 1918. Variation in labial characters in the nymph of *Gomphus spicatus* (Odonata). *Entomological News* 29: 68-70.

St. Quentin, D. 1939. Die systematische Stellung der Unterfamilie der Corduliinae Selys (Ordnung Odonata). *Verhandlungen der VII. Kongress von Entomologie*, Berlin 1: 345-360.

———. 1967. Die Gattung *Gomphoides* Selys (Ordnung Odonata) und ihre Verwandten in der neotropischen Region. *Beiträge zur Neotropischen Fauna* 5(2): 132-152.

Svihla, A. 1958. Nymph of *Tanypteryx hageni* Selys (Odonata). *Entomological News* 69(10): 261-266.

———. 1960a. Notes on *Phenes raptor* (Petaluridae). *Tombo* 3(3/4): 23-24.

———. 1960b. A comparison of the habitats of the nymphs of *Tanypteryx* (Petaluridae). *Tombo* 3(3/4): 24-25.

Switzer, P.V. 1997a. Past reproductive success affects future habitat selection. *Behavioral Ecology and Sociobiology* 40: 307-312.

———. 1997b. Factors affecting site fidelity in a territorial animal, *Perithemis tenera*. *Animal Behaviour* 53(4): 865-877.

———. 2002a. Individual variation in the duration of territory occupation by males of the dragonfly *Perithemis tenera* (Odonata: Libellulidae). *Annals of the Entomological Society of America* 95(5): 628-636.

———. 2002b. Territory quality, habitat selection, and competition in the amberwing dragonfly, *Perithemis tenera* (Say) (Odonata: Libellulidae): population patterns as a consequence of individual behavior. *Journal of the Kansas Entomological Society* 75(3): 145-157.

Switzer, P.V., and P.K. Eason. 2000. Proximate constraints on intruder detection in the dragonfly *Perithemis tenera* (Odonata: Libellulidae): effects of angle of approach and background. *Annals of the Entomological Society of America* 93(2): 333-339.

Switzer, P.V., and J.K. Schultz. 2000. The male-male tandem: a novel form of mate guarding in *Perithemis tenera* (Say) (Anisoptera: Libellulidae). *Odonatologica* 29(2): 157-161.

Switzer, P.V., and W. Walters. 1999. Choice of lookout posts by territorial amberwing dragonflies, *Perithemis tenera* (Anisoptera: Libellulidae). *Journal of Insect Behavior* 12(3): 385-398.

Tennessen, K.J. 1975. Description of the nymph of *Somatochlora provocans* Calvert (Odonata: Corduliidae). *Florida Entomologist* 58(2): 105-110.

———. 1977. Rediscovery of *Epitheca costalis* (Odonata: Corduliidae). *Annals of the Entomological Society of America* 70(2): 267-273.

———. 1979. Distance traveled by transforming nymphs of *Tetragoneuria* at Marion County Lake, Alabama, United States (Anisoptera: Corduliidae). *Notulae odonatologicae* 1(4): 63-65.

———. 1993. The larva of *Progomphus bellei* Knopf and Tennessen (Anisoptera: Gomphidae). *Odonatologica* 22(3): 373-378.

———. 1994. Description of the nymph of *Epitheca (Tetragoneuria) spinosa* (Hagen) (Odonata: Corduliidae). *Bulletin of American Odonatology* 2(2): 9-14.

———. 2001. *Coryphaeschna huaorania* spec. nov. from central Ecuador with keys to all species in the genus (Odonata: Aeshnidae). *International Journal of Odonatology* 4(1): 71-81.

———. 2004. *Cordulegaster talaria*, n. sp. (Odonata: Cordulegastridae) from west-central Arkansas. *Proceedings of the Entomological Society of Washington* 106(4): 830-839.

Tennessen, K.J., and J. Louton. 1984. The true nymph of *Gomphus (Gomphurus) crassus* Hagen (Odonata: Gomphidae), with notes on adults. *Proceedings of the Entomological Society of Washington* 86(1): 223-227.

Tennessen, K.J., and S.A. Murray. 1978. Diel periodicity in hatching of *Epitheca cynosura* (Say) eggs (Anisoptera: Corduliidae). *Odonatologica* 7(1): 59-65.

Theischinger, G., and J.A.L. Watson. 1984. Larvae of Australian Gomphomacromiinae, and their bearing on the status of the *Synthemis* group of Genera (Odonata: Corduliidae). *Australian Journal of Zoology* 32: 67-95.

Tillyard, R. 1917. The biology of dragonflies (Odonata or Paraneuroptera). Cambridge Univ. Press xii + 396 pp.

Tillyard, R., and F.C. Fraser. 1938-1940. A reclassification of the order Odonata. *Australian Zoologist* 9: 125-169, 195-221, 359-396.

Trapero Quintana, A., and C. Naranjo López. 2001. New locality reports for *Crocothemis servilia* (Drury, 1773) (Odonata: Libellulidae) in Cuba. *Argia* 13(2): 3.

————. 2003. Revision of the order Odonata in Cuba. *Bulletin of American Odonatology* 7(2): 23-40.

Trueman, J.W.H. 1996. A preliminary cladistic analysis of odonate wing venation. *Odonatologica* 25(1): 59-72.

Tsuda, S. 2000. A Distributional List of World Odonata. Osaka, 430 pp.

Upson, S. 2000. Life history observations on *Sympetrum signiferum* in Arizona. *Argia* 12(4): 5-6.

USFWS (United States Fish and Wildlife Service). 2001. Hine's emerald dragonfly (*Somatochlora hineana* Williamson) recovery plan. Great Lakes-Big Rivers Region, Fort Snelling, Minnesota, 120 pp.

Van Buskirk, J. 1986. Establishment and organization of territories in the dragonfly *Sympetrum rubicundulum*. *Animal Behaviour* 34: 1781-1790.

Young, A.M. 1980. Observations on feeding aggregations of *Orthemis ferruginea* (Fabricius) in Costa Rica (Anisoptera: Libellulidae). *Odonatologica* 9(4): 325-328.

Vogt, T.E., and W.A. Smith. 1993. *Ophiogomphus susbehcha* spec. nov. from North central United States (Anisoptera: Gomphidae). *Odonatologica* 22(4): 503-509.

von Ellenrieder, N. 1999. Description of the last larval instar of *Aeshna (Hesperaeschna) cornigera planaltica* Calvert, 1952 (Odonata: Aeshnidae). *Revista de la Sociedad Entomológica Argentina* 58(3-4): 151-156.

————. 2000. Additions to the description of *Gomphomacromia nodisticta* Ris, 1928 (Anisoptera: Corduliidae). *Bulletin of American Odonatology* 6 (1): 7-11.

————. 2001. The Larvae of Patagonian *Aeshna* Fabricius species (Anisoptera: Aeshnidae). *Odonatologica* 30(4): 423-434.

————. 2002. A phylogenetic analysis of the extant Aeshnidae (Odonata: Anisoptera). *Systematic Entomology* (London) 27: 437-467.

————. 2003. A synopsis of the Neotropical species of '*Aeshna*' Fabricius: The genus *Rhionaeschna* Förster (Odonata: Aeshnidae). *Tijdschrift voor Entomologie* 146: 67-207.

————. 2005. Taxonomy of the South American genus *Phyllopetalia* (Odonata: Austropetaliidae). *International Journal of Odonatology* 8(2): 311-352.

von Ellenrieder, N., and J. M. Costa. 2002. A new species of *Aeshna*, *A. brasiliensis* (Odonata, Aeshnidae) from South and Southeastern Brazil, with a redescription of its larva. *Neotropical Entomology* 31(3): 369-376.

von Ellenrieder, N., and R.W. Garrison. 2003. A synopsis of the genus *Triacanthagyna* (Odonata: Aeshnidae). *International Journal of Odonatology* 6(2): 147-184.

————. 2005a. A synopsis of the South American genus *Gomphomacromia* (Odonata: Libellulidae). *International Journal of Odonatology* 8(1): 83-98.

————. 2005b. Case 3294: *Triacanthagyna* Selys, 1883 and *Gynacantha* Rambur, 1842 (Insecta, Odonata): proposed conservation of usage by designation of *Gynacantha nervosa* Rambur, 1842 as type species of *Gynacantha*. *Bulletin of Zoological Nomenclature* 62(1): 14-17.

von Ellenrieder, N., and J. Muzón. 1999. The Argentinean species of the genus *Perithemis* Hagen (Anisoptera: Libellulidae). *Odonatologica* 28(4): 385-398.

————. 2000. Description of the last larval instar of *Erythrodiplax nigricans* (Rambur) (Anisoptera: Libellulidae). *Odonatologica* 29(3): 267-272.

————. 2003. Description of the last larval instar of *Aeshna (Marmaraeschna) pallipes* Fraser, 1947 (Anisoptera: Aeshnidae). *Odonatologica* 32(1): 95-98.

Walker, E.M. 1912. The North American Dragonflies of the Genus *Aeshna*. *University of Toronto Studies*, Biological Series, pp.1-213.

————. 1913. New nymphs of Canadian Odonata. *Canadian Entomologist* 45(6): 161-170.

———. 1914. New and little-known nymphs of Canadian Odonata. *Canadian Entomologist* 46: 369-377.

———. 1915. Notes on *Staurophlebia reticulata* Burm. *Canadian Entomologist* 47: 387-395.

———. 1916. The nymphs of N. American species of *Leucorrhinia*. *Canadian Entomologist* 48: 414-422.

———. 1917. The known nymphs of the North American species of *Sympetrum* (Odonata). *Canadian Entomologist* 49(12): 409-418.

———. 1925. The North American dragonflies of the genus *Somatochlora*. *University of Toronto Studies*, Biological Series 26: 1-202.

———. 1928. The nymphs of the *Stylurus* group of the genus *Gomphus* with notes on the distribution of this group in Canada (Odonata). *Canadian Entomologist* 60: 79-88.

———. 1932. The nymph of *Gomphus quadricolor* Walsh (Odonata). *Canadian Entomologist* 54: 270-273.

———. 1933. The nymphs of the Canadian species of *Ophiogomphus* (Odonata, Gomphidae). *Canadian Entomologist* 65: 217-229.

———. 1937. A new *Macromia* from British Columbia (Odon. Corduliidae). *Canadian Entomologist* 69(1): 5-13.

———. 1941. The nymph of *Somatochlora walshii* Scudder. *Canadian Entomologist* 73: 203-205.

———. 1958. The Odonata of Canada and Alaska, vol. 2. Part III: The Anisoptera, four families. University of Toronto Press, Toronto, xi + 318 pp.

———. 1966. On the generic status of *Tetragoneuria* and *Epicordulia* (Odonata: Corduliidae. *Canadian Entomologist* 98(9): 897-902.

Walker, E.M., and P.S. Corbet. 1975. The Odonata of Canada and Alaska. Vol. 3. Univ. of Toronto Press, Toronto 15 + 307 pp.

Walsh, B.D. 1862. List of the pseudoneuroptera of Illinois contained in the cabinet of the writer, with descriptions of over forty new species, and notes on their structural affinities. *Proceedings of the Academy of Natural Sciences of Philadelphia* 14: 361-402.

Weith, R., and J.G Needham. 1901. The life history of *Nannothemis bella*. *Canadian Entomologist* 33: 252-255.

Westfall, Jr., M.J. 1950. Nymphs of three species of *Gomphus* (Odonata). *Florida Entomologist* 33(1): 33-39.

———. 1951. Notes on *Tetragoneuria sepia* Gloyd, with descriptions of the female and nymph. *Florida Entomologist* 34(1): 9-14.

———. 1953. The nymph of *Miathyria marcella* Selys (Odonata). *Florida Entomologist* 36: 21-25.

———. 1965. Confusion among species of *Gomphus*. *Quarterly Journal of the Florida Academy of Sciences* 28(3): 245-254.

———. 1974. A critical study of *Gomphus modestus* Needham, 1942, with notes on related species (Anisoptera: Gomphidae). *Odonatologica* 3(1): 63-73.

———. 1988. *Elasmothemis* gen. nov., a new genus related to *Dythemis* (Anisoptera: Libellulidae). *Odonatologica* 17(4): 419-428.

———. 1989. The larvae of *Desmogomphus paucinervis* (Selys, 1873) and *Perigomphus pallidistylus* (Belle, 1972) (Anisoptera: Gomphidae). *Odonatologica* 18(1): 99-106.

———. 1992. Notes on *Micrathyria*, with descriptions of *M. pseudeximia* sp. n., *M. occipita* sp. n., *M. dunklei* sp. n. and *M. divergens* sp. n. (Anisoptera: Libellulidae). *Odonatologica* 21(2): 203-218.

———. 1998. Description of the true larva of *Tauriphila australis* (Hagen, 1867) from Limoncocha, Ecuador (Anisoptera: Libellulidae). *Odonatologica* 27(4): 491-512.

Westfall, Jr., M.J., and M.L. May. 1996. Damselflies of North America. Scientific Publishers, Inc., Gainesville, FL, vii + 649 pp.

Westfall, Jr., M.J., and K.J. Tennessen. 1979. Taxonomic clarification within the genus *Dromogomphus* Selys (Odonata: Gomphidae). *Florida Entomologist* 62(4): 266-273.

———. 1996. Odonata, pp. 164-211. *In*: Merritt, R.W., and K.W. Cummins (eds.), An introduction to the Aquatic Insects of North America. 3rd edition. Dubuque, Kendall Hunt.

Westfall, Jr., M.J., and R.P. Trogdon. 1962. The true *Gomphus consanguis* Selys (Odonata: Gomphidae). *Florida Entomologist* 45: 29-41.

Westman, A., F. Johansson, and A.N. Nilsson. 2000. The phylogeny of the genus *Leucorrhinia* and the evolution of larval spines (Anisoptera: Libellulidae). *Odonatologica* 29(2): 129-136.

Westwood, J.O. 1840. An introduction to the modern classification of Insects; founded on the natural habits and corresponding organization of different families. Vol. II. London, Longman, Orme, Brown, Green, and Longmans, 158 pp.

White, H.B., III, and R.A. Raff. 1970. The nymph of *Williamsonia lintneri* (Hagen) (Odonata: Corduliidae). *Psyche* 77(2): 252-257.

Whitehouse, F.C. 1941. British Columbia dragonflies (Odonata), with notes on distribution and habits. *American Midland Naturalist* 26(3): 488-557.

Wildermuth, H. 1991. Behaviour of *Perithemis mooma* Kirby at the oviposition site (Anisoptera: Libellulidae). *Odonatologica* 20(4): 471-478.

————. 1992. Visual and tactile stimuli in choice of oviposition substrates by the dragonfly *Perithemis mooma* Kirby (Anisoptera: Libellulidae). *Odonatologica* 21(3): 309-321.

————. 1993. Habitat selection and oviposition site recognition by the dragonfly *Aeshna juncea* (L.): an experimental approach in natural habitats (Anisoptera: Aeshnidae). *Odonatologica* 22(1): 27-44.

————. 1994. Reproductive behaviour of *Diastatops intensa* Montgomery (Anisoptera: Libellulidae). *Odonatologica* 23(2): 183-191.

Williams, C.E. 1977. Courtship display in *Belonia croceipennis* (Selys), with notes on copulation and oviposition (Anisoptera: Libellulidae). *Odonatologica* 6(4): 283-287.

Williams, C.E., and S. Dunkle. 1976. The larva of *Neurocordulia xanthosoma* (Odonata: Corduliidae). *Florida Entomologist* 59(4): 429-433.

Williams, F.X. 1937. Notes on the biology of *Gynacantha nervosa* Rambur (Aeschnidae), a crepuscular dragonfly in Guatemala. *Pan-Pacific Entomologist* 13(1-2): 1-8.

Williamson, E.B. 1901. On the manner of oviposition and on the nymph of *Tachopteryx thoreyi* (Order Odonata). *Entomological News* 12: 1-4.

————. 1903. A proposed new genus of Odonata (dragonflies) of the subfamily Aeschninae, group Aeschna. *Entomological News* 14(1): 2-10.

————. 1905. Oviposition of *Tetragoneuria* (Odonata). *Entomological News* 16: 255-256.

————. 1908. A new dragonfly (Odonata) belonging to the Cordulinae, and a revision of the classification of the subfamily. *Entomological News* 19(9): 428-434.

————. 1918. Results of the University of Michigan-Williamson expedition to Colombia 1916-17. I. Two interesting new Colombian Gomphines (Odonata). *Occasional Papers of the Museum of Zoology, University of Michigan* 52: 1-14.

————. 1919. Results of the University of Michigan-Williamson expedition to Colombia 1916-17. III. *Archaeogomphus*, a new genus of dragon-flies (Odonata) *Occasional Papers of the Museum of Zoology, University of Michigan* 63: 1-8.

————. 1920. A new Gomphine genus from British Guiana with a note on the classification of the subfamily (order Odonata). *Occasional Papers of the Museum of Zoology, University of Michigan* 80:1-11.

————. 1922. Notes on *Celithemis* with descriptions of two new species (Odonata). *Occasional Papers of the Museum of Zoology, University of Michigan* 108: 1-22.

————. 1923a. Notes on American species of *Triacanthagyna* and *Gynacantha*. *Miscellaneous Publications of the Museum of Zoology University of Michigan* 9: 1-67.

————. 1923b. Notes on the genus *Erythemis* with a description of a new species (Odonata). *Miscellaneous Publications of the Museum of Zoology University of Michigan* 11: 3-18.

Wolf, L.L., and E.C. Waltz. 1984. Dominions and site-fixed aggressive behavior in breeding male *Leucorrhinia intacta* (Odonata: Libellulidae). *Behavioural Ecology and Sociobiology* 14: 107-115.

Wright, M. 1946a. A description of the nymph of *Sympetrum ambiguum* (Rambur), with habitat notes. *Journal of the Tennessee Academy of Sciences* 21(1): 133-138.

————. 1946b. Notes on nymphs of the dragonfly genus *Tarnetrum*. *Journal of the Tennessee Academy of Sciences* 21(2): 198-200.

————. 1946c. Notes on nymphs of the dragonfly genus *Helocordulia* Needham. *Ohio Journal of Science* 46(6): 337-339.

————. 1949. Notes on nymphs of the dragonfly genus *Boyeria*. *Journal of the Tennessee Academy of Sciences* 24(3): 213-215.

Young, A.M. 1965. Some observations on territoriality and oviposition in *Anax junius* (Odonata: Aeshnidae). *Annals of the Entomological Society of America* 58(5): 767-768.

————. 1980. Observations on feeding aggregations

of *Orthemis ferruginea* (Fabricius) in Costa Rica (Anisoptera: Libellulidae). *Odonatologica* 9(4): 325-328.

Young, W.C., and C.W. Bayer. 1979. The dragonfly nymphs (Odonata: Anisoptera) of the Guadalupe River Basin, Texas. *Texas Journal of Science* 31: 85-97.

Yousuf, M., and H. Yunus. 1974. A new subfamily, Kuldanagasterinae of the family Cordulegastridae (Odonata: Anisoptera). *Pakistan Journal of Zoology* 6(1-2): 141-146.

Zimsen, E., 1964. *The type material of I.C. Fabricius*. Munksgaard, Copenhagen 656 pp.

Distribution tables

Distribution of genera by country (based on examined material and reliable literature records).
CA: Canada, US: United States of America, MX: Mexico, CU: Cuba; BA: Bahamas, HI: Hispaniola (Dominican Republic and Haiti), JA: Jamaica, PR: Puerto Rico, LA: Lesser Antilles, GU: Guatemala, BE: Belize, ES: El Salvador, HO: Honduras, NI: Nicaragua, CR: Costa Rica.

PETALURIDAE	CA	US	MX	GU	CU	BA	HI	JA	PR	LA	BE	ES	HO	NI	CR
Phenes															
Tachopteryx		■													
Tanypteryx	■														
Subtotal	1	2	0	0	0	0	0	0	0	0	0	0	0	0	0

AUSTROPETALIIDAE	CA	US	MX	GU	CU	BA	HI	JA	PR	LA	BE	ES	HO	NI	CR
Hypopetalia															
Phyllopetalia															
Subtotal	0	0	0	0	0	0	0	0	0	0	0	0	0	0	0

AESHNIDAE	CA	US	MX	GU	CU	BA	HI	JA	PR	LA	BE	ES	HO	NI	CR
Aeshna	■	■	■												
'A'.williamsoniana			■								■				■
Allopetalia															
Anax	■	■	■	■	■	■	■	■	■	■		■	■	■	■
Andaeschna															
Basiaeschna	■	■													
Boyeria	■	■													
Castoraeschna															
Coryphaeschna		■	■	■	■	■	■	■	■		■	■	■	■	■
Epiaeschna	■	■													
Gomphaeschna	■	■													
Gynacantha		■	■	■	■	■	■	■	■	■	■	■	■	■	■
Limnetron															
Nasiaeschna	■	■													
Neuraeschna													■		■
Oplonaeschna		■	■									■			
Racenaeschna															
Remartinia		■	■										■	■	■
Rhionaeschna	■	■	■	■	■						■	■	■	■	■
Staurophlebia			■	■							■				
Triacanthagyna		■	■	■											
Subtotal	8	13	10	7	6	4	5	5	5	4	7	5	8	6	9

GOMPHIDAE	CA	US	MX	GU	CU	BA	HI	JA	PR	LA	BE	ES	HO	NI	CR
Agriogomphus			■		■										■
Anomalophlebia															
Aphylla			■	■			■				■			■	
Archaeogomphus															
Arigomphus		■	■												
Brasiliogomphus															
Cacoides															
Cyanogomphus															
Desmogomphus														■	
Diaphlebia															
Dromogomphus	■	■	■												
Ebegomphus															
Epigomphus			■	■	■						■	■	■	■	■
Erpetogomphus		■	■							■					

312

Distribution of genera by country (based on examined material and reliable literature records).
PA: Panama, TR: Trinidad/Tobago, VE: Venezuela, CO: Colombia, EC: Ecuador, PE: Peru, BO: Bolivia, GY: Guiana, SU: Surinam, FR: French Guiana, BR: Brazil, PY: Paraguay, UR: Uruguay, AR: Argentina, CH: Chile.

PETALURIDAE	PA	TR	VE	CO	EC	PE	BO	GY	SU	FR	BR	PY	AR	UR	CH
Phenes													■		■
Tachopteryx															
Tanypteryx															
Subtotal	0	0	0	0	0	0	0	0	0	0	0	0	1	0	1

AUSTROPETALIIDAE	PA	TR	VE	CO	EC	PE	BO	GY	SU	FR	BR	PY	AR	UR	CH
Hypopetalia															■
Phyllopetalia													■		■
Subtotal	0	0	0	0	0	0	0	0	0	0	0	0	1	0	2

AESHNIDAE	PA	TR	VE	CO	EC	PE	BO	GY	SU	FR	BR	PY	AR	UR	CH
Aeshna															
'A'.williamsoniana	■														
Allopetalia			■	■	■	■									■
Anax	■	■	■	■	■	■	■	■	■	■	■	■	■	■	
Andaeshna			■	■	■										
Basiaeschna															
Boyeria															
Castoraeschna			■	■	■	■	■		■		■		■		
Coryphaeschna	■	■	■	■	■	■	■	■	■	■	■	■	■	■	
Epiaeschna															
Gomphaeschna															
Gynacantha			■	■	■	■	■	■	■	■	■	■	■		
Limnetron						■	■					■	■		
Nasiaeschna															
Neuraeschna			■	■	■	■	■		■	■	■				
Oplonaeschna															
Racenaeschna			■	■											
Remartinia	■	■	■	■	■	■					■	■	■	■	
Rhionaeschna	■		■	■	■	■	■						■	■	
Staurophlebia	■		■	■	■	■	■	■	■	■	■				
Triacanthagyna	■	■	■	■	■	■	■	■	■	■	■	■			
Subtotal	8	5	12	10	9	12	9	6	7	6	10	9	10	5	2

GOMPHIDAE	PA	TR	VE	CO	EC	PE	BO	GY	SU	FR	BR	PY	AR	UR	CH
Agriogomphus	■		■	■			■		■		■				
Anomalophlebia															
Aphylla			■	■	■	■	■	■	■	■	■	■	■		
Archaeogomphus			■			■		■			■				
Arigomphus															
Brasiliogomphus											■				
Cacoides			■	■				■	■						
Cyanogomphus											■		■		
Desmogomphus	■		■	■			■								
Diaphlebia															
Dromogomphus															
Ebegomphus			■	■											
Epigomphus	■				■				■						
Erpetogomphus	■														

Distribution of genera by country (based on examined material and reliable literature records).
CA: Canada, US: United States of America, MX: Mexico, CU: Cuba; BA: Bahamas, HI: Hispaniola (Dominican Republic and Haiti), JA: Jamaica, PR: Puerto Rico, LA: Lesser Antilles, GU: Guatemala, BE: Belize, ES: El Salvador, HO: Honduras, NI: Nicaragua, CR: Costa Rica.

GOMPHIDAE	CA	US	MX	GU	CU	BA	HI	JA	PR	LA	BE	ES	HO	NI	CR
Gomphoides															
Gomphus	■	■	■												
Hagenius	■	■													
Idiogomphoides															
Lanthus	■	■													
Melanocacus															
Mitragomphus															
Neogomphus															
Octogomphus	■	■	■												
Ophiogomphus	■	■	■												
Perigomphus															■
Peruviogomphus															
Phyllocycla			■	■							■			■	■
Phyllogomphoides		■	■	■									■		
Praeviogomphus															
Progomphus	■	■	■		■		■	■			■	■	■	■	■
Stylogomphus	■	■													
Stylurus	■	■	■												
Tibiagomphus															
Zonophora															
Subtotal	10	13	13	7	2	0	2	1	0	0	7	4	4	7	10

CORDULEGASTRIDAE	CA	US	MX	GU	CU	BA	HI	JA	PR	LA	BE	ES	HO	NI	CR
Cordulegaster	■	■	■	■											■
Subtotal	1	1	1	1	0	0	0	0	0	0	0	0	0	0	1

NEOPETALIIDAE	CA	US	MX	GU	CU	BA	HI	JA	PR	LA	BE	ES	HO	NI	CR
Neopetalia															
Subtotal	0	0	0	0	0	0	0	0	0	0	0	0	0	0	0

MACROMIINAE	CA	US	MX	GU	CU	BA	HI	JA	PR	LA	BE	ES	HO	NI	CR
Didymops	■	■													
Macromia	■	■	■												
Subtotal	2	2	1	0	0	0	0	0	0	0	0	0	0	0	0

CORDULIINAE	CA	US	MX	GU	CU	BA	HI	JA	PR	LA	BE	ES	HO	NI	CR
Aeschnosoma															
Cordulia	■	■													
Dorocordulia	■	■													
Epitheca	■	■	■												
Gomphomacromia															
Helocordulia	■	■													
Lauromacromia															
Navicordulia															
Neocordulia			■	■										■	■
Neurocordulia	■	■													
Paracordulia															
Rialla															
Santosia															
Somatochlora	■	■													
Williamsonia	■	■													
Subtotal	7	7	2	1	0	0	0	0	0	0	0	0	0	1	1

Distribution of genera by country (based on examined material and reliable literature records).
PA: Panama, TR: Trinidad/Tobago, VE: Venezuela, CO: Colombia, EC: Ecuador, PE: Peru, BO: Bolivia, GY: Guiana, SU: Surinam, FR: French Guiana, BR: Brazil, PY: Paraguay, UR: Uruguay, AR: Argentina, CH: Chile.

GOMPHIDAE	PA	TR	VE	CO	EC	PE	BO	GY	SU	FR	BR	PY	AR	UR	CH
Gomphoides											■	■	■		
Gomphus															
Hagenius															
Idiogomphoides									■	■	■				
Lanthus															
Melanocacus			■						■						
Mitragomphus															
Neogomphus														■	■
Octogomphus															
Ophiogomphus															
Perigomphus	■														
Peruviogomphus				■	■						■				
Phyllocycla			■	■	■	■	■	■	■	■	■	■	■		
Phyllogomphoides			■	■	■	■	■	■	■	■	■	■			
Praeviogomphus															
Progomphus			■	■	■	■	■	■	■	■	■	■	■	■	■
Stylogomphus															
Stylurus															
Tibiagomphus											■	■	■		
Zonophora			■	■	■	■	■	■	■	■	■	■	■		
Subtotal	9	4	14	11	8	9	6	9	13	8	19	8	11	4	2

CORDULEGASTRIDAE	PA	TR	VE	CO	EC	PE	BO	GY	SU	FR	BR	PY	AR	UR	CH
Cordulegaster	■														
Subtotal	1	0	0	0	0	0	0	0	0	0	0	0	0	0	0

NEOPETALIIDAE	PA	TR	VE	CO	EC	PE	BO	GY	SU	FR	BR	PY	AR	UR	CH
Neopetalia													■		■
Subtotal	0	0	0	0	0	0	0	0	0	0	0	0	1	0	1

MACROMIINAE	PA	TR	VE	CO	EC	PE	BO	GY	SU	FR	BR	PY	AR	UR	CH
Didymops															
Macromia															
Subtotal	0	0	0	0	0	0	0	0	0	0	0	0	0	0	0

CORDULIINAE	PA	TR	VE	CO	EC	PE	BO	GY	SU	FR	BR	PY	AR	UR	CH
Aeschnosoma			■	■							■				
Cordulia															
Dorocordulia															
Epitheca															
Gomphomacromia					■	■	■						■		■
Helocordulia															
Lauromacromia			■							■					
Navicordulia			■							■	■		■		
Neocordulia	■		■							■	■		■		
Neurocordulia															
Paracordulia			■			■			■	■					
Rialla													■		■
Santosia											■				
Somatochlora															
Williamsonia															
Subtotal	1	0	5	1	2	4	1	1	2	3	6	2	3	0	2

Distribution of genera by country (based on examined material and reliable literature records).
CA: Canada, US: United States of America, ME: Mexico, CU: Cuba; BA: Bahamas, HI: Hispaniola (Dominican Republic and Haiti), JA: Jamaica, PR: Puerto Rico, LA: Lesser Antilles, GU: Guatemala, BE: Belize, ES: El Salvador, HO: Honduras, NI: Nicaragua, CR: Costa Rica.

LIBELLULINAE	CA	US	MX	GU	CU	BA	HI	JA	PR	LA	BE	ES	HO	NI	CR
Anatya															
Antidythemis															
Argyrothemis															
Brachymesia															
Brechmorhoga															
Cannaphila															
Celithemis															
Crocothemis															
Dasythemis															
Diastatops															
Dythemis															
Edonis															
Elasmothemis															
Elga															
Erythemis															
Erythrodiplax															
Fylgia															
Gynothemis															
Idiataphe															
Leucorrhinia															
Libellula															
Macrodiplax															
Macrothemis															
Miathyria															
Micrathyria															
Misagria															
Nannothemis															
Nephepeltia															
Nothodiplax															
Oligoclada															
Orthemis															
Pachydiplax															
Paltothemis															
Pantala															
Perithemis															
Planiplax															
Pseudoleon															
Rhodopygia															
Scapanea															
Sympetrum															
Tauriphila															
Tholymis															
Tramea															
Uracis															
Ypirangathemis															
Zenithoptera															
Subtotal	11	26	28	27	22	13	18	17	15	12	27	20	25	23	27

ANISOPTERA	CA	US	MX	GU	CU	BA	HI	JA	PR	LA	BE	ES	HO	NI	CR
Total number	40	64	55	43	30	17	25	23	20	16	41	29	37	37	48

Distribution of genera by country (based on examined material and reliable literature records).
PA: Panama, TR: Trinidad/Tobago, VE: Venezuela, CO: Colombia, EC: Ecuador, PE: Peru, BO: Bolivia, GY: Guiana, SU: Surinam, FR: French Guiana, BR: Brazil, PY: Paraguay, UR: Uruguay, AR: Argentina, CH: Chile.

LIBELLULINAE	PA	TR	VE	CO	EC	PE	BO	GY	SU	FR	BR	PY	AR	UR	CH
Anatya	■	■	■	■	■	■	■	■	■	■	■		■		
Antidythemis			■					■	■	■	■				
Argyrothemis			■			■	■	■	■	■	■				
Brachymesia	■	■	■	■	■	■		■	■	■	■	■		■	
Brechmorhoga	■		■	■	■	■	■				■	■	■		
Cannaphila	■	■	■	■	■	■		■	■	■	■				
Celithemis															
Crocothemis															
Dasythemis			■	■	■	■	■	■	■	■	■	■	■		
Diastatops			■	■	■	■	■	■	■	■	■	■	■		
Dythemis	■		■	■	■	■	■	■	■	■	■	■	■		■
Edonis			■	■	■	■	■	■	■	■	■		■		
Elasmothemis	■	■	■	■	■	■	■	■	■	■	■	■	■		
Elga			■	■	■	■	■		■	■	■	■	■		
Erythemis	■	■	■	■	■	■	■	■	■	■	■	■	■	■	
Erythrodiplax	■	■	■	■	■	■	■	■	■	■	■	■	■	■	■
Fylgia				■	■		■		■	■	■				
Gynothemis		■	■		■		■	■	■	■	■				
Idiataphe	■		■	■	■	■	■	■	■	■	■		■		
Leucorrhinia															
Libellula	■		■	■	■	■	■	■		■	■		■		
Macrodiplax	■	■	■	■	■	■	■	■	■	■	■	■	■		
Macrothemis	■	■	■	■	■	■	■			■	■	■	■		
Miathyria	■	■	■	■	■	■	■	■	■	■	■	■	■	■	
Micrathyria	■	■	■	■	■	■	■	■	■	■	■	■	■	■	
Misagria		■	■		■	■	■	■	■	■	■				
Nannothemis															
Nephepeltia	■	■	■	■	■	■	■	■	■	■	■	■			
Nothodiplax			■	■	■	■	■	■		■	■				
Oligoclada	■	■	■	■	■	■	■	■	■	■	■	■			
Orthemis	■	■	■	■	■	■	■	■	■	■	■	■	■	■	
Pachydiplax															
Paltothemis															
Pantala	■	■	■	■	■	■	■	■	■	■	■	■	■	■	
Perithemis	■	■	■	■	■	■	■	■	■	■	■	■	■		
Planiplax		■	■	■		■	■			■	■	■	■		
Pseudoleon															
Rhodopygia	■	■	■	■	■	■	■	■	■	■	■				
Scapanea															
Sympetrum	■		■	■	■	■		■		■	■		■	■	■
Tauriphila	■	■	■	■	■	■	■	■	■	■	■	■	■		
Tholymis	■	■	■	■	■	■	■	■	■	■	■	■	■	■	
Tramea	■	■	■	■	■	■	■	■	■	■	■	■	■		
Uracis	■	■	■	■	■	■	■	■	■	■	■	■			
Ypirangathemis			■		■		■		■	■	■	■			
Zenithoptera		■	■	■	■	■	■	■	■	■	■				
Subtotal	26	27	36	30	30	33	29	31	31	27	35	23	26	13	6

ANISOPTERA	PA	TR	VE	CO	EC	PE	BO	GY	SU	FR	BR	PY	AR	UR	CH
Total number	45	36	67	52	49	58	45	47	53	44	70	42	53	22	16

List of Figures

The following list, arranged numerically, provides structure, genus and/or species, brief locality data, repository, and illustrator in brackets for each drawing, in this order: *Figure, structure, view – taxon sex, locality, specimen location (if not RWG) [illustrator]*. Most illustrations are based on material from the personal collection of Rosser W. Garrison (RWG); specimens from other sources are briefly noted. A few of the illustrations were taken or modified from other published accounts and are acknowledged. All other illustrations were done by Rosser Garrison (RWG) or Natalia von Ellenrieder (NVE). Wings were scanned and edited digitally except as noted.

Index of Taxa

Note: Entries are listed alphabetically, and include generic and infrageneric (subgeneric, specific, and subspecific) epithets. Page numbers in **boldface** indicate starting page of generic account; page numbers followed by k, indicate location of name in a generic key. Valid genera are set in ***BOLD CAPITAL ITALIC TYPE***, generic synonyms in *CAPITAL ITALIC TYPE*, valid specific names in ***bold italic type***, and specific synonyms in *italic type*. Infraspecific entries are listed under their parent species and separately with their respective parenthetical notation (ssp. = subspecies, forma = form, var. = variety) following each entry. *Nomina nuda* are followed by the parenthetical expression (n.n.).